International Conference on Transport and Infrastructure of the Siberian Region (SibTrans 2020)

AA007091

IOP Conference Series: Materials Science and Engineering
Volume 1151

Irkutsk, Russia
11 - 13 November 2020

ISBN: 978-1-7138-3421-2
ISSN: 1757-8981

Printed from e-media with permission by:

Curran Associates, Inc.
57 Morehouse Lane
Red Hook, NY 12571

Some format issues inherent in the e-media version may also appear in this print version.

This work is licensed under a Creative Commons Attribution 3.0 International Licence.
Licence details: http://creativecommons.org/licenses/by/3.0/.

No changes have been made to the content of these proceedings. There may be changes to pagination and minor adjustments for aesthetics.

Printed with permission by Curran Associates, Inc. (2021)

For permission requests, please contact the Institute of Physics at the address below.

Institute of Physics
Dirac House, Temple Back
Bristol BS1 6BE UK

Phone: 44 1 17 929 7481
Fax: 44 1 17 920 0979

techtracking@iop.org

Additional copies of this publication are available from:

Curran Associates, Inc.
57 Morehouse Lane
Red Hook, NY 12571 USA
Phone: 845-758-0400
Fax: 845-758-2633
Email: curran@proceedings.com
Web: www.proceedings.com

TABLE OF CONTENTS

IOP Conf. Series: Materials Science and Engineering 1151 (2021) 011001 doi:10.1088/1757-899X/1151/1/011001

PREFACE

The 2020 International Conference on Transport and Infrastructure of the Siberian Region (SibTrans 2020) has successfully taken place at Irkutsk State Transport University on November 11–13, 2020. SibTrans 2020 continues to be a forum mainly for the Commonwealth of Independent States professional and academic community, engaged in locomotive engineering, railway infrastructure development and related areas of engineering.

The major goal of SibTrans 2020 was to bring together scientists and practitioners, engineers and industry researchers to share their experience and novations. The four-day conference provided a great opportunity to initiate discussions on topical issues of instrumentation for testing and maintain railways and locomotives, machine-building technologies and machinery, and discuss the challenges encountered and the solutions adopted.

Scientists and practitioners, mostly from Russia, submitted over ten dozen papers. All the papers were reviewed by the editors and invited experts. The most up-to-date and comprehensive manuscripts were selected for this volume. We sincerely hope that the SibTrans 2020 proceedings will show the readers a broad overview of the latest research results on the field of transportation.

At last, we would like to express our deep appreciation and gratitude to all the members of the SibTrans 2020 scientific committee for their strenuous efforts. Without their assistance and dedication, it would be impossible to hold the discussion at such a high level and to collect quality papers for compiling the conference proceedings. Finally, we want to show appreciation of all the authors, participants and editors of the proceedings for their critical contribution to the work, organization, and holding of the conference and to the preparation of the papers.

With kindest regards,

Dmitriy I. Sachkov,
Anton P. Kutsiy

Content from this work may be used under the terms of the Creative Commons Attribution 3.0 licence. Any further distribution of this work must maintain attribution to the author(s) and the title of the work, journal citation and DOI.
Published under licence by IOP Publishing Ltd

IOP Conf. Series: Materials Science and Engineering 1151 (2021) 011001 doi:10.1088/1757-899X/1151/1/011001

Programm Committee:

S. K. Kargapol'tsev, Doctor of Engineering Science, the Rector of IrGUPS.

A. P. Khomenko, Doctor of Engineering Science, the President of IrGUPS.

Yufeng Luo, Professor, the President of East China Jiaotong University (ECJTU).

Park Joon Hoon, the President of the Korea National University of Transportation.

Yu. I. Kulazhenko, Professor, the Rector of BelSU.

I. V. Bychkov, a member of the Russian Academy of Sciences, Research Superviser of the Presidium of the Irkutsk Scientific Center of the Siberian Branch of the Russian Academy of Sciences

V. F. Frolov, Head of the East-Siberian Railway.

D. Zhigzhidnyamaa, Head of the Ulan-Bator Railway.

V. Yu. Dorofeev, First Deputy Governor of the Irkutsk Region.

A. M. Suleimenov, Minister of Housing Policy, Energy and Transport of the Irkutsk Region.

Organizational Committee

A. I. Orlenko, Ph.D. in Engineering Science, Director of KrIZhT (Krasnoyarsk Institute of Railway Transport)

A. N. Saveliev, Acting Director of UUIZhT (Ulan-Ude Institute of Railway Transport)

S. I. Noskov, Doctor of Engineering Science, Professor of the Subdepartment of Information Systems and Information Security

M. V. Malova, Ph.D. in Engineering Science, Dean of the Department of Transportation Systems

O. I. Rusakova, Ph.D. in Engineering Science, Dean of the Department of Economics and Finance

E. A. Rusch, Doctor of Engineering Science, Head of the Subdepartment of Technosphere Safety

A. V. Kryukov, Doctor of Engineering Science, Professor of the Subdepartment of Electric Power Engineering of Transport

N.P. Shustov, Ph.D. in Engineering Science, Dean of the Department of Transport Support Systems

A.V. Dimov, Ph.D. in Engineering Science, Dean of the Department of Transport Management and Information Technologies

V. A. Bubnov, Ph.D. in Economic Science, Dean of the Department of Management, Logistics and Customs Affairs

IOP Conf. Series: Materials Science and Engineering 1151 (2021) 011001 doi:10.1088/1757-899X/1151/1/011001

E. V. Filatov, Ph.D. in Engineering Science, Dean of the Department of Construction of Railways

V. V. Tretyakov, Ph.D. in Historical Science, Associate Professor of the Subdepartment of Philosophy and Socio-Humanitarian Sciences

V. N. Zheleznyak, Ph.D. in Engineering Science, head of the Subdepartment of Cars and Rolling Stock

O. V. Goreva, Ph.D. in Physics and Mathematics, Vice-Rector for Information Technology and Communication with Branches

A. A. Pykhalov, Doctor of Engineering Science, Director of Educational Center of CTEA (Computer Technologies of Engineering Analysis)

D. I. Sachkov, Ph.D. in Economic Science, Head of the Research and Development Department

M. A. Savosteeva, Head of the Organization of Scientific and Contractual Work, Researach and Development Department

Peer review declaration

All papers published in this volume of *IOP Conference Series: Materials Science and Engineering* have been peer reviewed through processes administered by the Editors. Reviews were conducted by expert referees to the professional and scientific standards expected of a proceedings journal published by IOP Publishing.

- **Type of peer review:** Single-blind peer review by two independent referees
- **Conference submission management system:** submissions were received and handled via e-mail
- **Number of submissions received:** 82
- **Number of submissions sent for review:** 60
- **Number of submissions accepted:** 45
- **Acceptance Rate (Number of Submissions Accepted / Number of Submissions Received X 100):** 55%
- **Average number of reviews per paper:** 2
- **Total number of reviewers involved:** 16
- **Any additional info on review process:** the review was organized in two stages.

Prior to the conference sessions chairs (or their affiliates) made an entry check of the suggested manuscripts. The authors of all qualified manuscripts got an invitation to the report.

All reported (and several not actually presented) studies were submitted to the review to ensure their scientific quality. The most papers had reviewed by the single external expert; however, in some cases two referees were involved.

All revised and accepted papers were checked and reviewed by the editorial team of three members. At this stage, style and language checks were applied.

- **Contact person for queries:** Dmitriy Sachkov, chief editor, Cand. of Eco., head of the Research Management Department, Irkutsk State Transport University, Irkutsk, Russia, leninb@yandex.ru

Content from this work may be used under the terms of the Creative Commons Attribution 3.0 licence. Any further distribution of this work must maintain attribution to the author(s) and the title of the work, journal citation and DOI.
Published under licence by IOP Publishing Ltd

IOP Conf. Series: Materials Science and Engineering 1151 (2021) 012001 doi:10.1088/1757-899X/1151/1/012001

Proper use of technical risk assessment tools as an element of the railway safety culture

O I Verevkina[1] and I A Veprinyak[2]

[1] Rostov State Transport University, 2, Lenina Str., Rostov-on-Don, 344038, Russia
[2] Military Academy of logistics named after General of the army A.V. Khrulev, 1, Suvorovskaya str., Peterhof, St. Petersburg, 198504, Russia

E-mail: ov18111966@mail.ru

Abstract. The work is aimed at developing ideas and improving the technical means of risk assessment – the risk matrix. A typical risk matrix defined by regulatory sources is considered. An original approach to evaluating the effectiveness of the risk matrix with associated scales is proposed. The definition of the risk matrix with minimum scale spacing is introduced and the existence of this matrix is proved. It is shown by example that a matrix with minimum scale spacing is a more accurate tool for determining the risk category compared to matrices with wider scale spacing. The problem of finding the parameters of such a matrix from the initial data in a mathematical statement is formulated. One of the branches of the developed simplified algorithm for constructing such a matrix is given.

1. Introduction

The current time in railway transport is a time of increasing the culture of traffic safety management. One of the important elements of traffic safety management is risk assessment, and the technical tool used for risk assessment is the risk matrix. The use of this tool for interval risk assessment is widespread in railway transport [1-5]. However, the issue of correctly setting the problem of determining the parameters of a typical risk matrix, as well as correctly calculating these parameters, has not been raised before. Below, the problem is formulated in a strict mathematical statement and compared with the solution in the normative source.

2. Preliminary statement of the problem

The problem statement is given for a typical risk matrix with the size of 6×4.

2.1 Definition 1

The risk plane (Fig. 2) is the I-th quadrant of the plane in which the axis scale is set in an exponential manner:

$$f = f_н \cdot K^y \quad (1)$$

$$c = c_н \cdot K^x \quad (2)$$

where $f_н$, $c_н$, K – positive constants,

x, y – non-negative coordinates (x – abscissa, y – ordinate).

Content from this work may be used under the terms of the Creative Commons Attribution 3.0 licence. Any further distribution of this work must maintain attribution to the author(s) and the title of the work, journal citation and DOI.
Published under licence by IOP Publishing Ltd

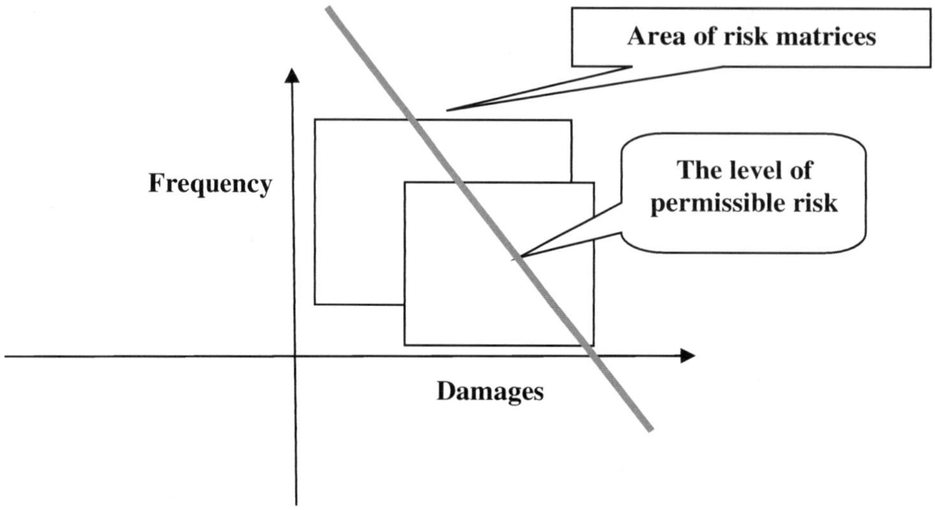

Figure 1. Risk Matrix – as a part of the risk plane

2.2 Definition 2

A risk matrix is a rectangular region in the plane of risk, with the coordinates of the vertices: (c_{min}, $f_{\min}$), (c_{max}, $f_{\min}$), (c_{min}, $f_{\max}$,), (c_{max}, $f_{\max}$) satisfying the conditions (3-4):

$$\left(\frac{f_{\max}}{f_{\min}}\right)^{1/2} = \left(\frac{c_{\max}}{c_{\min}}\right)^{3/8} \quad (3)$$

$$f_{\max} \cdot (c_{\min} \cdot c_{\max})^{1/2} = R \quad (4)$$

where R – positive number assigned by the level of "permissible risk", the left and right part of the ratio (3) is called the scale spacing K of the risk plane:

$$K = \left(\frac{f_{\max}}{f_{\min}}\right)^{1/2} = \left(\frac{c_{\max}}{c_{\min}}\right)^{3/8}. \quad (5)$$

The problem in this statement is formulated in [3, 4]

3. The problem of determining the parameters of the risk matrix

Let us set the frequency limits ($a_{\min}$, $a_{\max}$) of adverse events, and the limits of consequences ($b_{\min}$, $b_{\max}$), as well as the level of permissible risk: R_{perm}.

Based on the data presented above, it is necessary to find such a risk matrix that in addition to condition (3), the following conditions are met:

$$f_{\max} \cdot (c_{\min} \cdot c_{\max})^{1/2} = R_{perm} \quad (6)$$

$$f_{\min} \leq a_{min} \quad (7)$$

$$f_{\max} \geq a_{max} \quad (8)$$

$$c_{min} \leq b_{min} \quad (9)$$

$$c_{max} \geq b_{max}. \quad (10)$$

We note that there are an infinite number of risk matrices that solve this problem, which differ in the magnitude of the scale spacing, as well as the coordinates that define the rectangle of the risk matrix, for the same level of permissible risk R_{perm}. This is due to the fact that 2 (two) equations and 4 (four) constraints on the unknowns are used to determine the 4 unknowns. Such problems are called undefined (not fully defined) and have an infinite set of solutions.

We will prove the following statement: there is the smallest of all K belonging to the set {K} of all risk matrices defined by the data: a_{min}, a_{max}, b_{min}, b_{max}, R_{perm}. (11)

4. Evidence

Since, according to (5) $K = \left(\frac{c_{max}}{c_{min}}\right)^{3/8}$, and according to the requirements for the risk matrix (9)

$c_{max} \geq b_{max}$, then:

$$\left(\frac{c_{max}}{c_{min}}\right)^{3/8} \geq \left(\frac{b_{max}}{c_{min}}\right)^{3/8}. \tag{12}$$

At the same time, according to (9) $c_{min} \leq b_{min}$:

$$\left(\frac{b_{max}}{c_{min}}\right)^{3/8} \geq \left(\frac{b_{max}}{b_{min}}\right)^{3/8}. \tag{13}$$

Combining (11) and (12) we get: $K \geq \left(\frac{b_{max}}{b_{min}}\right)^{3/8}$. (14)

Similarly, using (7)-(8), we obtain: $K \geq \left(\frac{a_{max}}{a_{min}}\right)^{1/2}$. (15)

Choosing the minimum of the two values on the right side (13) and (14):

$$K_{min} = \max\left\{\left(\frac{b_{max}}{b_{min}}\right)^{3/8}, \left(\frac{a_{max}}{a_{min}}\right)^{1/2}\right\} > 0,$$

we get that for any K:

$$K \geq K_{min}.$$

Since the set of all K is a limit from below, according to the theorem [6, 7] on the lower bound of the set, the set {K} has a lower limit that we will consider the smallest value, or the smallest spacing of the risk matrix defined by fixed initial data, and thus the statement is proved.

Since it is known [4] that the accuracy of estimation using the risk matrix decreases with increasing K, it is natural to consider the risk matrices with the smallest scale spacing as the best one (we can hypothesize that these risk matrices are real).

Here is an example of the fact that when manipulating an arbitrary extension of the initial data (that is, essentially increasing K), errors may occur in determining the risk category: undesirable, permissible, not taken into account. As an example, let us take the data from [4]:

$a_{min} = 293$ 1/ year; $a_{max} = 1031$ 1/year;

$b_{min} = 0{,}081$ train-hour; $b_{max} = 0{,}727$ train-hour.

$R_{perm} = 150$ train-hour/year4.

Figures 2 and 3 below show 2 risk matrices, the parameters of the first one are calculated based on the minimum scale spacing requirement (without expanding the initial data), and the second one is calculated taking into account the initial scale expansion, which is a part of the calculation algorithm presented in [4]. The sloping lines are lines of the equal level of risk R_{perm}, R_{perm}/K, R_{perm}/K^2 respectively.

The risk category is determined not by belonging to a rectangle of a particular color, but by its location relative to these lines, according to table 1. Such matrices are called refined according to [4].

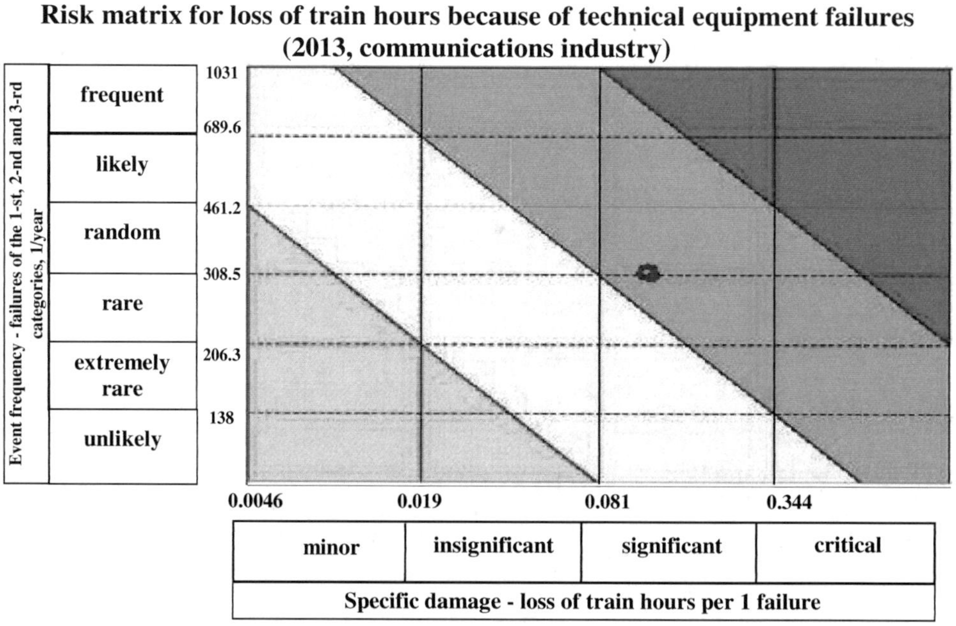

Figure 2. The risk matrix with the minimum scale spacing

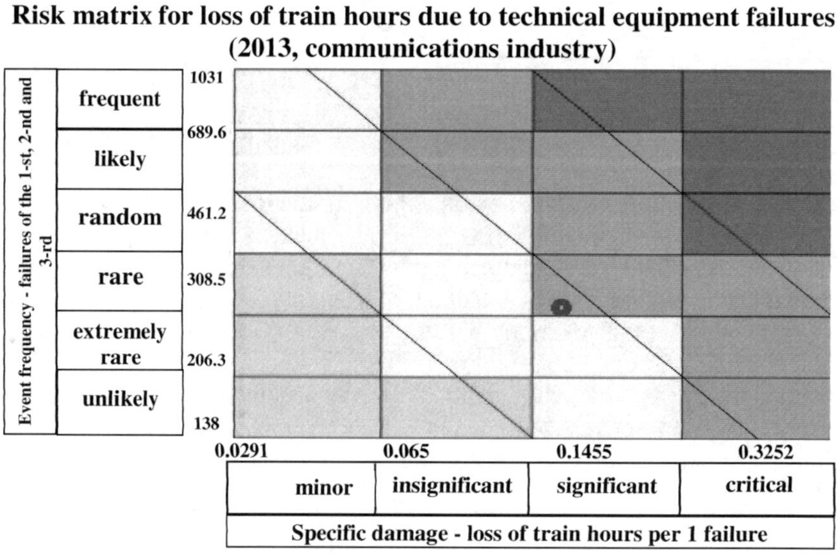

Figure 3. Risk matrix with an expanded scale spacing

Let us consider the differences in the possibilities of the criterion risk assessment, taking into account that the risk category is assigned in accordance with Table 1 [3.4].

We will estimate the category of hypothetical risk consisting in the fact that there are 210 failures with the damage of 0.163 train-hours per year. To do this we shall note the coordinates of the point on both risk matrices. Obviously, whilst assessing risk with a minimum scale spacing matrix (Figure 2), the risk category is permissible. When assessing risk with an expanded scale spacing matrix, the same risk falls into the category undesirable. Calculated category boundaries for both risk matrices are presented in Table 1.

Table 1. The boundaries of the criterion values

Criterion	$R \le R_{perm}K^2$	$R_{perm}/K^2 < R \le R_{perm}/K$	$R_{perm}/K < R \le R_{perm}$	$R \ge R_{perm}$
risk category	risk is not taken into account	permissible risk	undesirable risk	impermissible level of risk
matrix with a scale spacing of 3.34	$R \le 13.4$	$13.4 < R \le 44.9$	$44.9 < R \le 150$	$R > 150$
matrix with a scale spacing of 8.68	$R \le 2.0$	$2.0 < R \le 17.3$	$17.3 < R \le 150$	$R > 150$

Let us consider the differences in the possibilities of the criterion risk assessment, taking into account that the risk category is assigned in accordance with Table 1 [3].

It is obvious that the second matrix misjudges the "permissible and undesirable risk" categories for all risks at the boundary, (17.3; 44.9). The 34.6 risk we are considering falls within this interval. The length of the interval is 27.6, which is about 25% of the length of the interval in the "undesirable risk" category.

The analysis clearly and numerically shows the significance of the fact that risk matrix should be derived from initial data without any preliminary arbitrary scale changes. After all, classifying a risk as "undesirable" means the need for increased attention to risk and the development of measures to reduce the risk to the "permissible" category in many cases.

All of the above forces us to formulate the problem of finding the parameters of the risk matrix in the following statement, which provides the minimum scale spacing for the initial data:

We find such a risk matrix that converting a rectangle with borders (b_{min}, a_{max}), (b_{max}, a_{min}), (b_{min}, a_{max}), (b_{max}, c_{max}) to a rectangle of the risk matrix provides minimal changes in coordinates.

Mathematical record of the last condition:

$$\left|1-\frac{f_{\min}}{a_{\min}}\right|+\left|1-\frac{c_{\min}}{b_{\min}}\right|+\left|1-\frac{f_{\max}}{a_{\max}}\right|+\left|1-\frac{c_{\max}}{b_{\max}}\right| \to \min. \tag{16}$$

In addition to condition (3), the above conditions (6)-(10) must be met. Thus, we come to the problem of minimizing a certain function, which is reduced to the problem of linear programming [8-11] when a number of restrictions are met.

The problem stated above can also be solved using a constructive algorithm. Next we will give one of its branches that works under the condition:

$$R_{perm} > f_{\max} \cdot (c_{\min} \cdot c_{\max})^{1/2}. \tag{17}$$

Table 2 shows the actions necessary to be performed depending on the ratio of parameters. Condition (16) is fulfilled in this case due to minimal changes in the parameters of the initial data rectangle, when it is reduced to the risk matrix.

Symbols used in calculating the parameters of the risk matrix are:

S_a – scale spacing of the initial data by frequency, calculated by the formula:

$$S_a = \left(\frac{a_{\max}}{a_{\min}}\right)^{1/2}; \tag{18}$$

S_b – scale spacing of initial data on damages, calculated by the formula:

$$S_b = \left(\frac{b_{\max}}{b_{\min}}\right)^{3/8}; \tag{19}$$

$S_b^{н}$ – new spacing in the damage scale;

$S_b^{н1}$ – final spacing of the damage scale;

b_0 – center of the initial damage data scale: $b_0 = (b_{min} \cdot b_{max})^{1/2}$;

$b_{\max}^{н}$ – new right limit of the damage scale;

$b_0^{н}$ – new center of the damage scale;

α_1 – coefficient of increase of the right limit of the damage scale;
α_2 – coefficient of increase in the damage scale in both directions;
β_1 – coefficient of reduction of the lower limit of the frequency scale;
β_2 – coefficient of increasing the upper limit of the frequency scale.

Table 2. Branch of the algorithm for calculating the parameters of the risk matrix

Ratio of scale spacings	Action (step 1) Fulfilling the condition (6)	Checking the ratio	Action (step 2): Equalizing of the spacing by scales	The result – the parameters of the risk matrix:
1. $S_a \geq S_b$	$\alpha_1 = \left(\frac{R}{b_0 a_{\max}}\right)^2$ $b_{\max}^{н} = \alpha_1 b_{\max}$ $b_0^{н} = \sqrt{\alpha_1} b_0$ $S_b^{н} = \alpha_1^{3/8} S_b$; explanation: by expanding the right border of the damage range, we achieve equality: $\frac{R}{b_0^{н}} = a_{\max}$	1.1 $S_a > S_b^{н}$	the left border of the damage range decreases, and the right border of the damage range increases proportionally: $y\alpha_2 = \left(\frac{S_a}{S_b^{н}}\right)^{4/3}$; $b_{\max}^{н1} = \alpha_2 b_{\max}^{н}$ $b_{\min}^{н} = \frac{b_{\min}}{\alpha_2}$; $S_b^{н1} = \alpha_2^{3/4} S_b^{н}$	Scale spacing: $K = S_a = S_b^{н1}$; the borders: a_{min}, a_{max}; $b_{\max}^{н1}, b_{\min}^{н}$;
		1.2 $S_a < S_b^{н}$	the lower border of the frequency range decreases: $\beta_1 = \left(\frac{S_b^{н}}{S_a}\right)^2$ $a_{\min}^{н} = a_{\min} / \beta_1$;	scale spacing: $K = S_b^{н}$; the borders: $a_{\min}^{н}$, a_{max}; $b_{\max}^{н}, b_{\min}$;
		1.3 $S_a = S_b^{н}$	no action;	scale spacing: $K = S_b^{н}$; the borders: $a_{\min}$, a_{max}; $b_{\max}^{н}, b_{\min}$;

$S_a < S_b$	$\beta_2 = \frac{R}{b_0 a_{max}}$, $a^{н}_{max} = \beta_2 a_{max}$ $a^{н}_0 = \sqrt{\beta_2} a_0$ $S^{н}_a = \beta_2^{1/2} S_a$; explanation: by expanding the right border of the frequency range, we achieve equality: $\frac{R}{b_0} = a_{max}$	$S^{н}_a > S_b$	the right border of the damage range increases, and the left border decreases: $\alpha_1 = \left(\frac{S^{н}_a}{S_b}\right)^{4/3}$; $b^{н}_{max} = \alpha_1 b_{max}$ $b^{н}_{min} = \frac{b_{min}}{\alpha_1}$; $S^{н}_b = \alpha_1^{3/4} S_b$;	scale spacing: $K = S^{н}_a$; the borders: a_{min}, $a^{н}_{max}$; $b^{н}_{max}$, $b^{н}_{min}$;
		$S^{н}_a = S_b$	no action;	step of the scale: $K = S^{н}_f$; the borders: a_{min}, $a^{н}_{max}$; b_{max}, b_{min}
		$S^{н}_a < S_b$	the left (lower) border of the frequency range decreases: $\beta_1 = \left(\frac{S_b}{S^{н}_a}\right)^2$ $a^{н}_{min} = a_{min} / \beta_1$	scale spacing: $K = S_b$ the borders: $a^{н}_{min}$, $a^{н}_{max}$; b_{max}, b_{min}

In conclusion it should be noted that rejection of the above-mentioned features of the problem statement, careless development of algorithms can result in errors not only in the assessment of significant risks, but also in the regulatory literature.

For example, despite the significance, usefulness and achievements of the document [4], the calculation of the risk matrix parameters was done with mistakes: the ratio (6) was not fulfilled (error – 250%), as a result, the risk related to the "undesirable" category was evaluated as "permissible" by this matrix. Below there is a matrix based on the minimum spacing rule with same initial data that shows that the risk is in the "undesirable" category. By the way, the spacing for the matrix is calculated correctly, if the risk category was evaluated according to Table 1, the result would be correct, hence the risk of 67.1 train hours / year belongs, according to table 1, to the "undesirable" category.

The conducted research allows us to draw the following conclusions.

The definition of a risk matrix with minimum scale spacing is introduced and the example shows that the matrix with minimum scale spacing is a more accurate tool for determining the risk category.

It is shown by examples and numerically that the matrix with the smallest spacing gives better results in comparison with matrices constructed using an arbitrary scale extension.

The problem of finding the parameters of such a matrix from the initial data in a mathematical statement is formulated.

One of the branches of the developed simplified algorithm for constructing such a matrix is given.

Risk matrix for loss of train hours due to technical equipment failures (2013, communications industry)

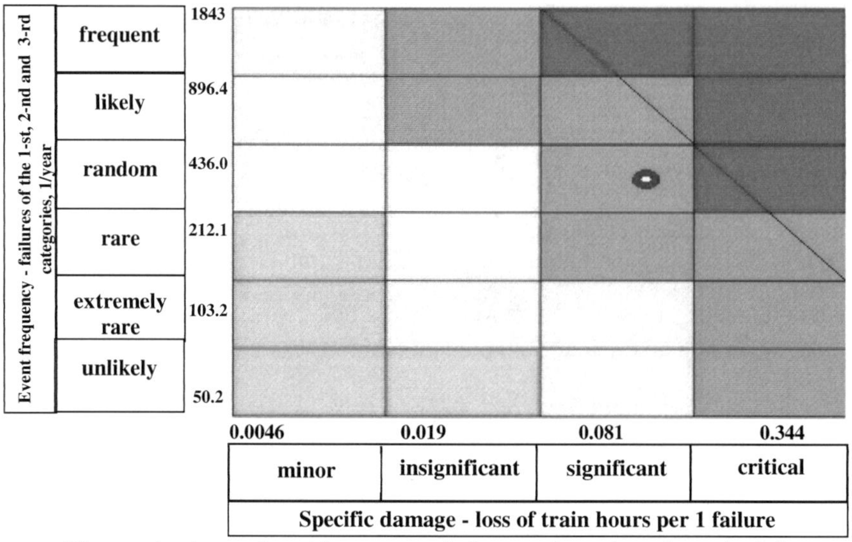

Figure 4. A matrix with an extended range based on the principle of the minimum scale spacing

5. Conclusion

This article reflects the development of ideas about the risk matrix as a tool for interval risk assessment for assigning it to a particular category.

The possibility of using the risk matrix is related to how effectively this tool will work. Therefore, the given method of matrix constructing that most accurately determines the risk category; the constructed algorithm for calculating parameters is relevant. Another aspect of the relevance of the work is the upcoming digitalization of risk assessments in railway transport with the resulting need to have effective algorithms for calculating the parameters of optimal risk matrices.

Reference

[1] Order of JSC «Russian Railways» No. 4246r. 2016 «Methodology for identifying risks in the field of functional safety of train traffic in on the infrastructure of JSC «Russian Railways» for the management of tracks and structures»

[2] *Risk assessment in audit planning. A guide for auditors on how best to assess risks when planning audit work. Internal Audit Community of Practice*, available at: http://www.pempal. org/data/upload/files/2014/09/rap_guide_eng.pdf

[3] 2014 *Methodology of traffic safety risk rationing depending on operational characteristics of dispatching areas: approved by JSC «RZhD»* (Moscow)

[4] Order of JSC «RZhD» No. 1946 2016 *Methodological recommendations for building a risk matrix* (Moscow)

[5] Verevkina O I 2020 Assessment of contribution of human factor and factors of material-technical supply to safety risks due to poor repair and technology *International Conference on Transport and Infrastructure of the Siberian Region (SibTrans-2019) IOP Conference Series: Materials Science and Engineering,* 760, 012065. DOI: 10.1088/1757-899X/760/1/012065

[6] Kudryavtsev L D 2012 *Course in mathematical analysis* (Moscow: Vysshaya shkola)

[7] Ilyin V A Poznyak E G 2007 Fundamentals of mathematical analysis (Moscow: RSHU)

[8] Bazara M, Shetty K 1982 Nonlinear programming. Theory and algorithms (Moscow: Mir)

IOP Conf. Series: Materials Science and Engineering 1151 (2021) 012001 doi:10.1088/1757-899X/1151/1/012001

[9] Lupashko S G 2012 *Solving the problem of linear programming in Microsoft Excel* (Kurgan: Kurgan State University)
[10] Yudin D B, Holstein E G 2010 *Problems and methods of linear programming. Transport-type tasks* (Moscow: Librocom)
[11] Paliy I A 2008 Linear programming (Moscow: Eksmo)

Improving the electric power quality in the networks adjacent to the railroad traction substations

Yu N Bulatov[1], **A V Kryukov**[2,3], **A V Cherepanov**[2]

[1] Bratsk State University, 40, Makarenko st., Bratsk, Irkutsk region, 665709, Russia
[2] Irkutsk State Transport University, 15, Chernyshevskogo st., Irkutsk, 664074, Russia
[3] Irkutsk National Research Technical University, 83, Lermontova st., Irkutsk, 664074, Russia

E-mail: bulatovyura@yandex.ru

Abstract. A significant deterioration in the electric power quality indicators (EPQI) is observed in the 110-220 kV electrical networks, which are directly adjacent to the traction substations of the Trans-Siberian Railway. Such situations are typical of many regions of Siberia and the Far East. The article presents the results of research aimed at improving the electric power quality in the networks adjacent to the traction substations. The simulation algorithms used were based on application of phase coordinates and lattice equivalent circuits. Software tools Matlab and AnyLogic were used to model transition processes and evaluate the effectiveness of the concept of an intelligent electricity quality controlling system. On the basis of the results, technical solutions have been proposed to bring the electric power quality indicators to standard values. It is shown that a comprehensive approach, including the following segments, is needed to address the problem of reduced electric power quality: the use of technology to improve electric power quality indicators; reinforcing 110-220 kV networks that feed traction substations; technologies of smart customization and control of technical means.

1. Introduction.

In the electrical networks that feed traction substations (TS) of the main AC railway lines, the problem of reduced quality of electric power (EP) has become particularly critical. There is a significant deterioration in electric power quality in Siberia and the Far East, because in some regions of these areas the main networks 110-220 kV are directly adjacent to the substations of the Trans-Siberian Railway [1–3]. Abruptly variable, non-linear and single-phase traction loads, as well as reduced short-circuit (SC) capacity levels at the points of connection of individual traction substations, cause the electric power quality indicators in these networks to go far beyond acceptable limits.

The results of the instrumental measurements of the electric power qialuty indicators show that there are significant violations of voltage symmetry and sinusoidality on high voltage buses of Transsib traction substations [2, 3]. The average per day of the voltage unbalance ratios in reverse sequence k_{2U} 9 regularly reach 3... 4%, and the maximum reach 6... 7 %; similar data on the total ratios of harmonic constituents k_U are appropriately equal to: 5... 6 and 8... 10 %.

Electric power quality indicators that go beyond regulatory limits can lead to the following negative consequences: disruption of technological processes; reducing product quality; reducing the performance of machines and mechanisms; increasing electric power losses; reducing the credibility of

Content from this work may be used under the terms of the Creative Commons Attribution 3.0 licence. Any further distribution of this work must maintain attribution to the author(s) and the title of the work, journal citation and DOI.
Published under licence by IOP Publishing Ltd

electric power accounting; reduced electric power reliability due to accelerated aging of insulation and malfunctioning of relay protection and automation devices [4, 5].

In recent years, the Irkutsk State Transport University has conducted comprehensive research aimed at developing multifunctional modelling methods for electric power systems (EPS) and railway electric power supply systems (REPSS) [6]. A unified methodological approach to the construction of models of elements of EPS and REPSS, implemented by lattice equivalent circuits, has been proposed. On the basis of this approach, methods and computer technologies have been implemented, the distinctive features of which are [6]:

- multiphasal nature, providing simulations of single-phase, three-phase, four-phase, six-phase systems, etc.;
- multi-mode nature, which is to be able to determine a wide range of electric power system and railway electric power supply system modes: normal, emergency, asymmetrical and non-sinusoidal, as well as limiting modes on static aperiotic resistance;
- multitasking, consisting of a number of additional tasks, such as the definition of induced voltages on adjacent electric power transmission lines (EPTL) and communication lines; calculating the strengths of the electromagnetic field created by electric power transmission lines and traction networks; parametric identification of electric power transmission lines and transformers based on measurements from PMU WAMS devices; taking into account the active elements of the smart grid when modeling electric power systems and railway electric power supply systems; correct accounting of the surface effect, as well as the proximity effect when modeling current wires with solid buses.

The article presents the results of research aimed at solving problems of improving the electric power quality in the networks adjacent to the traction substations. To improve electric power quality indicators, one can use intelligent electrical network technologies [7–11] and multi-agent control systems [12, 13]. On the basis of the results, technical solutions have been proposed to bring the electric power quality indicators to standard values.

2. The simulation technique.

Railway electric power supply systems are a combination of complex subsystems: the part of the electric power systems adjacent to traction substations; traction power supply system (TPSS); electric power supply regions (EPSR) of non-traction and non-transport consumers. Simulation methods are used to model railway electric power supply system modes; in this case, the study time interval is broken down into small intervals, within which the parameters of the railway electric power supply systems are considered unchanged. A comparison of measurements on real-world objects and computer simulation results [6] shows that the accepted assumption does not make a noticeable error in the results of the calculations.

The algorithm of simulation of the railway electric power supply system is based on the application of phase coordinates and lattice equivalent circuits with fully meshed topology. Details of the simulation of electricity sources, electrical loads, and elements used to regulate the modes of the electric power systems are discussed in work [6]. To model transient processes and form the concept of an intelligent electric power quality controlling system, extra software tools were used, such as Matlab and AnyLogic.

3. Methods for improving electric power quality.

The problem of improving the quality of electric power in the networks adjacent to the traction substations can be solved by building a control system for the railway electric power supply system modes. The global goal of such controlling is uninterrupted energy-efficient and high-quality supply of electric power to electric rolling stock (ERS) and non-traction consumers. To implement it, one needs to solve the following problems:

- ensuring voltage levels at electric rolling stock current collectors required by regulations;
- reducing equalizing currents and losses of electricity in traction networks and EPSR to optimal values;
- compliance of electric power quality indicators at the railway electric power supply systems

and the supply electric power system with regulatory requirements.

To solve the problem of the reduced electric power quality, one can use technical means presented in table 1. It is necessary to note the versatility of the presented technical devices.

Table 1 The main and additional functions of technical means to improve the electric power quality

Technical equipment group	Tasks solved				
	Stabilization of voltage levels in the traction network	Reducing voltage deviations on buses 6 (10, 35) – 110–220 kV traction network	Reducing voltage asymmetry on buses 6 (10, 35) –110–220 kV traction network	Reducing harmonic distortions on buses 6 (10, 35) – 110–220 kV traction network	Flicker in 0.4 kV EPSR networks in the presence of distributed generation plants
Main technical equipment	Controllable single-phase reactive power sources in traction substations	Energy storages [14], reactive power sources controlled phase-by-phase [1]	Single-phase reactive power sources in traction substations and reactive power sources controlled phase-by-phase on buses 6 (10, 35)-110-220 kV of traction substations, DC inserts, energy routers	Active harmonic conditioners [17, 18], passive filters of higher harmonics	Concordantly adjusted regulators of excitation and speed of the generator's rotor
Additional technical equipment	Energy storages, distributed generation plants	Distributed generation plants, DC inserts, energy routers [15]	Distributed generation plants, unbalance-to-balance transformer, balancing devices based on the Steinmetz scheme [16]	DC inserts, energy routers	Energy routers

Voltage fluctuations can be significantly reduced by the use of regulated reactive power sources (RPS). Reducing voltage deviations and fluctuations can be obtained with energy storage that allow improving the non-stationary performance of the traction power supply system mode.

The use of phase-controlled reactive power source makes it possible to lower the voltage asymmetry ratios k_{2U} on the buses of the power voltage of traction substations. The degree of lowering depends on the volume of the train traffic, the short circuit capacity and the condition of the track profile.

One of the most effective means of reducing harmonic distortions are active harmonic conditioners (AHC). The results of simulation of real electricity systems of the main railway lines showed that on the basis of the AHC, installed on the power supply arms of the traction network, levels of non-sinusoidality on the buses of 110-220 kV traction substations can be reduced to standard limits.

Modelling of the prospective modes of the electric power supply system (EPSS) of one of the main railway lines of Eastern Siberia has shown that without the use of means of improving the electric power quality, the maximum values of harmonic components ratios k_U are up to 23%. When placing

IOP Conf. Series: Materials Science and Engineering 1151 (2021) 012002 doi:10.1088/1757-899X/1151/1/012002

blocks of single-phase active harmonic conditioners on the inputs of 27.5 kV traction substations, the maximum of k_U on 220 kV buses did not exceed 2%.

To improve the quality of electric power and to develop the concept of intelligent electrical networks, it is necessary to apply energy routers (ER) in EPSS. These energy routers should be constructed with the use of high-frequency solid-state transformers (Figure 1), which are single-phase high-frequency power transformers in mono-block design, equipped with two active semiconductor bridges.

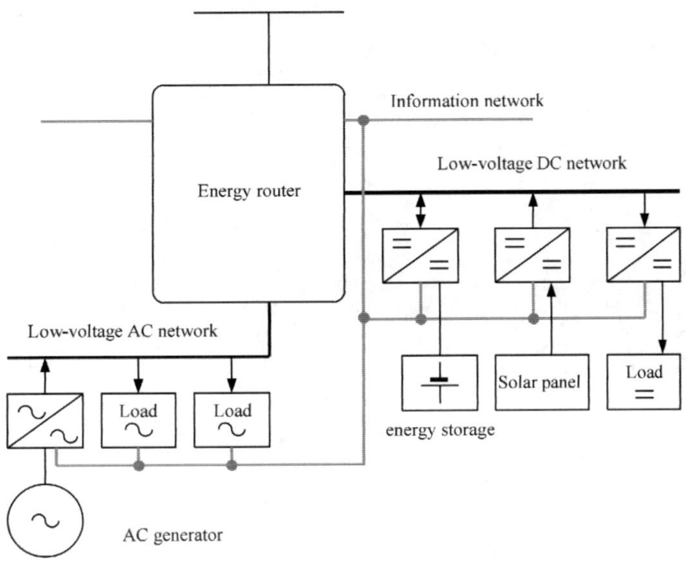

a)

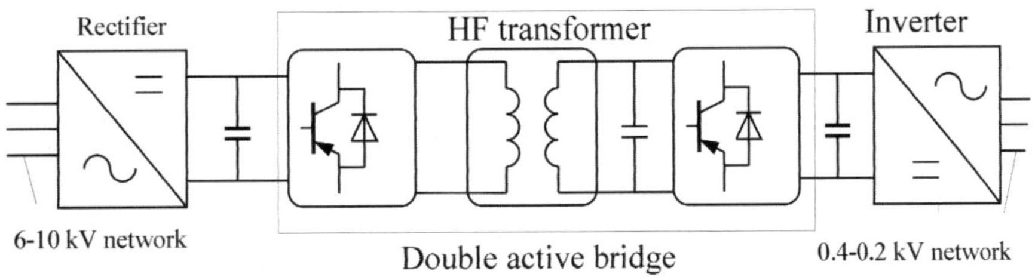

b)

Figure 1. The electric power supply scheme with an energy router (a) and the router scheme built on the basis of a high-frequency transformer (b)

In order to be able to regulate voltage, the inverter, working on the 0.4 kV network, should be provided with an automatic control system. The consumer electric power quality depends on its correct operation. Taking into account the requirements described in the MATLAB system, Simulink and SimPowerSystems developed an energy router model with a voltage regulator. The results of the studies carried out on the railway electric power supply model with distributed generation plants connected through the energy router to the electric power supply region of non-traction consumers showed that when using the energy router, consumers received a quality electric power (Figure 2 b). It lacked the negative characteristics inherent in the networks of railway non-traction consumers. For the example considered on the 10 kV buses of the energy router, the maximum values of ratios k_U, k_{2U} reached 10.7 and 4.2% respectively. On 0.4 kV buses, these values were 0.1 and 0.05 percent (table 2).

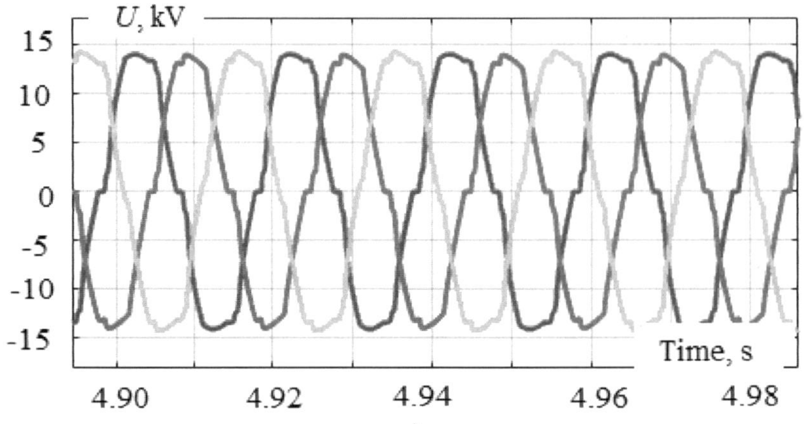

a)

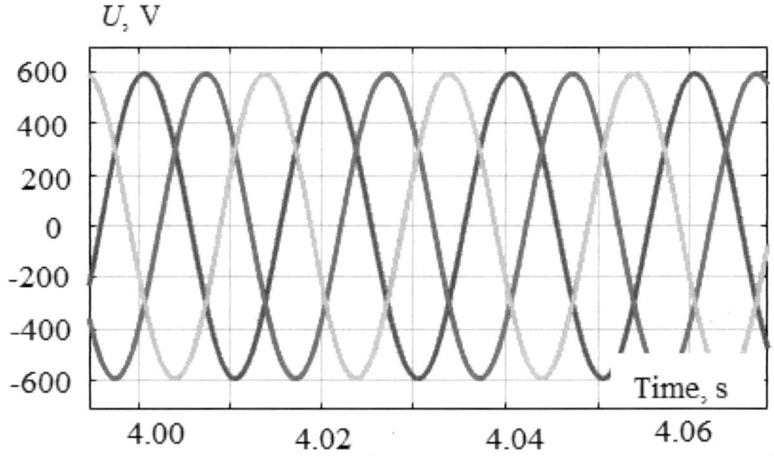

b)

Figure 2. 10 kV bus (a) and non-traction consumer 0.4 kV bus voltage oscillograms in the presence of an energy router (b)

Table 2 Voltage harmonics ratios and asymmetry ratios according to the reverse sequence, %

	Place of measurement	k_{Uab}	k_{Ubc}	k_{Uca}	k_{2U}
1	On 10 kV buses	10.66	9.11	8.74	4.24
2	On 0.4 kV buses	0.09	0.09	0.09	0.05

The simulation results also showed that the ER makes it possible to effectively integrate distributed generation plants into EPSS, including those based on renewable energy sources.

The use of distributed generation plants allows one to unload networks, which helps to reduce electric power losses, improve the reliability of electric power supply to consumers and electric power quality. However, low-power generators can cause voltage fluctuations that lead to a flicker [19–21]. This effect usually manifests itself with a sharp decrease in voltage in the connection node of the distributed generation plant generator [22] and is associated with the interaction of equipment and the dynamic behavior of machines. Studies carried out in the MATLAB system on the electric power supply system model with distributed generation plants have shown that when turbine generators with mismatched excitation and frequency controllers operate, oscillations may occur, leading to flicker in the connection node of the distributed generation plant. The corresponding oscillograms of frequency and voltage, as well as flickermeter readings, are presented in figures 3 and 4.

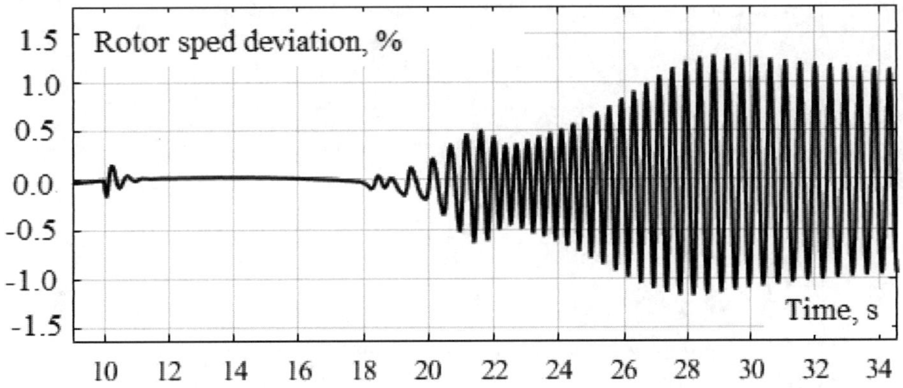

a)

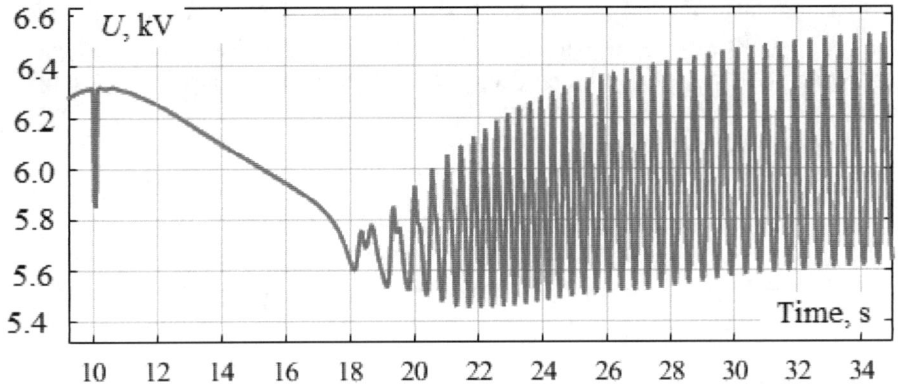

b)

Figure 3. Fluctuations in the rotation rate of the generator rotor (a) and the voltage in the distributed generation plant (b) connecting node when a powerful load is turned on and off

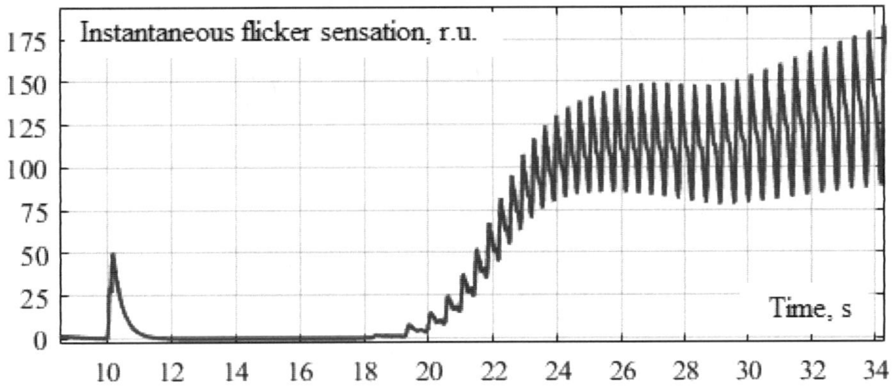

Figure 4. Flickermeter reading

Studies have also shown that the use of predictive control algorithms in harmonized regulators of distributed generation excitation and frequency allows one to eliminate flicker. The corresponding flickermeter readings are presented in Figure 5.

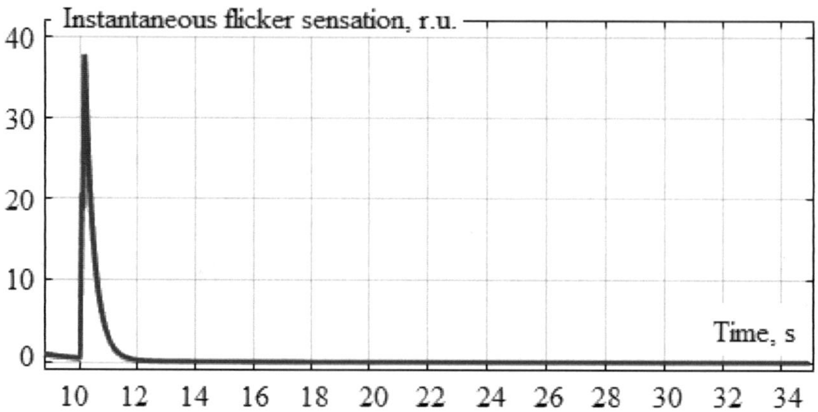

Figure 5. A reading of the flickermeter (a dose of flicker) when enabling and disabling a powerful load using predictive coordinated regulators of distributed generation plant generators

The problem of reduced electric power quality in the railway electric power quality systems can also be effectively solved by applying the concept of intelligent electrical networks (smart grid) and multi-agent technologies [23, 24]. The structural diagram of the multi-agent control system (MACS) by the railway electric power supply system modes is presented in Figure 6.

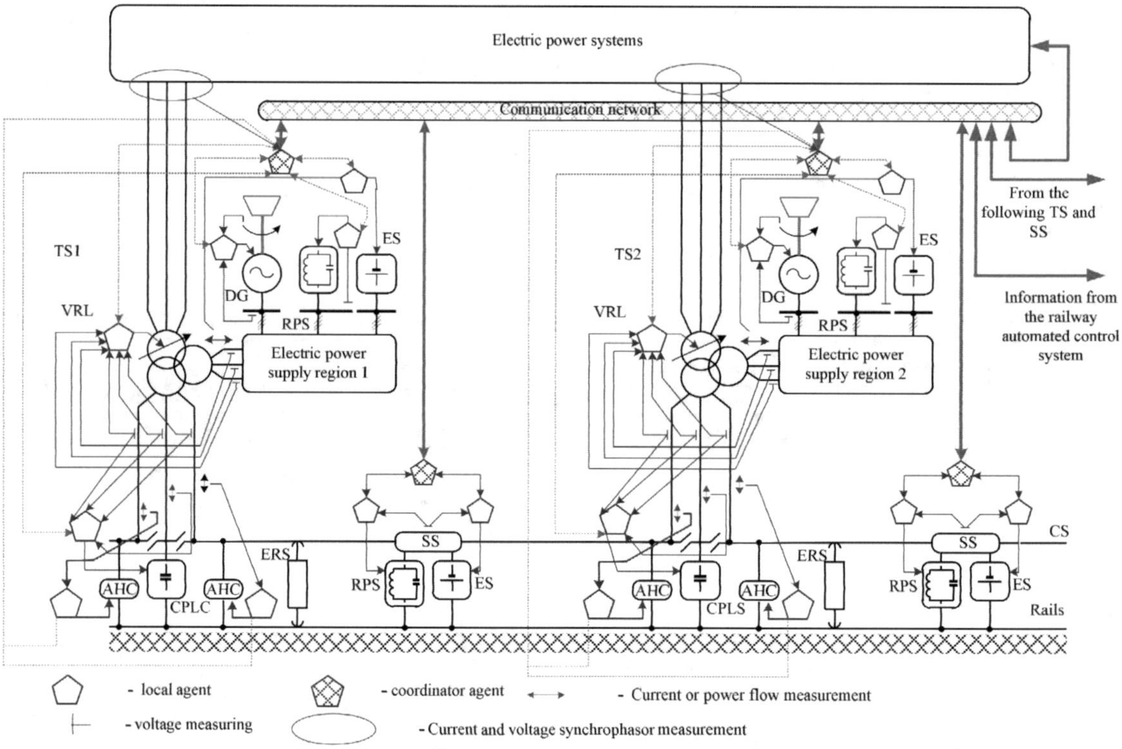

Figure 6. The structural scheme of MACS by the system of railway electric power supply system modes: CS – contact system; ES – energy storage; SS – sectioning station; VRL – a voltage regulator under load; ERS – electric rolling stock; CPLC – controlled plant of longitudinal compensation; RPS – reactive power source; DG – distributed generation

The autonomous MACS agents interacting with each other are partially independent and have limited representation, as each agent does not have complete information about the object of controlling and external environment. The concept of decentralization is also being used; there are no agents who fully control the entire object.

Work [24] provide the results of a simulation of a multi-agent system for controlling distributed generation plants based on synchronous generators, in which agents had the following basic set of properties:

- activity – the ability to organize and implement impacts on the controlled object, as well as on other agents;
- reactivity – perception of an object's condition through sensors and messages from other agents;
- autonomy – relative independence from the environment, as well as the presence of "freedom of will" in some agents, conditioning their own behavior;
- the sociability provided by advanced communication protocols that enables an individual agent to solve its problems together with others;
- purposefulness, which implies the presence of one's own sources of motivation.

The results of MACS simulations of distributed generation plants in electric power supply region, obtained in the AnyLogic environment, showed a 91% reduction in voltage and frequency regulation time and a 10% overregulation, as well as an improvement in electric power quality indicators and transient process quality [24].

4. Conclusion

The problem of reduced quality of electric power in the railway electric power supply systems can be effectively solved on the basis of a comprehensive approach, which includes the following segments:

- the use of technical equipment to improve electric power quality indicators;
- reinforcing of 110-220 kV networks that feed traction substations connected to nodes with reduced capacity of short circuit;
- applying intellectual technologies of customization and control of technical means.

5. Acknowledgments

The research was carried out with the financial support on a grant provided by the Russian Ministry of Science themed as "Improving the quality of electric energy and electromagnetic safety in the electricity supply railway systems equipped with Smart Grid devices by applying methods and means of mathematical modeling based on phase coordinates".

References

[1] Kuro J 2005 Modern technologies for improving the quality of electricity during its transmission and distribution *News of electrical engineering* **1(31)**
[2] Smirnov S S 2010 *Higher harmonics in high voltage networks* (Novosibirsk)
[3] Mustafa G M, Gusev S I, Ershov A M et al. 2015 Calculation of the power of an active filter balancing device to normalize the voltage on the buses of the 220 kV Skovorodino substation *Electric power plants* **3** 46-53
[4] Kartashev I I, Tulsky V N, Shamonov R G et al 2006 *Electricity quality control* (Moscow)
[5] Zhezhelenko I V, Shidlovsky A K, Pivnyak G G et al 2012 *Electromagnetic compatibility of consumers* (Moscow)
[6] Zakaryukin V P and Kryukov A V 2011 *Methods of joint modeling of traction and external power supply systems for AC railways* (Irkutsk)
[7] Wang R, Wang P and Xiao G 2018 *Intelligent Microgrid Management and EV Control Under Uncertainties in Smart Grid* (Springer)
[8] Suslov K, Solonina N and Stepanov V 2015 A principle of power quality control in the intelligent distribution networks, *International symposium on smart electric distribution systems*

and technologies EDST 2015 Proceedings, pp. 260-264.
[9] Wang J, Huang A Q, Sung W, Liu Y and Baliga B J 2009 Smart Grid Technologies *IEEE Industrial Electronics Magazine* **2 (3)** 16-23
[10] Mohsen F N, Amin M S and Hashim H 2013 Application of smart power grid in developing countries. *Proc. 2013 IEEE 7th International Power Engineering and Optimization Conference (PEOCO)*, pp. 427-431.
[11] Buchholz B M and Styczynski Z A 2014 *Smart Grids – Fundamentals and Technologies in Electricity Networks* (Berlin Heidelberg: Springer-Verlag)
[12] DeLoach S A 2009 Moving multi-agent systems from research to practice *Int. J. Agent-Oriented Software Engineering* **3(4)** pp. 378-382.
[13] Miiller J and Fisher K 2013 Application Impact of Multiagent Systems and Technologies. *Springer* pp. 1-26.
[14] Lombardi P, Styczynski Z A, Sokolnikova T and Suslov K 2014 Use of energy storage in isolated micro grids *Power Systems Computation Conference (PSCC)* 1-6
[15] Tiefu Zhao, Liyu Yang, Jun Wang and Huang A Q 2007 270 kVA Solid State Transformer Based on 10 kV SiC Power Devices, *Electric Ship Technologies Symposium, ESTS '07. IEEE* 145-149
[16] Jordi O, Sainz L and Chindris M 2002 Steinmetz system design under unbalanced conditions. *European Transactions on Electrical Power* **12(4)** 283–290.
[17] Zamula K V, Sokolov Yu V and Karmanov A V 2018 Active harmonic filter as a means of improving the quality of electrical energy *Energy of the Unified Network* **2 (37)** 22-32
[18] Boyarskaya N P, Derbenev A M and Dovgun V P 2011 Adaptive control signal generation system for active harmonic filters *Polzunovsky vestnik* **2-1** 25-29
[19] Voropai N I *Distributed Generation in Electricity Systems,* available at: http://www.combienergy.ru /stat/983- Raspredelennaya-generaciya-v-elektroenergeticheskih-sistemah
[20] Barker Ph P and De Mello R W 2000 Determining the Impact of Distributed Generation on Power Systems: Part 1 - Radial Distribution Systems *2000 IEEE PES Summer Meeting,* Seattle, WA, USA. July 11-15, pp. 222-233.
[21] Jenkins N, Allan R, Grossley P, Kirschen D, Strbac G 2000 *Embedded Generation* (London: IEEE)
[22] Sikorski T and Rezmer J 2015 Distributed Generation and Its Impact on Power Quality in Low-Voltage Distribution Networks, Power Quality Issues in Distributed Generation, *IntechOpen*, DOI: 10.5772/61172, available at: https://www.intechopen.com/books/power-quality-issues-in-distributed-generation/distributed-generation-and-its-impact-on-power-quality-in-low-voltage-distribution-networks.
[23] Skobelev P O 2002 Open Multi-Agent Systems for On-line Data Processing in the Course of Decision Making *Avtometriya* **6** 45-61
[24] Bulatov Yu N, Kryukov A V and Suslov K V 2017 Multi-agent technologies for control of distributed generation plants in the isolated power systems *Far East Journal of Electronics and Communications* **17(5)** 1197-1212

Complex use of smart grid technology in railway traction networks

A V Kryukov[1,2], **A V Cherepanov**[1], **I A Lyubchenko**[1]

[1] Irkutsk State Transport University, 15, Chernyshevskogo st., Irkutsk, 664074, Russia
[2] Irkutsk National Research Technical University, 83, Lermontova st., Irkutsk, 664074, Russia

E-mail: and_kryukov@mail.ru

Abstract. The goal is to develop a methodology for modeling electric power supply systems for railways equipped with a complex of devices implemented using smart grid technologies. The results obtained showed that a reliable and high-quality electric power supply for traction of trains and non-traction consumers could be implemented on the basis of the integrated use of Smart Grid active elements, such as a phase converter, an active harmonic conditioner, a controlled reactive power source and a distributed generation plant. On the basis of computer simulation, it has been established that in the absence of reactive power sources, noticeable voltage fluctuations are observed on the 10 kV buses of a non-traction consumer. The asymmetry approaches the limit of normally permissible values; turning off the active filter leads to an increase in the total voltage harmonic distortion up to 16%; in the presence of the whole complex of active devices, high quality of electricity is achieved. The phase number converter is robust and has low sensitivity to parameter setting errors; voltage deviations caused by a limited range of changes in reactive power in its source are short-term and do not exceed values acceptable for practice. Thus, on the basis of Smart Grid technologies, it is possible to implement the connection of distributed generation plants directly to the traction network using a device for transducing the number of phases, formed according to the inverted Steinmetz's circuit. Elimination of harmonic distortions created by rectifier electric locomotives is carried out by means of an active conditioner of higher harmonics. A controlled reactive power source can be used to maintain voltage levels.

1. Introduction

Reliable and high-quality electric power supply of railway transport [1] in modern conditions cannot be provided without the use of solutions based on the concept of intelligent electric technologies (Smart Grid) [2–7]. These technologies are based on the use of active elements, which include the following devices: controlled reactive power sources (RPS), active harmonic conditioners (AHC), distributed generation plants (DGP), phase converters (PhC), etc. Smart Grid technologies can be used to improve the efficiency of power supply for traction of trains and railway infrastructure facilities, especially the most critical consumers [8–19], for example, stations of electrical and dispatch centralization, the operation disruption of which can lead to significant damage.

In the context of digitalization of the transport electric power industry, to use the described technologies in practice, it is necessary to develop computer models of railway electric power supply systems (REPSS), equipped with active Smart Grid elements. These models can be implemented on

Content from this work may be used under the terms of the Creative Commons Attribution 3.0 licence. Any further distribution of this work must maintain attribution to the author(s) and the title of the work, journal citation and DOI.
Published under licence by IOP Publishing Ltd

the basis of methods and means of modeling the railway electric power supply (REPS) modes in phase coordinates, developed at Irkutsk State Transport University [20, 21].

2. Problem formulation

The article presents the results of computer studies that have confirmed the possibility of implementing a railway electric power supply system (REPSS) equipped with a complex of active Smart Grid elements (Fig. 1). To obtain a symmetric three-phase system of voltages, a phase converter was used on the basis of the inverted Steinmetz's circuit (Fig. 2). The reduction of harmonic distortions generated by electric locomotives was carried out using active harmonic conditioner [20, 22]. The controlled reactive power source with reactive power regulation limits of –10 ... + 10 MVAr maintained a given voltage level of 28 kV at the point of connection of the wire-to-ground line with a rated voltage of 25 kV. A distributed generation plant with a capacity of 2 MW was connected to the 10 kV phase converter buses.

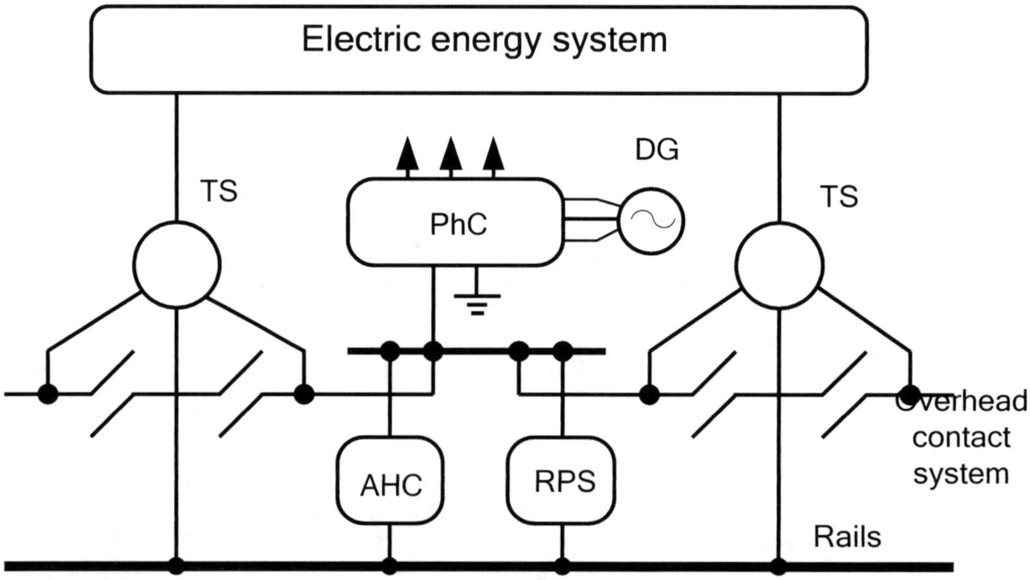

Figure 1. Railway electric power supply circuit: TS - traction substations

The possibility of obtaining a symmetrical three-phase voltage system with only a single-phase source is shown for the circuit shown in Figure 2 a. If the voltage angles are measured in relation to the line voltage between the phases AB, then the following relationships can be written:

$\dot{U}_{AB} = Ue^{j0°}$;

$\dot{U}_{BC} = \underline{a}^2 U$;

$\dot{U}_{CA} = \underline{a}\, U$;

$\dot{U}_{AC} = -\underline{a}\, U$,

where $\underline{a} = e^{j120°} = -0{,}5 + \dfrac{j\sqrt{3}}{2}$.

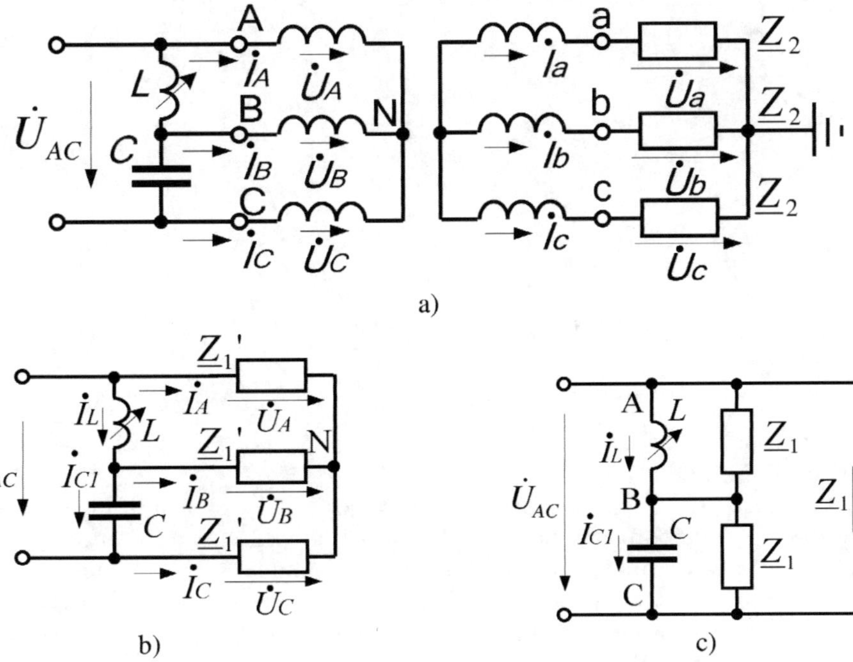

Figure 2. Steinmetz's inverted circuit: a – original; b – equivalent circuit; c – transformed equivalent circuit

Using the transformation ratio kT, you can recalculate the load resistances $\underline{Z}_2 = R_2 + jX_2$ on the primary circuit (Fig. 2 b):

$\underline{Z}_1' = k_T^{\,2}\underline{Z}_2 = R_1' + jX_1'$.

To simplify the notation, the load resistance star should be converted to a triangle (Fig. 2 c):

$\underline{Z}_1 = 3\underline{Z}_1' = R_1 + jX_1$.

Having performed the transformations corresponding to the parallel connection of elements, one can write:

$$\underline{Z}_{AB} = \frac{jX_L\underline{Z}_1}{jX_L + \underline{Z}_1};\ \underline{Z}_{BC} = \frac{-jX_C\underline{Z}_1}{-jX_C + \underline{Z}_1}.$$

The supply voltage determines the currents of phases AB and BC, which are offset relative to each other by an angle of 120º:

$$\dot{I}_{AB} = \dot{I}_{BC} = \frac{-\underline{a}\,U}{\underline{Z}_{AB} + \underline{Z}_{BC}};\ \dot{I}_{AB}\underline{Z}_{AB} = \underline{a}\,\dot{I}_{BC}\underline{Z}_{BC};$$

$$\frac{jX_L}{jX_L + \underline{Z}_1} = \left(-0{,}5 + \frac{j\sqrt{3}}{2}\right)\frac{-jX_C}{-jX_C + \underline{Z}_1}.$$

After separating the real and imaginary parts, one can write:

$$\frac{X_L}{R_1} = \frac{b\sqrt{3} - (b-2)\mathrm{tg}\varphi}{3b};\ X_C = bX_L,$$

where $b = \dfrac{\sqrt{3} - \mathrm{tg}\varphi}{\sqrt{3} + \mathrm{tg}\varphi}$.

The resulting ratios can be represented in terms of powers:

$$P_H = \frac{U^2}{R_1(1+\mathrm{tg}^2\varphi)};$$

$$Q_L = \frac{U^2}{X_L};\ Q_C = \frac{U^2}{X_C};$$

$$\mathrm{tg}\varphi = \frac{Q_H}{P_H};$$

$$\frac{Q_L}{P_H} = \frac{3b(1+\mathrm{tg}^2\varphi)}{b\sqrt{3}-(b-2)\mathrm{tg}\varphi};$$

$$\frac{Q_C}{Q_L} = \frac{X_L}{X_C} = \frac{1}{b}.$$

3. The simulation technique.

Computer simulation of the railway line electric power supply, shown in Figure 1, is performed on the basis of the Fazonord software package [21], which provides simulation of a series of modes formed on the basis of a preset train schedule (Fig. 4 a). The model was built in relation to a double-track section of the main railway. Electric power supply to the traction network was carried out from five traction substations. The electric power supply diagram of a non-traction consumer was presented in detail, including a SWER single-wire power line and a single-phase-three-phase converter (Fig. 3).

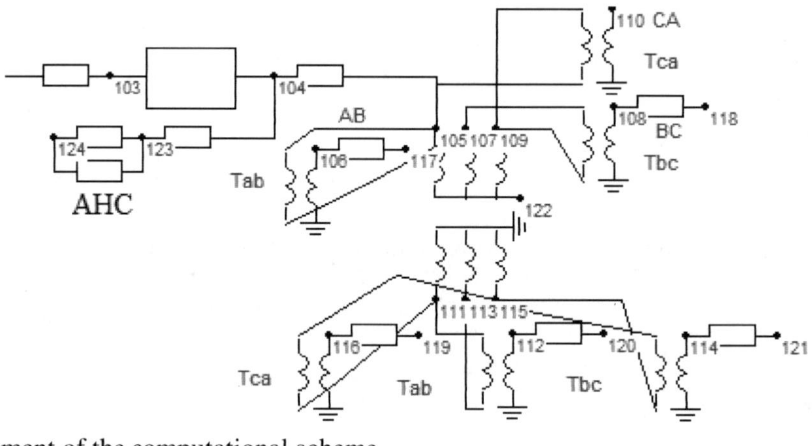

Figure 3. Fragment of the computational scheme

The external network supplying the traction substations is made of 110 kV electric power lines connected at regional substations with a 220 kV system-forming network, which was characterized by the presence of significant generating capacities. The considered railway section corresponds to an increased electric power consumption on ascents and significant recuperation when moving downhill, which is associated with a complex mountain-pass profile, the height differences of which reach 17 ‰.

The electric power supply of the non-traction consumer was carried out from the overhead contact system (CS) via a single-wire line 5 km long, made with an AC-300 wire and connected to the contact system at the parallel connection point. During the simulation, the movement of 40 trains weighing from 2700 to 6000 tons in up and down directions was considered (Fig. 4 a). For example, Fig. 4 b shows the current profile of an up train weighing 6000 t.

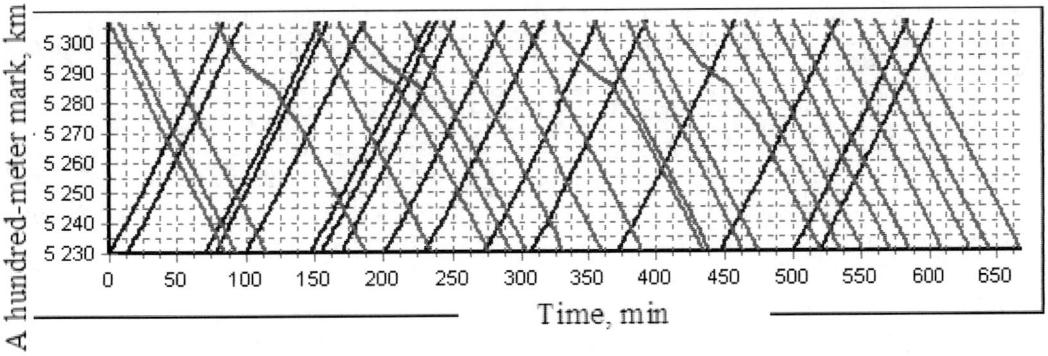

a)

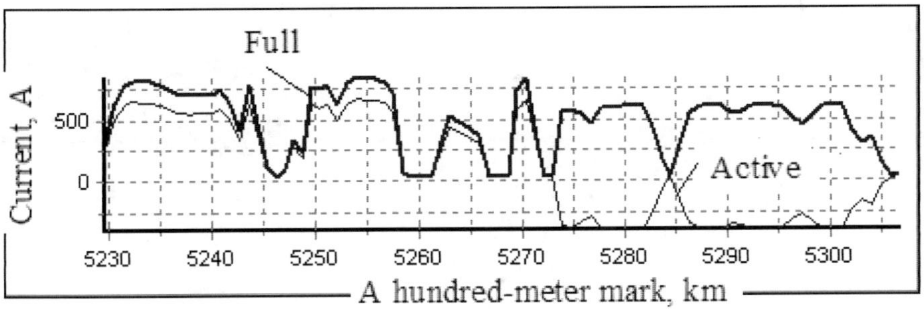

b)

Figure 4. Schedule (a) and current profile (b) for a locomotive weighing 6000 t

Isolation transformers with transformation ratios equal to unity made it possible to obtain the values of line voltages and connect loads in the nodes of the secondary windings.

4. Modeling results.

The circuit shown in Fig. 1 can operate in two modes:

1. Power delivery to the electric traction network, when the total load at the output of the PhC P_H is less than the power P_{DG} developed by the DGP installation.

2. Power consumption from the traction network provided that: $P_{DG} < P_H$.

Below are the simulation results (Tables 1–3, Fig. 5–9) corresponding to the first mode.

Table 1. Negative sequence imbalance coefficient values k_{2U}, %

Mode	$k_{2U\text{mid}}$, %	$k_{2U\text{max}}$, %
Reactive power source is off	1.14	1.92
Reactive power source is on	1.11	1.33
Difference, %	2.53	30.73

Figure 5 illustrates the dynamics of changes in phase voltages on the 10 kV side of the converter, it can be seen from this figure that in the absence of reactive power source, there are voltage fluctuations in the range of 9.5 ... 12.3 kV. When the reactive power source is turned on, the voltages are stabilized. Due to the limited regulating range of the reactive power source, short-term voltage drops are observed, the relative values of which do not go beyond the acceptable limits in practice.

Figure 6 shows the dynamics of voltage changes on the current collector of an up train weighing

6000 tons with the reactive power source switched on and off. The time dependences shown in this figure indicate that due to the inclusion of a controlled reactive power source, a noticeable stabilization of voltage levels is achieved.

Figure 7 shows the graphs of changes in the imbalance coefficient k_{2U} on the 10 kV side of the converter. When the reactive power source is on, the maximum of the coefficient k_{2U} decreases by 30% (see Table 1). Fig. 8 shows similar graphs for the total harmonic distortion. The dependencies shown in Figure 7 are constructed for two options: with the switched on and switched off reactive power source. Similarly, the graphs k_U are presented with the active harmonic conditioner enabled and disabled. The total data for k_U, given in table 2, show that when the active harmonic conditioner is switched on, there are practically no harmonic distortions.

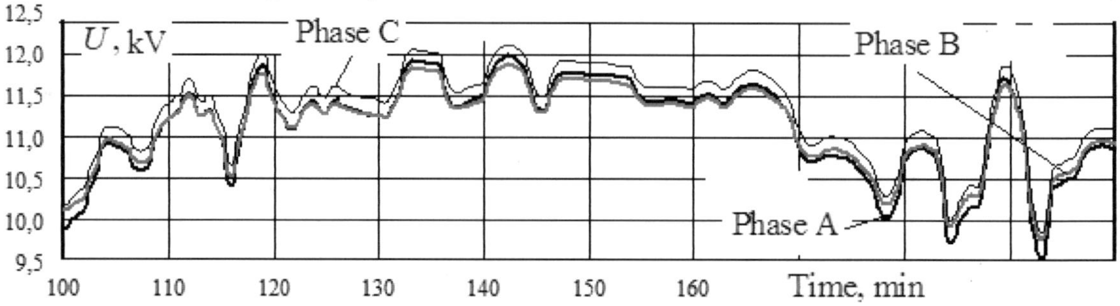

a)

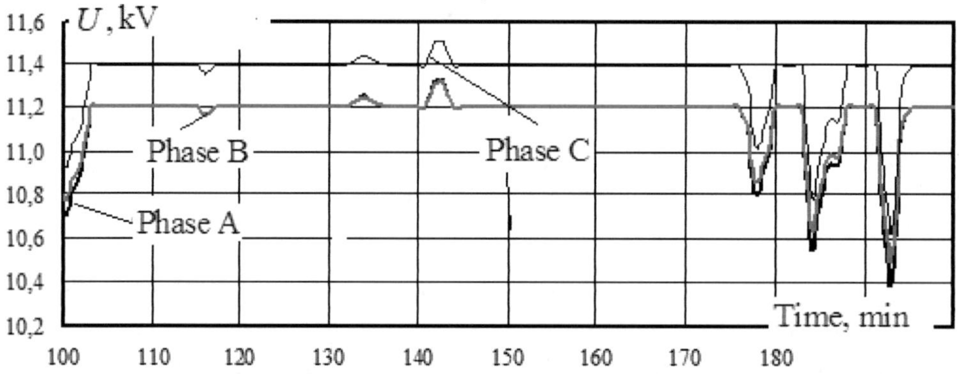

b)

Figure 5. Dynamics of voltage changes on the 10 kV side of Steinmetz's transducer: a – the reactive power source is off; b – the reactive power source is on

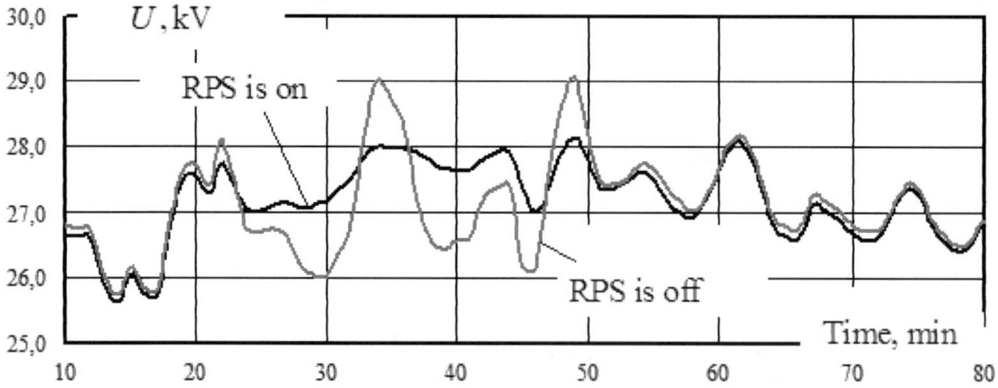

Figure 6. Dynamics of voltage changes at the current collector of an up train weighing 6000 t

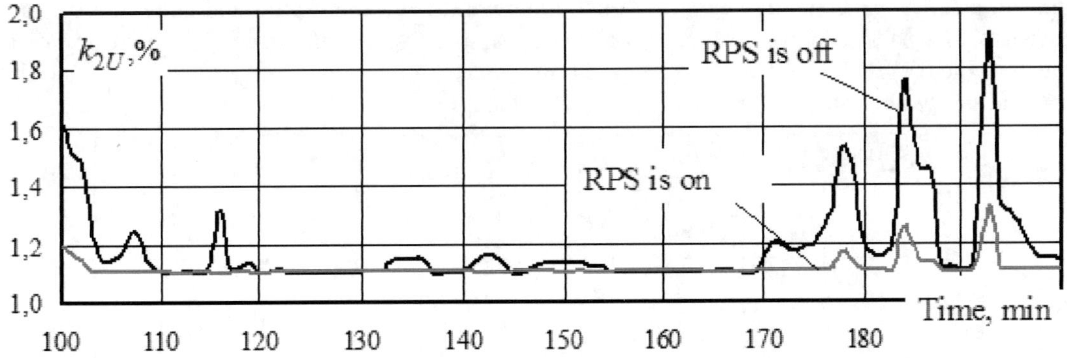

Figure 7. Dynamics of changes in the imbalance coefficient k_{2U} on the 10 kV side of Steinmetz's transducer

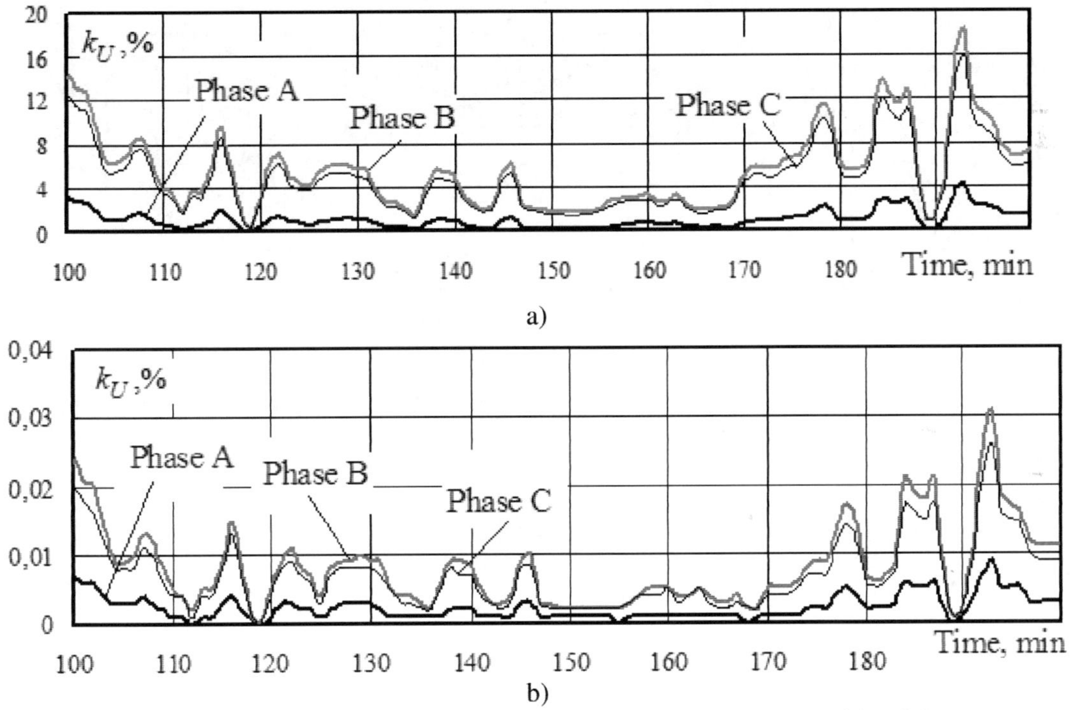

Figure 8. Dynamics of changes in the total harmonic distortion on the 10 kV side of the converter: a – active harmonic conditioner is off; b – active harmonic conditioner is on

Table 2. Values of total harmonic distortions k_U, %

Mode	Indicator	Phase		
		A	B	C
Active harmonic conditioner is on	Average value	0.001	0.005	0.004
	Maximum	0.009	0.031	0.026
Active harmonic conditioner is off	Average value	0.64	3.42	3.00
	Maximum	4.18	18.40	15.88

Figure 9 shows the dependences of the transmitted powers and losses in the SWER line [23]

connecting the phase converter with the traction network; from this figure and table 3, it can be seen that when the reactive power source is on, these parameters are stabilized. It should be noted that instead of the SWER line, a two-wire electric power line can be used if, according to the conditions of electrical safety, the use of SWER technology is limited.

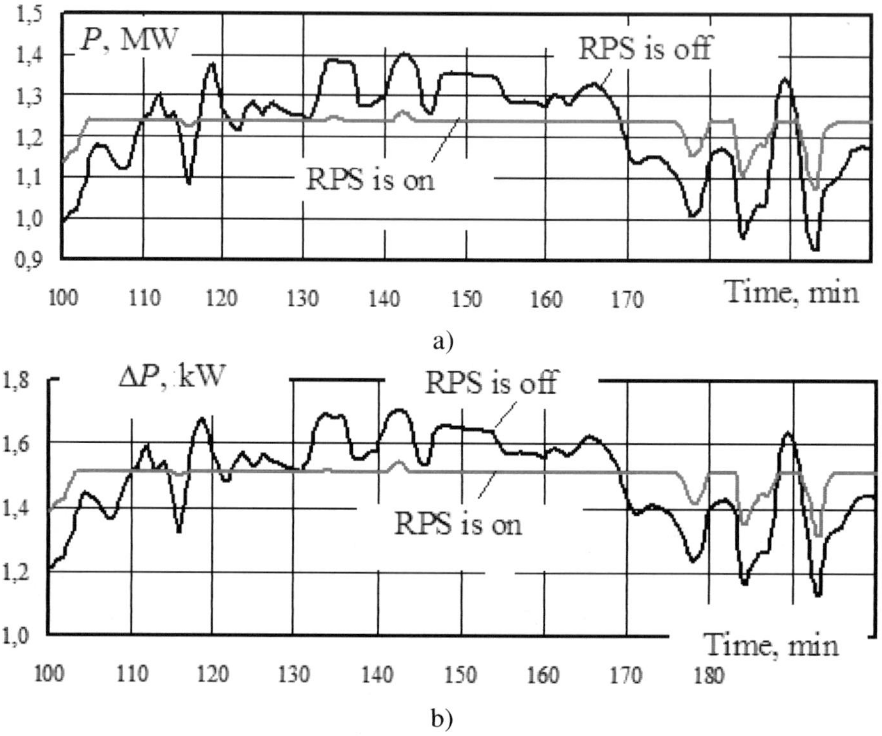

Figure 9. Time dependences of the transmitted power and losses in the wire-to-ground electric power line: a – transmitted power; b – losses

Table 3. Losses in the wire-to-ground electric power line, kW

Indicator	Mode		Difference, %
	Reactive power source is off	Reactive power source is on	
Average value	1.56	1.51	3.73
Maximum	1.71	1.55	10.87

The presented results allow us to formulate the following conclusions:

1. Reliable and high-quality electric power supply of critical railway transport facilities can be implemented through the complex use of active elements, such as a phase converter, active harmonic conditioners and reactive power source; in the absence of reactive power source, there are significant voltage fluctuations on the buses of the substation of a non-traction consumer, as well as asymmetry approaching the limit of the normally permissible value of 2 %; the total voltage harmonic distortion in the absence of an active filter exceeds 18%, and the voltage waveform deviates significantly from the sinusoid.

2. With the presence of the entire system of devices, high quality of electric power is achieved at the consumer's substation.

3. The phase transducer is robust, as it has a low sensitivity to errors in setting parameters.

4. Voltage deviations (see Fig. 5) caused by the limited range of available reactive power of the reactive power source (–10 ... + 10 Mvar) are short-term and do not exceed the limits permissible in

practice.

5. The presence of the reactive power source ensures the maintenance of the required voltage level at the locomotive current collectors

6. The transmission of electricity using a single-wire SWER line is possible over a considerable distance, up to 25 km.

5. Conclusion

On the basis of the complex use of active Smart Grid elements, such as low-power generators controlled by the reactive power source, phase converters and active harmonic conditioners, a reliable and high-quality electric power supply of critical railway transport facilities can be organized. Distributed generation plants and three-phase consumers can be connected to a single-phase traction network using a SWER single-wire electric power line. The controlled source of reactive power ensures the maintenance of the required voltage levels on the current collectors of locomotives and the buses of substations for non-traction consumers. An active harmonic conditioner eliminates harmonic distortions generated by electric rectifier locomotives. Unconventional sources of electricity, for example, wind turbines and micro-hydroelectric power plants, can be used for the actuator of distributed generation plants.

6. Acknowledgments

The research was carried out with the financial support of a grant of the state assignment of the Ministry of Science and Higher Education of Russia (project No. 0667-2020-0039).

References

[1] Dobrynin E V and Efremova I A 2018 Controlling the objects of the traction electricity system in the concept of digital railway *Science and education for transport* **1** 237-240

[2] Huang, Xie G H, Kong W Z, Li Q H 2012 Study on Smart Grid and key technology system to promote the development of distributed generation *IEEE PES Innovative Smart Grid Technologies* https://doi.org/10.1109/ISGT-Asia.2012.6303265

[3] Buchholz B M, Styczynski Z 2014 *Smart Grids – fundamentals and technologies in electricity networks* (Berlin Heidelberg: Springer-Verlag) https://doi.org/10.1007/978-3-642-45120-1

[4] Nejad M F, Saberian A M, Hizam H, Radzi M A M, Ab Kadir M Z A 2013 Application of smart power grid in developing countries *IEEE 7th International Power Engineering and Optimization Conference* https://doi.org/10.1109/PEOCO.2013.6564586

[5] Agüero J R 2012 Applications of Smart Grid technologies on power distribution systems *IEEE PES Innovative Smart Grid Technologies* https://doi.org/10.1109/ISGT.2012.6175536

[6] Wang Jun, Huang Alex Q, Sung Woongje, Liu Yu, Baliga B J 2009 Smart Grid technologies. *IEEE Industrial Electronics Magazine* **3(2)** 16–23

[7] Suslov K V, Stepanov V S, Solonina N N 2013 Smart Grid: effect of high harmonics on electricity consumers in distribution networks *IEEE International Symposium on Electromagnetic Compatibility* (Brugge: Institute of Electrical and Electronics Engineers) 841–845

[8] Tretyakov E A 2014 Researching the components of the intelligent electricity supply system for non-traction consumers *Innovative projects and technologies in education, industry and transport.* Omsk, pp. 76-83.

[9] Tretyakov E A 2014 Multi-agent controlling of the distribution of electric energy in the electricity supply system of non-traction railway consumers. *Science and education in the 21st century: theory, practice, innovation.* Moscow, pp. 45–48.

[10] Tretyakov E A *Improving the efficiency and throughput capacity of the electric power supply system through the controlling of modes and the introduction of modern equipment and materials* **3-1** 139–142 (Belgorod)

[11] Tretyakov E A 2015 Regulation of the mode parameters in the system of electricity supply to non-traction railway consumers *Omsk scientific herald. Series: Society. History. Modern times* **2** 155–

159
[12] Tretyakov E A 2013 Efficiency of electricity storage in railway distribution networks. Y*oung Russia: promoting advanced technologies into industry* **2** 347-349 (Omsk)
[13] Tretyakov E A and Malysheva N N 2010 Modeling the established modes of the electric power supply system of non-traction consumers *Improving electromechanical energy converters* pp. 54–60 (Omsk)
[14] Valyakhmetova V K, Nikolaev V L and Vlasova V A 2018 Improving the reliability of electric power supply to non-traction railway consumers *Modern innovations in science, education and technology,* pp. 27-29 (Moscow)
[15] Shelomentsev A O and Kosyakov A A 2012 Improving the organization of the designing of electricity systems for non-traction consumers in railway transport and distribution networks of public service. *Innovative transport,* **5,** 15-19
[16] Drobov A V, Galushko V N, Alferov A A and Alferova T V 2017 Determining the energy efficiency of electrical equipment of non-traction railway consumers by simulation modeling during designing *Herald of Sukhoy Homel State Technical University* **2** 95–105
[17] Mukharyamov R I, Dobrynin E V and Okladov S A 2015 Automation of control of the current state of the electricity power supply system of non-traction railway consumers *Science and education for transport* **1** 136-138
[18] Ozhiganov N V 2012 Improving the quality of electricity for the railway automation and telemechanics *Automation, communication, computer science* **1** 22–26
[19] Avilov V D, Tretyakov E A and Krause A V 2013 Controlling electricity quality in the distribution networks of railway transport. *Omsk scientific herald. Series: Electrical engineering. Power industry* **1** 183–187
[20] Zakaryukin V P, Kryukov A V and Cherepanov A V 2014 Static models of active harmonic conditioners *Modern technologies. System analysis. Modeling* **3(43)** 48-55
[21] Zakaryukin V, Kryukov A and Cherepanov A 2017 Intelligent traction power supply system *International Scientific Conference Energy Management of Municipal Transportation Facilities and Transport EMMFT 2017. Advances in Intelligent Systems and Computing,* Vol. 692, Pp. 91–99 https://doi.org/10.1007/978-3-319-70987-1_10
[22] Ermolenko A V and Ermolenko D V 2009 Prospects of the use of modern active filtration devices to normalize the quality of electric energy in the AC electric traction system *Bulletin of Research Institute of Railway Transport* **5** 7-12
[23] Bakkabulindi G, Hesamzadeh M R, Amelin M, Da Silva I P and Lugujjo E 2011 A heuristic model for planning of single wire earth return power distribution *IASTED Power and Energy Systems and Applications: systems: Proceedings of International Conference* (Pittsburgh, 7–9 November) pp. 215–222

Simulation of induced voltages created by the 25 kV traction network across the pipeline in ice melting mode

A V Kryukov[1,2], A V Cherepanov[1], A E Kryukov[1], A D Stepanov[1]

[1] Irkutsk State Transport University, 15, Chernyshevskogo st., Irkutsk, 664074, Russia
[2] Irkutsk National Research Technical University, 83, Lermontova st., Irkutsk, 664074, Russia

E-mail: and_kryukov@mail.ru

Abstract. Long metal structures and facilities can be located along the routes of electrified railways with 25 kV traction networks. A typical example of such facilities is pipelines designed to transport liquid or gaseous products. Due to the electromagnetic effects of electric traction networks, induced voltages can occur on structural parts, which are dangerous for the operating personnel. A set of special measures is needed to improve electrical safety and protect people from the influence of induced voltages. In the context of digitalization of the electric power industry, the choice of such measures requires the use of computer technologies, which can be implemented on the basis of methods and tools for simulation of railway power supply systems developed in Irkutsk State Transport University. The article presents the results of research aimed at the development of computer models designed to determine the electromagnetic effects of electric traction networks on the main ground-laid pipeline in the modes of ice melting.

1. Introduction.

etal structures and facilities along electrified railway tracks can be located in the areas of noticeable electromagnetic influences of electric traction networks (ETN). An example of such facilities is pipelines [1-3] designed to transport liquid or gaseous products. Due to the electromagnetic influences of traction networks, induced voltages [4-17] may occur on the parts of their structures, which are dangerous for operating personnel.

Special measures are used to protect people from the effects of induced voltages. In the context of digitalization of the electric power industry, the choice of such measures should be based on computer technologies which can be implemented on the basis of methods and tools for simulation of railway power supply systems developed in Irkutsk State Transport University [18].

The induced voltages at certain points of the grounded metal structure are determined mainly by the mechanism of magnetic influence; at the same time, the voltage levels depend on currents flowing through the wires of the electric traction network. Significant currents in the wires of contact suspensions take place in short circuit modes (SC) and ice melting modes. Damaged sections with short circuits are quickly turned off by relay protection, so the degree of negative impact of voltages induced in such modes is small. Ice melting is carried out for a long time (up to several hours), so in these situations the probability of electrical injuries increases due to exposure to induced voltages.

Below are the results of research aimed at implementing the method of computer simulation of

Content from this work may be used under the terms of the Creative Commons Attribution 3.0 licence. Any further distribution of this work must maintain attribution to the author(s) and the title of the work, journal citation and DOI.
Published under licence by IOP Publishing Ltd

induced voltages on the ground-laid pipeline in the modes of melting ice on the wires of the electric traction network.

2. Simulation technique

The methods of modeling the electric traction network modes developed in Irkutsk State Transport University [18] allow one to implement a systematic approach to the modeling of the electromagnetic influences of electric traction networks on long metal structures. The consistency is ensured by the following circumstances:

1. Induced voltages are determined on the basis of calculating the mode of a complex power supply system, which includes a multi-wire electric traction network and a high-voltage network of the power supply system adjacent to traction substations (TS).

2. The simulation takes into account all significant factors that affect the levels of electromagnetic influences of the traction network: harmonic distortions in the currents and voltages of the traction network, conductivity of the ground, the presence of stationary and distributed groundings of the metal structure; in this case, the values of specific conductivity of the ground can be differently set on separate sections of the approach line of the traction network and the analyzed structure.

3. The algorithms used for calculating the mutual inductive resistances work correctly in the near, intermediate and far zones of Carson's integral [19].

4. The technique is universal and allows one to determine the induced voltages for traction networks of different constructive design.

5. The approach of the metal structure with the traction network can be carried out along parallel, non-parallel and complex trajectories.

3. Simulation results

The simulation was carried out in relation to the diagrams shown in Figure 1.

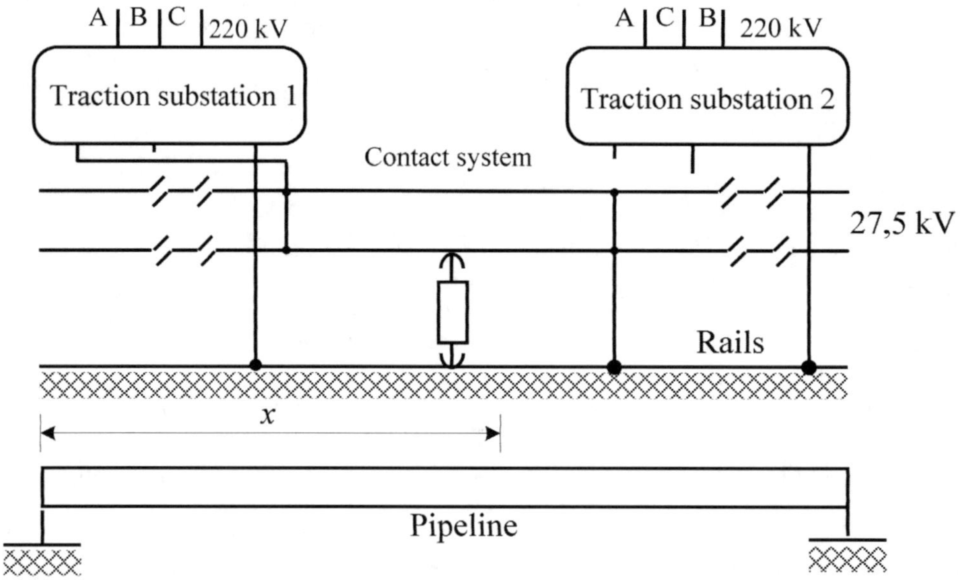

a)

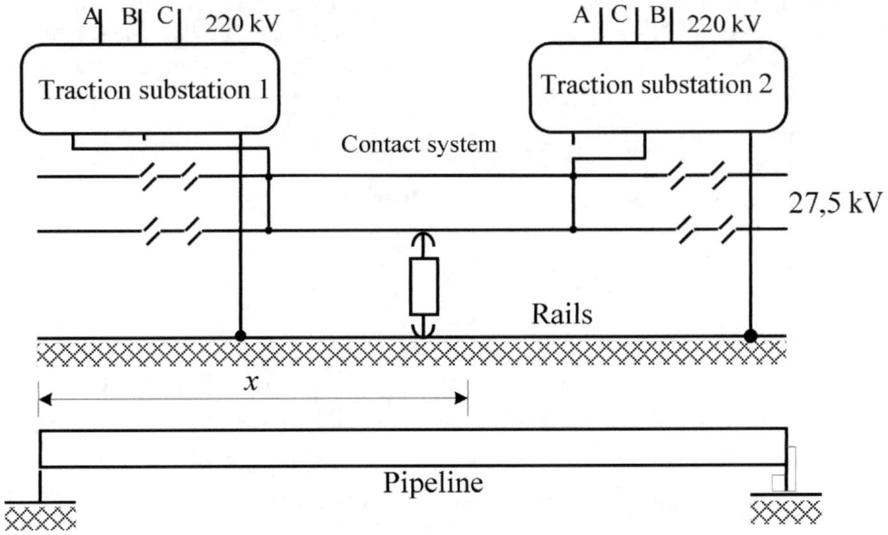

b)

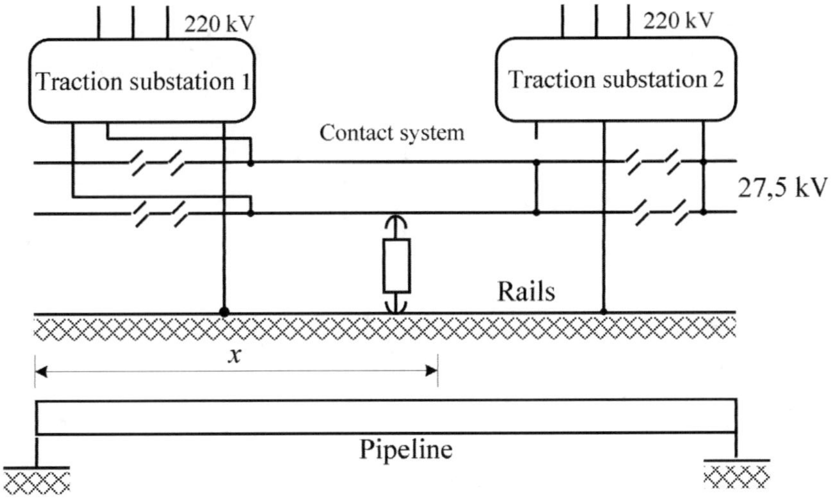

c)

Figure 1. Ice melting diagram: a – by artificial short circuit (SC); b – by feeding contact suspensions (CS) from different phases of adjacent traction networks; c – by feeding contact suspensions from different phases of the same traction network

During the simulation it was assumed that a surface steel pipeline with a pipe diameter of 250 mm was laid parallel to the railway track at a distance of 100 m. The coordinates of the conductive parts are shown in Figure 2. The simulation was carried out on the basis of the Fazonord software complex. Computational schemes, fragments of which are represented in Figure 3, included models of the following segments of the electric power supply system:

- a high-voltage network adjacent to traction substations, made by 220 kV electric power transmission lines;
- traction transformers with a rated power of 40 MVA;
- electric traction networks of two inter-substation zones (ISZ) with a length of 50 km.

The inter-substation zone model, which provided approach to the pipeline, was divided into five sections 10 kilometers long. At the ends of the pipeline, the presence of stationary groundings with a spreading resistance of 1 Ohm was assumed; in addition, the distributed grounding of the pipe with a specific conductivity of 0.05 reciprocal Ohm/km was taken into account. The specific conductivity of the ground on the approach line was set being equal to 0.01 reciprocal Ohm/m and was accepted as the same for all sections of the traction network.

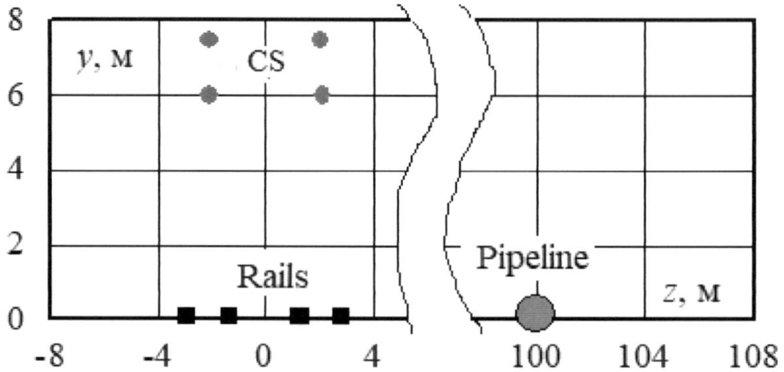

Figure 2. Coordinates of conductive parts

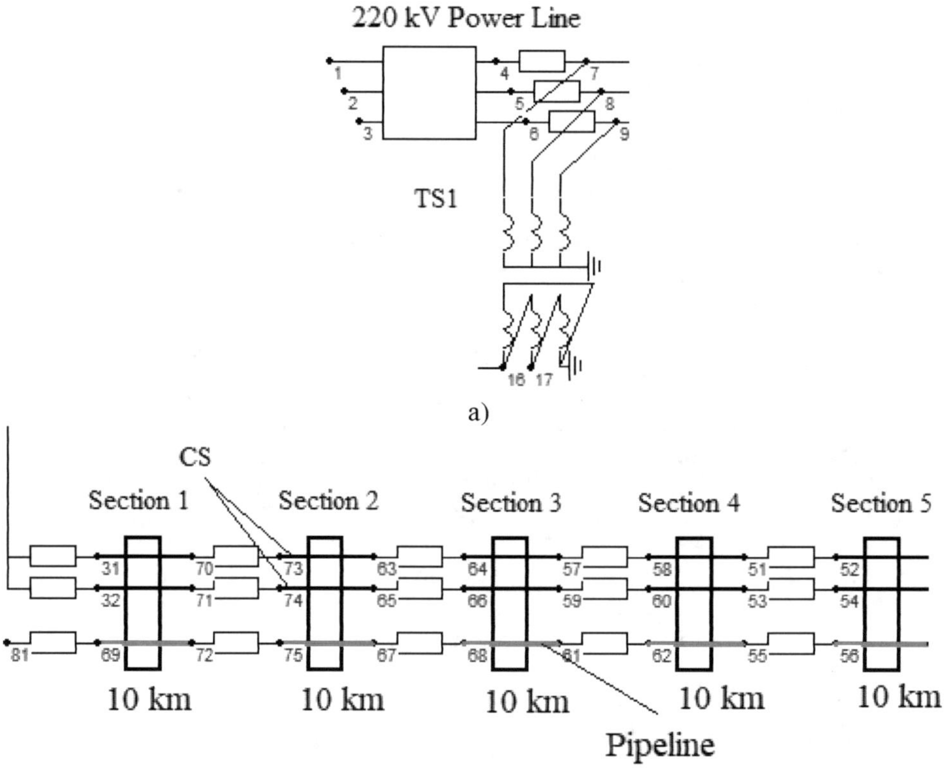

a)

b)

Figure 3. Fragments of diagrams of design models: a - external network and traction substation 1; b - overhead contact system and pipeline

The results of modeling are shown in Tables 1–3 and are illustrated in Figures 4-7. Due to artificial short circuit or multi-phase connection, significant currents (Table 1, Figure 4), sufficient for their preventive heating in order to avoid the formation of ice, flowed along the wires of contact suspensions. The currents that flowed in the rails are presented in Table 2 and are shown in the graph presented in Figure 5.

Table 1. Results of the calculation of currents and voltages

Scheme	Track	Conductive part	Current		Voltage	
			Module, A	Phase, deg	Module, A	Phase, deg
Artificial short circuit	Down	Cwire	465.54	−81.08	20.44	−5.86
		CC	251.55	−49.91	20.44	−5.86
	Up	Cwire	464.84	−81.10	20.44	−5.86
		CC	251.54	−49.86	20.44	−5.86
	–	Pipe	85.98	107.28	0.09	−72.72
Multi-phase power supply from two traction substations	Down	Cwire	393.58	−145.93	24.16	10.93
		CC	212.58	−114.78	24.16	10.93
	Up	Cwire	393.11	−145.98	24.16	10.93
		CC	212.65	−114.76	24.16	10.93
	–	Pipe	72.72	42.44	0.07	−137.56
Multi-phase power supply from one traction substation	Down	Cwire	493.91	164.92	26.20	−57.46
		CC	276.60	−161.46	26.20	−57.46
	Up	Cwire	496.09	−14.56	27.60	-4.43
		CC	277.48	18.94	27.60	−4.43
	–	Pipe	2.10	9.40	0.00	−170.60

Note: Cwire - contact wire; CC - carrying cable; the data correspond to the beginning of the first section of the traction network adjacent to traction substation 1.

Table 2. Results of calculation of currents flowing along the rails

Rail number	Artificial short circuit		Multi-phase power supply from two traction substations		Multiphase power supply from one traction substation	
	Current, A	Current, deg	Current, A	Current, deg	Current, A	Current, deg
1	234.16	114.49	197.98	49.61	23.69	−158.47
2	220.48	111.96	186.42	47.08	11.70	−160.50
3	220.38	111.94	186.33	47.07	9.71	11.36
4	233.88	114.46	197.73	49.58	21.65	17.58

Note: The data correspond to the beginning of the first section of the traction network adjacent to traction substation 1.

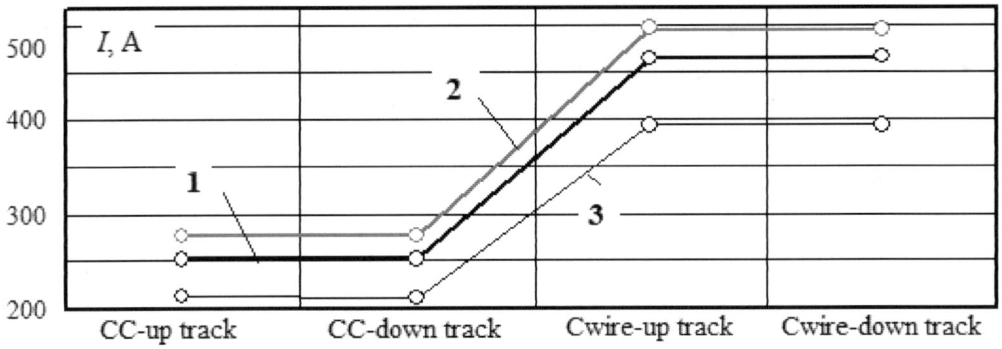

Figure 4. Contact suspension wire currents at the beginning of the first section of the traction network: 1 – artificial short circuit; 2 – multi-phase power supply from one traction substation; 3 – multi-phase power supply from two traction substations

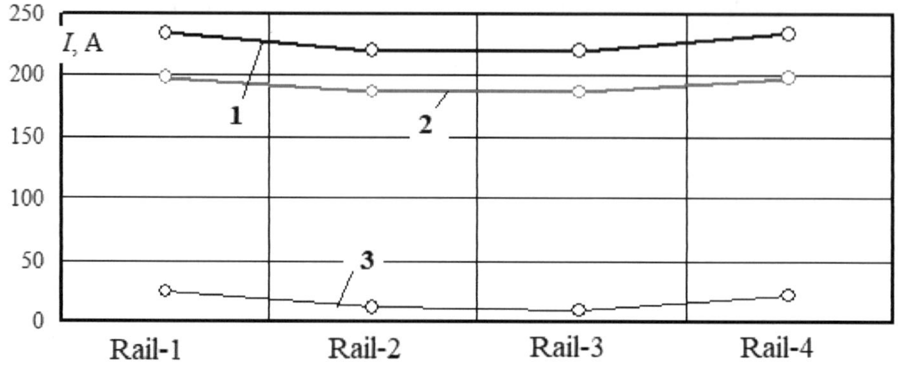

Figure 5. Rail currents at the beginning of the first section of the traction network: 1 – artificial short circuit; 2 – multi-phase power supply from two traction substations; 3 – multi-phase power supply from one traction substation

The results of determining the induced voltages are summarized in Table 3 and illustrated in the graphs in Figures 6 and 7.

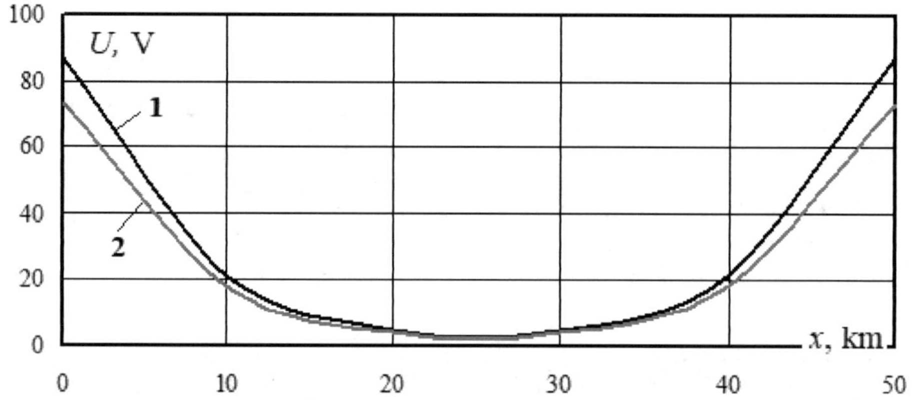

Figure. 6. Dependences of induced voltages on the coordinate x:
1 – artificial short circuit; 2 – multi-phase power supply from two traction substations

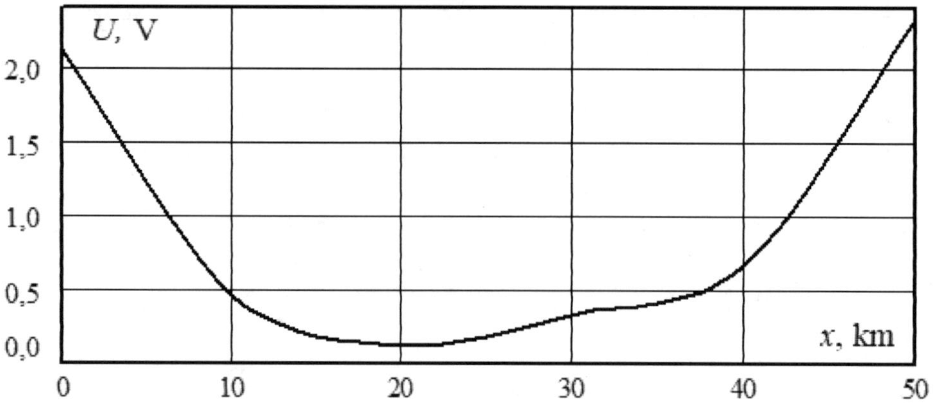

Figure. 7. Dependences of induced voltages on the coordinate x with multi-phase power supply from one traction substation

Table 3 Induced voltages across the pipeline

Scheme	Coordinate x, km					
	0	**10**	**20**	**30**	**40**	**50**
Artificial short circuit	86.8	21.2	4.9	4.6	21.7	87
Multi-phase power supply from one traction substation	2.12	0.46	0.12	0.33	0.67	2.34
Multi-phase power supply from two traction substations	73.5	17.7	3.9	4.1	18.5	73.5

The results presented above allow us to formulate the following conclusions:

1. In the considered modes, significant currents flow through the contact suspension wires; the highest currents of the contact wire, reaching 496 A, are observed with a multi-phase power supply from one traction substation; the current of the carrying cable in this situation is equal to 277 A (Table 1, Fig. 4); significantly lower currents take place in the mode of a multi-phase power supply from adjacent traction substations (394 A for Cwire and 213 A for CC).

2. In multi-phase power supply schemes, only induced currents flow in the rails; with that, in Fig. 1c these currents do not exceed 24 A, which is associated with the phase opposition of currents flowing through the contact suspension of separate tracks (Figure 5, Table 2).

3. The currents, induced due to magnetic influence in the pipeline, have the greatest values in the schemes of artificial short circuit and multi-phase power supply from adjacent traction substations: 86 and 73 A, respectively. During multi-phase power supply from one traction substation, the phases of currents of contact suspension of separate tracks vary by 180° (Table 1); therefore, the induced current in the pipeline is reduced to 2 A.

4. At the inputs of traction substations, voltages of 20 ... 28 kV are maintained (Table 1), which corresponds to the permissible limits.

5. Levels of induced voltages exceed the allowable values of 60 V [20] in the modes of artificial short circuit and multi-phase power supply from adjacent traction substations (Figure 6). Such voltages are observed at the ends of the simulated section of the structure. In the mode of multi-phase power supply from one traction substation due to the phase opposition of currents flowing through the contact suspension of separate tracks, the levels of induced voltages do not exceed 2.5 V.

4. Conclusion

The presented technique and developed computer models can be used in practice when planning measures to ensure electrical safety in ice melting modes on the wires of contact suspensions.

Acknowledgments

The research was carried out with the financial support on a grant of the Russian Ministry of Science themed as "Improving the quality of electric energy and electromagnetic safety in the electricity supply railway systems equipped with Smart Grid devices by applying methods and means of mathematical modeling based on phase coordinates".

References

[1] Borodavkin P P and Berezin V L 1987 *The construction of main pipelines* (Moscow)

[2] Zinevich A M, Glazkov V I and Kotik V G 1975 *Protection of pipelines and tanks against corrosion* (Moscow)

[3] Tretyakova M V and Furkin A V 2011 Assessment of the hazardous interference of currents on power lines on underground pipelines *Rassokhin readings.* (Ukhta) pp. 320-325.

[4] Yabluchansky A I 2008 Methods of assessing the dangerous effects of high-voltage electric power transmission lines on the gas pipeline currently being designed *Materials of the industry meeting on corrosion protection.* pp. 110-123 (Moscow)

[5] Zakharov D B, Piont D Yu and Yabluchansky P A 2018 The assessment of the impact of a high-voltage electric power transmission line on the underground pipeline and its protection from influence of the induced alternating current *Gas Industry* **9(774)** pp. 84-90

[6] Zakharov D B, Yabluchansky P A and Titov A V 2013 On the assessment of the corrosion impact of electric power transmission lines on an underground pipeline at their intersection *"NEFTEGAZ" Territory* **12** 68-74

[7] Kotelnikov A V and Kosarev A B 1992 Electromagnetic effect of AC electrical traction networks on metal structures *Electricity.* **9** 26-34

[8] Strizhevsky I V and Dmitriev V I 1967 *Theory and calculation of the influence of an electrified railway on underground metal structures* (Moscow)

[9] Cherepanov A V and Kryukov A E 2019 Determination of electromagnetic effects of electric traction networks on pipelines *IOP Conference Series: Materials Science and Engineering* **760** DOI: 10.1088/1757-899X/760/1/012014.

[10] Ratner M P 1966 *Inductive influence of electrified railways on electrical networks and pipelines* (Moscow)

[11] Konnova E I and Kosarev A B 1990 Calculation of the electromagnetic effect of AC networks on metal supply lines *Bulletin of Research Institute of Railway Transport* **2** 17-19

[12] Caulker D, Ahmad H and Mohamed Ali M S 2008 Effect of lightning induced voltages on gas pipelines using ATP-EMTP program *IEEE 2nd International Power and Energy Conference* (IEEE)

[13] Ahmed S. Al Shahri, Nirmal-Kumar C. Nair 2016 Overhead shielding wire and induced potential on pipeline located close to high voltage transmission *Australasian Universities Power Engineering Conference (AUPEC)* (IEEE)

[14] Chuco Paucar B, Roel Ortiz J L, Pereira Pinto J O, Koltermann P I 2007 Induced Voltage on Gas Pipeline With Angle Between a Transmission Line, *2007 IEEE Lausanne Power Tech* (IEEE)

[15] Ahmed S AlShahri, Minh Thi Nguyet Dinh, Nirmal Kumar C. Nair 2014 Induced voltage on pipeline located close to high voltage lines due to electromagnetic induction, *2014 Australasian Universities Power Engineering Conference (AUPEC)* (IEEE)

[16] El-Sayed Mohamed El-Refaie, AbdEl-latif Sayed Ahmed, Sabry Mousa Mohamed, Hossam Mohamed Gaber 2017 Electromagnetic field interference between high voltage transmission lines and nearby metallic gas pipelines *2017 Nineteenth International Middle East Power Systems Conference (MEPCON)* (IEEE)

[17] Nagat M K Abdel-Gawad, Essam M Shaalan, M M F Darwish, Mahmoud A M Basuny 2019 Influence of Fault Locations on the Pipeline Induced Voltages Near to Power Transmission Lines *21st International Middle East Power Systems Conference (MEPCON)* (IEEE)

[18] Zakaryukin V P and Kryukov A V 2015 The refined method of determining the mutual electromagnetic influences of adjacent electric power transmission lines P*roceeedings of universities. Problems of power economy* **3-4** 29-35

[19] Carson I R 1926 Wave propagation in overhead wires with ground return *Bell System Techn. J.* **5** 539–554

[20] Technische Richtlinien-71 (TRL-71). *EMR-Technic Kathodischer Korrosionsschutz fur Erdgasfernleitungen*

IOP Conf. Series: Materials Science and Engineering 1151 (2021) 012005 doi:10.1088/1757-899X/1151/1/012005

Improving the efficiency of the railway electric power system

T L Alekseeva[1], N L Ryabchyonok[1], L A Astrakhancev[1], N L Mihalchuk[2] and V A Tikhomirov[1]

[1] Irkutsk State Transport University, 15, Chernishevskogo ave., Irkutsk, 664074, Russia
[2] Center RZD, 2, Novaya Basmannaya ave., Moscow, 105066, Russia

E-mail: talecseeva843@gmail.com

Abstract. The railway productivity can be increased by increasing the efficiency of the electric power system, the main elements of which are traction substations, overhead contact systems and electric locomotives. The paper analyzes the technical infrastructure of the electric power system of the railway and evaluates the energy efficiency of the equipment. To assess the energy efficiency of the system elements, the revised law of conservation of energy in the electromagnetic field, spectral analysis of electrical quantities and mathematical computer modeling of electromagnetic processes in the system are used. The proposed energy characteristics of the elements of the electric power system made it possible to determine the reasons that hinder the further development of the throughput and carrying capacity of the railway. In order to eliminate the negative impact of inductive resistance of the system elements on the energy indicators of the traction supply and the AC electric rolling stock, the feasibility of the use of high voltage DC electric traction has been proved. Through the complete and continuous use of the voltage of the electricity system for electric traction, it is possible to ensure movement in the intersubstation zone of three linked multiple trains instead of one and increase the speed of trains by 20%. Using high-voltage direct current, the efficiency of electric traction of trains can be increased by 7% compared to electric traction of trains using alternating current by reducing the current by 40% in the overhead contact system.

1. Introduction

The throughput and carrying capacity of the railway largely depends on the energy supply system for train movement and electric traction rolling stock. From the analysis of the directions of research [1, 2], design and engineering work [3 – 5] and the production of equipment for electric traction of trains [6, 7], it follows that the problem of increasing the productivity of transport operations is urgent.

The tasks of energy-saving use of electric energy in the technological process of transport works and reducing the negative technogenic impact on the environment remain unresolved [8, 9]. On Russian railways, 3.3 kV DC voltage and 27.5 kV AC voltage with a frequency of 50 Hz are used for electric traction of trains. Three-phase transformers, twelve-, twenty-four pulse three-phase rectifier-inverter converters with supply and amplifying wires in the overhead contact system are used at DC traction substations. Electric rolling stock with collector traction electric motors and contactor power control are mainly used in a DC system. Three-phase transformers with voltage distribution over the feeder zones of the overhead contact system are mostly used at AC traction substations; to enhance the traction power supply, a 2 × 25 kV system with autotransformer points on the line is being introduced.

Content from this work may be used under the terms of the Creative Commons Attribution 3.0 licence. Any further distribution of this work must maintain attribution to the author(s) and the title of the work, journal citation and DOI.
Published under licence by IOP Publishing Ltd

Electric locomotives with collector traction electric motors and pulse-phase power control of a single-phase reversible converter are mainly used in the AC system. Two-system electric locomotives with three-phase asynchronou traction electric motors have been developed and tested, the power of which is regulated by a four-quadrant 4-qS input converter and three-phase autonomous inverters (AIV) with pulse-width modulation (PWM) of a frequency-controlled three-phase voltage on the stator windings of the traction electric motors. To increase the traction power of locomotives, the plants manufacture them in a three-section, four-section design and technology of driving linked multiple trains is introduced. The experience of operating three-sectional electric locomotives on separate railway sections is accompanied by the maximum permissible voltage deviations, slack of the contact wire and the decrease in the speed of trains. Our research [10] is confirmed by the results of the experimental driving of linked multiple trains, the results of acceptance tests of new traction rolling stock. Equipment for cleaning and monitoring systems, electric rolling stock with collector and three-phase asynchronous traction motors are used in the system of direct and alternating current of the national railways, and enterprises of the machine-building industry in Russia continue to manufacture them.

The theory of energy processes, which is used in educational institutions in Russia and abroad, has methodological contradictions with the theory of fundamental electrical engineering, power supply, electrical machines and electric drives. Future specialists are focused on improving the features that allow one to evaluate energy processes in the use of electrical energy, and thus develop, exploit and eliminate the consequences of unsatisfactory equipment operation [11].

2. Materials and methods

Insulating materials with increased dielectric and mechanical strength, made on the basis of modern nanotechnology, allow using high voltage in power equipment of the energy supply system for train movement and electric rolling stock. In power transformers, reactors, electrical machines and other electromagnetic devices with the use of hard magnetic, ferrite and other core materials instead of electrical steel, the efficiency of galvanic isolation and energy transfer increases, the ability to match the voltage of high voltage electrical circuits with the voltage of the traction networks of the energy supply system for train movement, with the voltage of the windings of the traction electric motor of the electric rolling stock. The use of modern electric energy storage devices gives new properties to semiconductor power regulators. Modern advances in digital electronics can be used to control power semiconductor devices and implement control algorithms. The above-mentioned equipment allows one to implement modern scientific developments to improve the efficiency of the railway electricity system.

Based on the specified law of conservation of energy in the electromagnetic field [12], this study uses new energy characteristics of elements of the electric power system [13], spectral FFT analysis of non-sinusoidal voltage and current to evaluate the efficiency of the electric power system. Calculations were performed using the MATLAB computer program in the Simulink environment. The error of the results of the study is assessed on the basis of the proposed power balance.

3. The study of electric power efficient use in transport

The power balance at the inlet of electrical plants with semiconductor converters [13] is obtained with the application of the second law of Kirchhoff and spectral analysis of voltage and current:

$$\sqrt{S^2 - \Delta S^2} = \sqrt{P^2 + Q^2}, \tag{1}$$

where S is the full power at the inlet of the electrical plant with semiconductor converters;

ΔS is the part of the full power at the inlet of the electrical plant during the non-conductive state of the converter power semiconductor devices;

P is the active power at the inlet of the electrical plant;

Q is the reactive power at the inlet of the electrical plant.

The balance of power constituents (1) is calculated using spectral voltage and current analysis at the inlet of the electrical plant.

Full power at the inlet of the electrical plant:

$$S = \sqrt{\sum_{k=0}^{n} U_k^2} \cdot \sqrt{\sum_{k=0}^{n} I_k^2} = U \cdot I, \quad (2)$$

where U_k is the RMS voltage of the k-th component of the Fourier series at the inlet of the electrical plant;

I_k is the RMS current of the k-th component of the Fourier series at the inlet of the electrical plant;

U is the RMS voltage at the inlet of the electrical plant;

I is the RMS current at the inlet of the electrical plant;

k is the number of the Fourier series component;

n is the number of the last harmonics to be considered.

The power at the inlet of the electrical plant during the non-conductive state of the converter power semiconductor devices:

$$\Delta S = \sqrt{\sum_{k=0}^{n} U_{pk}^2} \cdot \sqrt{\sum_{k=0}^{n} I_k^2} = U_P \cdot I, \quad (3)$$

where U_{pk} is the RMS voltage of the k-th component of the Fourier series at the inlet of the electrical plant during the non-conductive state of the converter power semiconductor devices;

where U_P is the RMS voltage at the inlet of the electrical plant during the non-conductive state of the converter power semiconductor devices.

Active power at the inlet of the electrical plant:

$$P = U_{C0} \cdot I_{C0} + \sum_{k=1}^{n} U_{Ck} \cdot I_k \cdot cos\varphi_k, \quad (4)$$

where U_{C0}, I_{C0} are the constant components of the Fourier series, respectively, voltage, current during the conducting state of the converter power semiconductor devices;

U_{Ck}, I_k are the RMS voltage, current of the k-th harmonics of the Fourier series at the inlet of the electric plant during the conducting state of the converter power semiconductor devices;

φ_k is the phase displacement angle of the k-th harmonic relative to the similarly named k-th voltage harmonic.

Reactive power at the inlet of the electrical plant:

$$Q = \sum_{k=1}^{n} U_{Ck} \cdot I_k \cdot sin\varphi_k. \quad (5)$$

The argument φ_Σ of the total inlet power and the inlet electrical resistance of the electrical plant can be calculated using the formula:

$$\varphi_\Sigma = arccos\frac{P}{S} = arccos\left[\frac{U_{C0} \cdot I_0 + \sum_{k=1}^{n} U_{Ck} \cdot I_k \cdot cos\varphi_k}{U \cdot I}\right]. \quad (6)$$

Using the proposed energy characteristics, the studies of the energy efficiency of a 12 kV DC traction power supply system and a three-section electric freight locomotive with a traction electric motor (figure 1) are carried out in work [14].

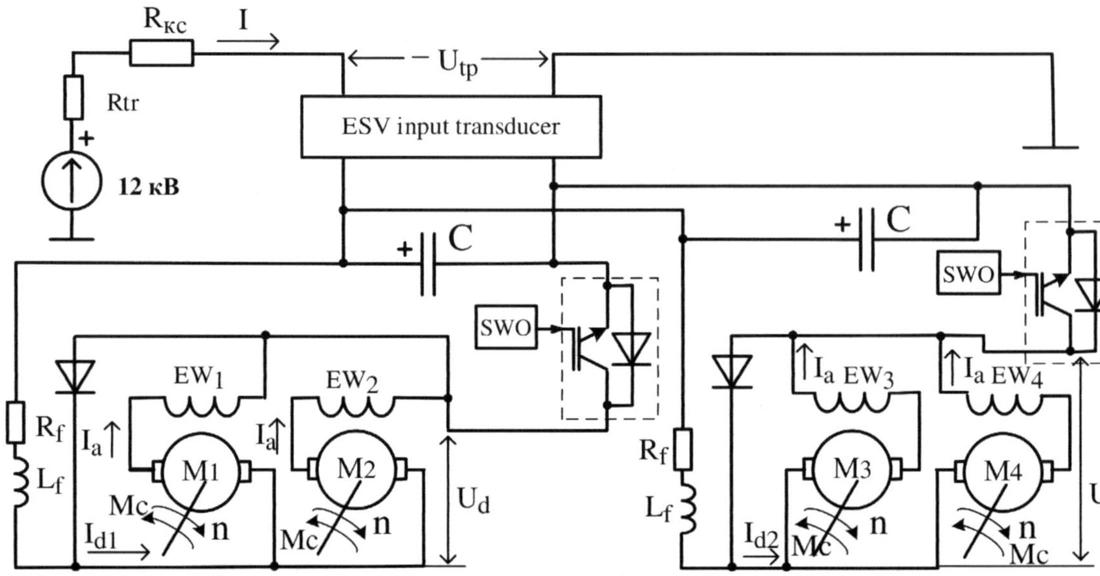

Figure 1. Electrical circuit of a 12 kV DC traction power supply system and a traction electric motor of one section of a freight electric locomotive.

In contrast to the AC electric power supply system, the inductivity of the substation's traction transformers, the overhead contact system and the inductivity of the traction transformers of electric rolling stock are eliminated, and the negative impact of inductive resistance on the efficiency of the railway electricity system is suppressed. The design of the traction electric motor of the electric rolling stock, thanks to the use of the High Voltage Electric Semiconductor Variator (ESV) input converter, allows one to change the way the three-section traction electric motor is linked in time and thus eliminate the grouping of the traction electric motor. Intermediate electrical energy storage devices C, IGBT transistor modules and reverse diodes provide practically constant current in the traction electric motor windings (current pulsation rate is no more than 2%) without the use of smoothing reactors. The energy of the electric field of C storage devices, the magnetic field of traction electric motor windings is used for the traction of the train and to compensate for the loss of active power with full and continuous use of voltage on the current collector.

If the loss of active power in the electrical circuits of the traction electric motor of the electric rolling stock is neglected, how many times the voltage of the Ud across the windings of the traction electric motor is less than the voltage across the current collector Utp. The current consumed by the electric rolling stock from the overhead contact system reduces by the same amount. Calculations performed with computer simulation and expressions (1–6) allow one to take into account the loss of effective power in the electrical traction system. Voltage across the current collector is Utp = 9221V. The current in the electric rolling stock current with the operation of 12 traction electric motors with a nominal load amounted to I = 1111 A, i.e. it exceeds the nominal current of one traction electric motor by 11%. The total current of 12 traction electric motors is 9723 A, which indicates that the power regulator has the property of an electric variator. The results of this study confirm the possibility of increasing the efficiency of the power regulator and traction electric motor of the electric rolling stock to 94.3%. The voltage on the current collector provides a nominal speed of the train movement. However, the efficiency of the DC overhead contact system with a voltage of 12 kV is 75.8%. In this regard, the total loss of active power in the AC electric traction system with a voltage of 27.5 kV and in the electrical traction system at a constant current of 12 kV is almost identical.

IOP Conf. Series: Materials Science and Engineering 1151 (2021) 012005 doi:10.1088/1757-899X/1151/1/012005

Three-phase six-pulse rectifiers assembled on diodes can be installed at the traction substations, which are equipped with traction three-phase transformers with a linear voltage of secondary windings of 27.5 kV. Voltage stabilization can be provided by regulators similar to ESV designs. The rectified voltage on the substation's DC buses is 37.1 kV. The power equipment of the traction transformer substation, the contact suspension and the equipment of the high-voltage electric rolling stock input converter must be manufactured with high dielectric, mechanical strength and through the use of modern technologies for the manufacture of materials for power equipment of the the energy supply system for train movement and electric rolling stock. The mathematical model of the 37.1 kV DC traction power supply system consists of an overhead contact system, which is applied at a variable voltage of 27.5 kV, and a three-section electric rolling stock with a high voltage input converter, ESV, and collector's traction electric motors. The results of the simulation were performed for the study of the start and acceleration to the 5 km/h steady-state speed of the train with a nominal moment of resistance on the shaft of the traction electric drice 7772 N·m in the middle of the intersubstation zone and are presented in Figures 2, 3. In the steady-state mode of electric traction of the train, electric rolling stock consumes the current $0I_{tp}$ = 75,51 A from the contact network. The voltage on the electric rolling stock collector is 36910 V. The voltage on the windings of the traction electric motor is U_d = 135,4 V, current I_a = 806,2 A flows along the windings of each traction electric motor, and the total current of 12 traction electric motor is equal to I_d = 9733 A (Figure 2). Active power at the inlet of the traction electric motor is 2787 kW. The efficiency of the traction electric motor is 69.8%, the ESV efficiency is 47.2% due to the low voltage mode on the traction electric motor windings. After 0.2 seconds from the beginning of oscillographing, the voltage of 37.1 kV is applied and the current I_{tp} = 1656 A flows along the electric rolling stock current collector to charge the intermediate electric energy storage device. At a moment of time of 0.25, the control pulses are coming in on the ESV IGBT transistors, so the current I_{dv} begins to flow in the winding of each traction electric motor. With the increase in the current of the traction electric motor to I_{dv} = 1000 A, the shaft of the electric motor begins to rotate, and the current on the electric rolling stock collector is reduced exponentially to I_{tp} = 600 A (Figure 3).

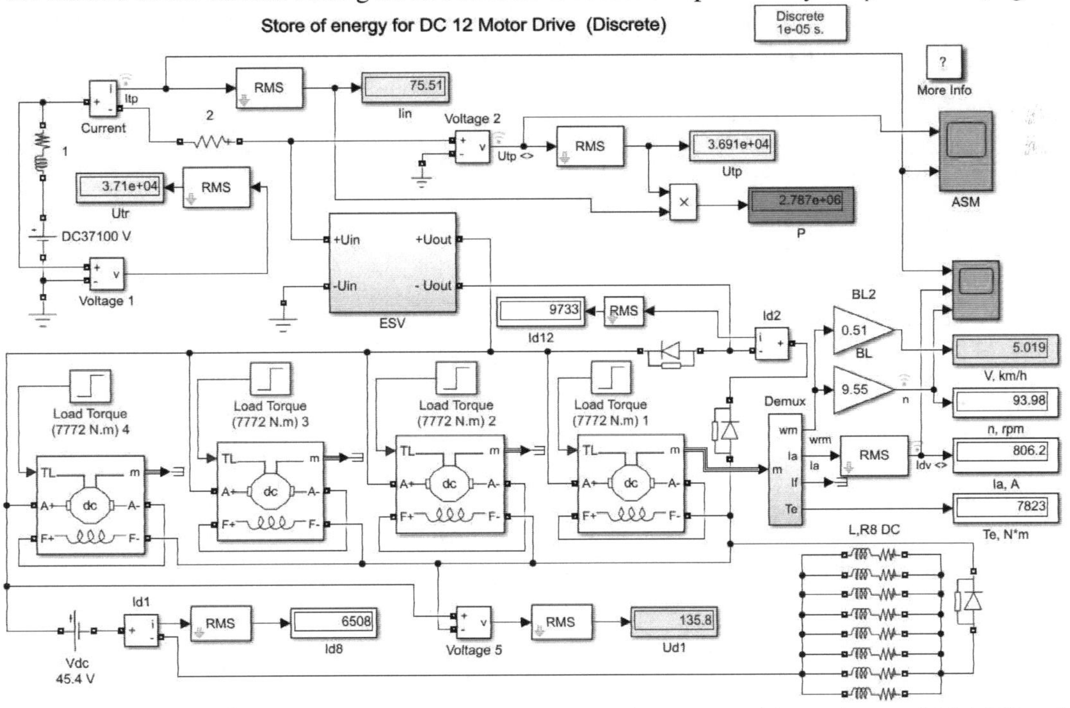

Figure 2. The mathematical model of the DC power supply system with a voltage of 37.1 kV and a traction electric motor of a three-section freight locomotive at a train speed of 5 km/h.

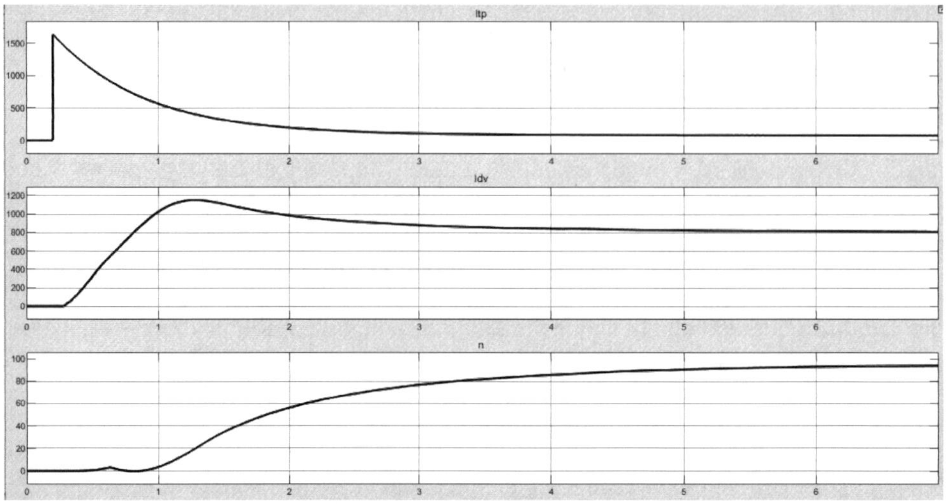

Figure 3. The current oscillograms across the current collector I_{tp}, current in the windings of the traction electric motor I_{dv} and the rotation speed of the shaft n in the traction electric motor of the electric rolling stock.

The maximum starting current in the windings of a single traction electric motor is I_{dv} = 1150 A, and the overhead contact system is loaded with a current of I_{tp} = 400 A (Figure 3). In the steady-state mode, currents are equal to: I_{dv} = 806 A, I_{tp} = 75 A, the rotation speed of the traction electric motor shaft n = 94 rpm.

The results of the calculation of the DC current traction power supply system with a voltage of 37.1 kV and a three-section electric rolling stock in the nominal mode of operation were obtained using a mathematical model presented in Figure 4.

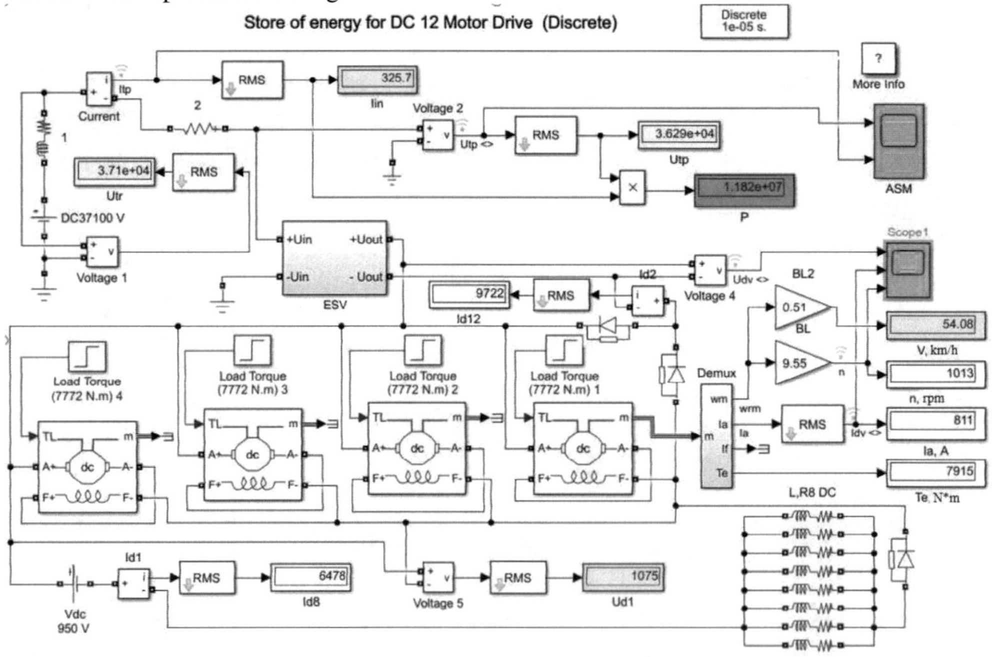

Figure 4. The mathematical model of the DC traction power supply system with a voltage of 37.1 kV and a traction electric motor of a three-sectional freight locomotive at a train speed of 54 km/h.

The results of the study of start and acceleration modes of electric rolling stock show a significant decrease in the current in the overhead contact system compared to the current in the windings of the traction electric motor, which is necessary for the traction and acceleration of the train. This result is achieved by the preliminary charge of the intermediate electric energy storage device and the properties of the electric variator.

The study found that with the transition to the DC energy supply system for train movement, together with increased voltage, eliminates inductive resistance of the overhead contact system and traction transformer of the electric rolling stock, and the speed and mass of trains can be increased. In a 27.5 kV overhead contact system, the RMS current is 1.4 times higher than the current in the same overhead contact system at a constant current of 37.1 kV. The 43 km/h speed of the train was obtained due to the voltage limit on the electric rolling stock current collector with reversible converter in the AC energy supply system for train movement due to the voltage drop on active and inductive resistances in the energy supply system for train movement. The voltage level across the ESV DC electric rolling stock allows one to drive three high-speed linked multiple trains when using the DC energy supply system for train movement.

Active power in the DC energy supply system for train movement and EPS with 11820 kW ESV is spent on train raction and the loss of active power in the power electrical circuit of the electric rolling stock. And the AC energy supply system for train movement and the electric rolling stock with reversible converter as part of the full power S_G = 11480 kVA contains the active capacities of the harmonic components P_1 =9773,1 kW, P_3 =98,4 kW, P_5 = 44,4 kW, P_{11} =2,9 kW, P_{15} = 1,14 kW, P_{19} = 1,16 kW. Reactive magnetization powers of harmonic components of the electric rolling stock traction transformer are Q_1 =4745,6 kVAr, Q_3 =350,6 kVAr, Q_5 = –86,7 kVAr, Q_7 =15,7 kVAr, Q_9 = –6,7 kVAr, Q_{11} =5,2 kVAr, Q_{13} = –3,3 kVAr, Q_{15} =1,2 kVAr, Q_{17} = –0,24 kVAr,Q_{19} = 0,022kVAr. Due to the operation of the traction transformer of the electric rolling stock with reversible converter in a short-circuit mode during the switching of current in the reversible converter tiristors, as part of the full-capacity S_G , the power is formed by the power ΔS = 3202. And the voltage across the AC electric rolling stock current collector and the train speed are reduced. The efficiency of AC energy supply system for train movement is 94.6% and the electric rolling stock efficiency with reversible converter is 79.3% due to significant losses of active power in power electrical circuits of the electric rolling stock.

The efficiency of DC energy supply system for train movement is 97.8% efficiency and the efficiency of the electric rolling stock traction electric motor with ESV is 83.7%, so the efficiency of DC electric traction of trains increases by 7% compared to the AC electric traction of trains.

In well-known scientific works, in addition to energy efficiency, other advantages and disadvantages of AC and DC electric traction are considered in sufficient detail. Once again, this study proves the feasibility of increasing voltage with an increase in the amount of work performed with electric energy.

4. Conclusion

Technical solutions have been developed on the basis of the refined law of energy conservation in the electromagnetic field. The energy characteristics of the elements of the railway's electricity system have been proposed. The application of the new power regulator ensures the full and continuous use of the voltage of the electric power supply system to perform transport work. In the high-voltage DC electric traction system, it is possible to ensure the movement of three linked multiple trains in the intersubstation zone of the railway with a modern contact suspension and increase the train speed by 20%. The RMS current in the overhead contact system is reduced by 40% due to the efficient use of voltage. The efficiency of high-voltage DC electric traction trains is 7% higher than the efficiency of AC electric traction trains.

Significant amounts of scientific and technical work are needed to implement electric traction at a constant high voltage current. In the development of design documentation, it is important to use modern insulation materials with high dielectric and mechanical strength. It is advisable to develop and manufacture electromagnetic machines, electric machines for electric traction by applying magnetic conductors with high magnetic permeability. Using the latest advances in power and information electronics, electric energy storage devices allow one to implement promising directions for further development of electric traction trains.

References

[1] Burkov A T 2015 *Electronics and converter equipment* (Moscow: Transport Publ.)

[2] Bose B K 2002 *Modern power electronics and AC converters* (Prentige Hall PTR)

[3] Tigelketter I and Schirenger D 2000 *Powerful converter on IGBT-transistors for application on railway rolling stock* (Minchen: Siemens AG)

[4] Kopanev A S and Khripkov P A 2010 Tests of the reactive power compensation system on the 3ES5K electric locomotive *Bulletin of the All-Russian Research and Design Institute of Electric Locomotive* **2(60)** 14–36

[5] Bakhvalov I A Buzalo G A 2006 *Dynamic processes in the asynchronous traction drive of the main electric locomotives* (Moscow: Marshrut Publ.)

[6] Sorin L N 2003 Future electric locomotives for the Russian railways *Railroads of the world* **8** 18–24

[7] Electro rectifier OAO 2008 *The reversible converter VIP-1000-U1: Operation manual IEAL.435411.046 RE* (Saransk)

[8] Akagi H and Watanabe E 2007 *Instantaneous power theory and applications to power conditioning* The power engineering (New York: Wiley)

[9] Macheret D A 2019 World of transport *Uncertainty of the future as a fundamental problem of long-term development of transport* **17** 6–19

[10] Vorotilkin A V and Mikhalchuk N L 2015 World of transport *Innovative perspectives of traction rolling stock* **13(6)** 62–76

[11] Ryabchenok N L and Alekseeva T L 2018 Materials of the III All-Russian scientific and practical. conf. with international participation *New opportunities for increasing the efficiency of the educational process in educational institutions of higher education* (Irkutsk: ISU) pp 431–434

[12] Ryabchenok N L and Alexeyeva T L 2015 Refined Energy Conservation Act *Materialy X1 Miedzynarodowej naukowi-praktycznej konferencji "Perspektywiczne opracowania sa nauka i technikami"* **11** (Przemysl (Poland): Nauka i studia) pp 63–74

[13] Alexeyeva T L and Ryabchenok N L 2020 Energy efficiency in electrical circuits with semiconductor devices *Bulletin of SUSU. Series: "Power engineering"* **20(2)** 89–99

[14] Ryabchenok N L and Alekseeva T L 2020 Energy efficiency of the traction drive of mainline electric locomotives *News of the Transsib* **1(41))** (Omsk: Omsk state transport university) 29–41

IOP Conf. Series: Materials Science and Engineering 1151 (2021) 012006 doi:10.1088/1757-899X/1151/1/012006

Electromagnetic influences of autotransformer overvoltage traction networks on pipelines

A V Cherepanov[1], **A E Kryukov**[1] **A V Kryukov**[1,2]

[1] Irkutsk State Transport University, 15, Chernyshevskogo st., Irkutsk, 664074, Russia
[2] Irkutsk National Research Technical University, 83, Lermontova st., Irkutsk, 664074, Russia

E-mail: and_kryukov@mail.ru

Abstract. The article presents the results of the simulation of the electromagnetic effects of an advanced overvoltage traction network on a steel pipeline with a diameter of 250 mm, laid on the earth's surface. Due to the electromagnetic imbalance of the traction network, at certain points of the structure, voltages with respect to the ground, dangerous for the maintenance personnel, may arise. In such situations, it is necessary to develop measures to improve electrical safety. In the conditions of digitalization of the electric power industry, the solution to this problem should be based on computer technologies that allow determining the induced stresses arising on the parts of pipelines due to the electromagnetic effect of the traction network. The simulation, performed with the help of the Fazonord software package, showed that at a distance of 100 meters from the railway route to the pipeline, the maximum induced voltage at individual points of the structure can reach 100 V, which is more than one and a half times the permissible value. This must be taken into account when designing advanced traction electric power supply systems in the areas where they are close to the main pipelines.

1. Introduction

To transport gas, oil and products of its processing in Russia, a large-scale network of pipelines has been created, the total operational length of which exceeds 250 thousand km [1]. Some of their sections can run along the lines of AC railways [2, 3]. Due to the electromagnetic imbalance of the traction network (TN) at certain points of the structure, voltages to earth, dangerous for the maintenance personnel, may arise [2–10]. In such situations, it is necessary to develop measures to improve electrical safety. In the context of digitalization of the electric power industry, the solution to this problem should be based on computer technology, which makes it possible to determine the induced voltages occurring on the parts of pipelines due to the electromagnetic influence of the traction network.

Such technologies can be implemented on the basis of methods and tools for simulating the modes of electric power systems in phase coordinates, proposed at Irkutsk State Transport University [11-12]. Below are the results of modeling the electromagnetic effects of an advanced overvoltage traction network [13, 14] on a ground surface laid pipeline.

2. The simulation technique.

The electromagnetic influences of electric traction networks on pipelines can be determined using the

Content from this work may be used under the terms of the Creative Commons Attribution 3.0 licence. Any further distribution of this work must maintain attribution to the author(s) and the title of the work, journal citation and DOI.
Published under licence by IOP Publishing Ltd

Fazonord software package [12]. Simulation of induced voltages can be based on the principles proposed in [6, 8].

One of the main principles is to correctly take into account the return of currents through the ground on the basis of calculating the resistances of the mutual inductive coupling, as well as the intrinsic resistances of the "wire – ground" circuits using the relations proposed in [8]; this provides an exact definition of the Carson integral [15] for the near, intermediate and far zones.

3. Simulation results.

The Fazonord software complex simulates the operating modes of an advanced overvoltage traction electric power supply system (TEPSS) [15] (Figure 1). Such a traction electric power supply system provides for the construction of base traction substations (TS) equipped with unbalance-to-balance transformers; at the same time, the distance between the base traction substations can reach 300 ... 350 km. Power supply to the traction network is carried out from single-phase transformers with the winding voltages of 93.9 / 27.5, located at distances of 30 ... 45 km from each other.

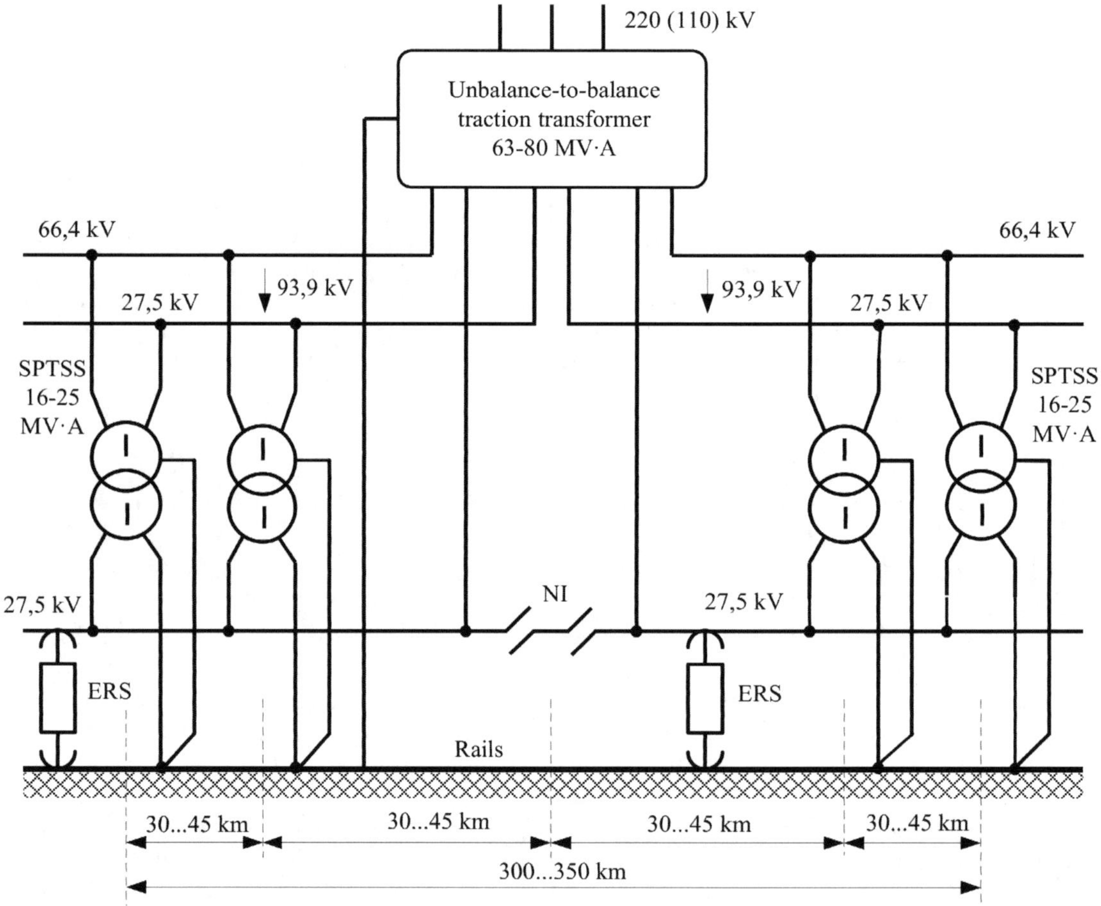

Figure 1. Traction electric power supply system (TEPSS) scheme of increased voltage with unbalance-to-balance transformers: SPTSS – single-phase traction substation; NI – neutral insert

A fragment of the computational model diagram is shown in Figure 2, and the train schedules – in Figure 3. The current profiles are shown in Figure. 4. The movement of 6 freight trains weighing 6000 tons in up and down directions was simulated at intervals of 40 minutes.

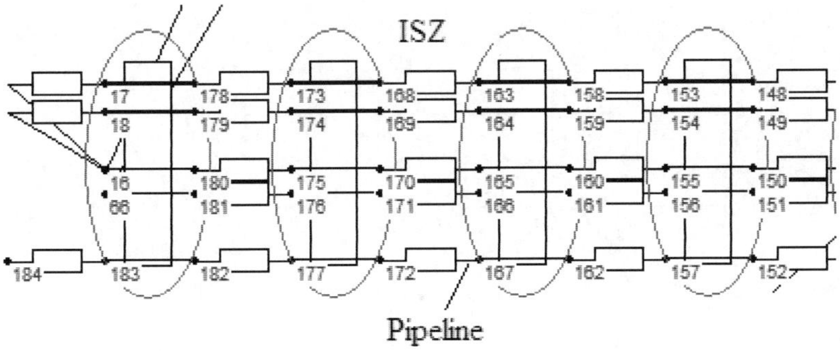

Figure 2. A fragment of the computational model scheme

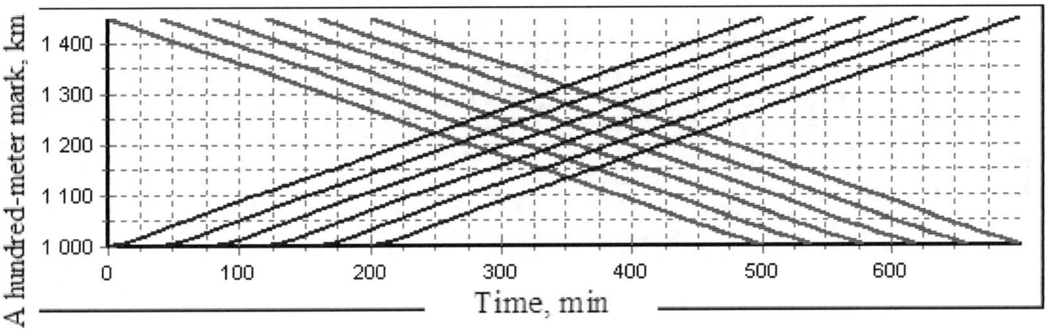

Figure 3. Train movement schedule

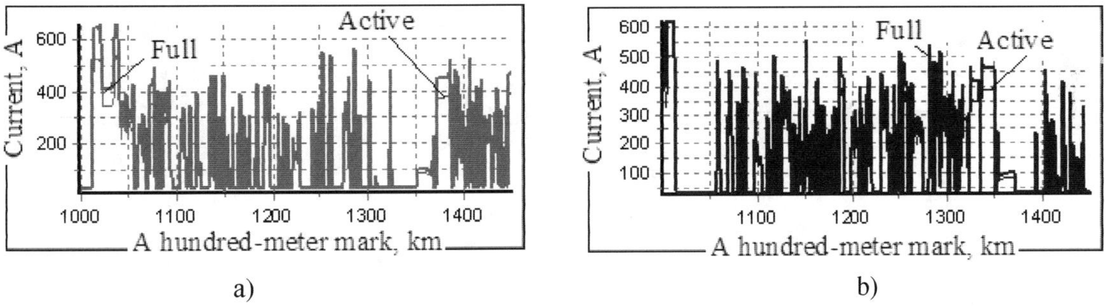

Figure 4. A fragment of the computational model scheme

The induced voltages were determined taking into account the influence of the higher harmonics of the current and voltage of the traction network. The results of modeling are shown in figures 5-10 and in table 1.

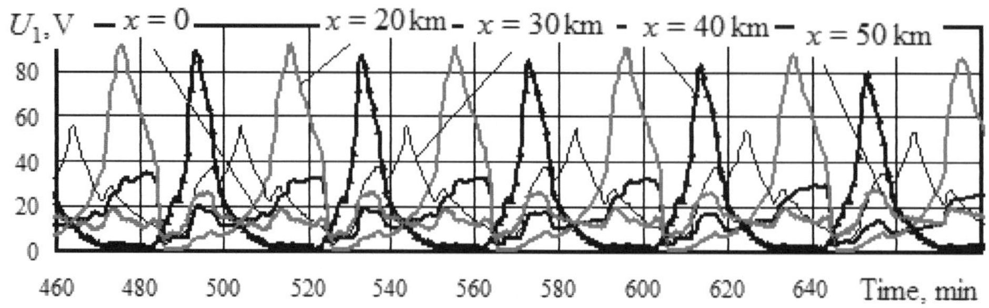

Figure 5. Dependences of the induced voltage U_1 of the fundamental frequency on time with the approach width a = 100 m: x – distance from the observation point to the left traction substation

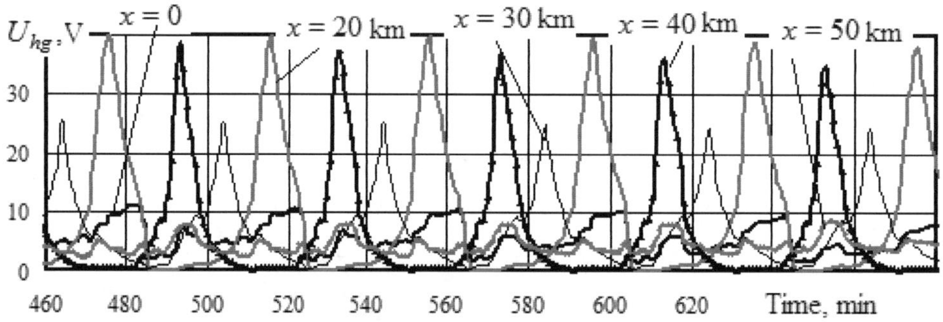

Figure 6. Dependences of the effective value U_{hg} of voltages of higher harmonics on time with the approach width a = 100 m

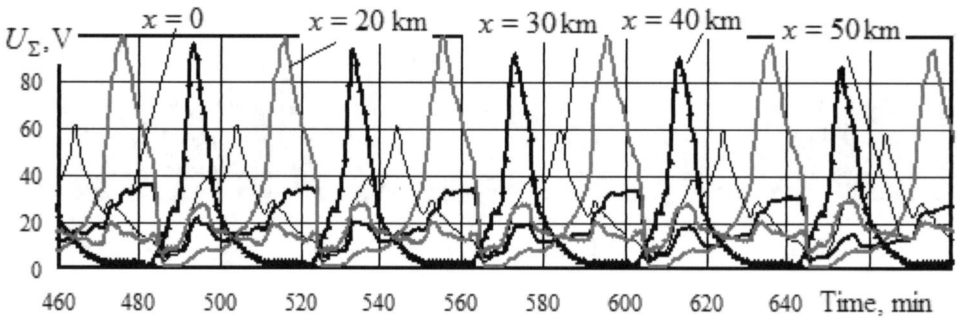

Figure 7. Dependences of the resulting induced voltage $U_\Sigma = U_1\sqrt{1+k_U^2}$ on time with the approach width a = 100 m: k_U is the total harmonic distortion

IOP Conf. Series: Materials Science and Engineering 1151 (2021) 012006 doi:10.1088/1757-899X/1151/1/012006

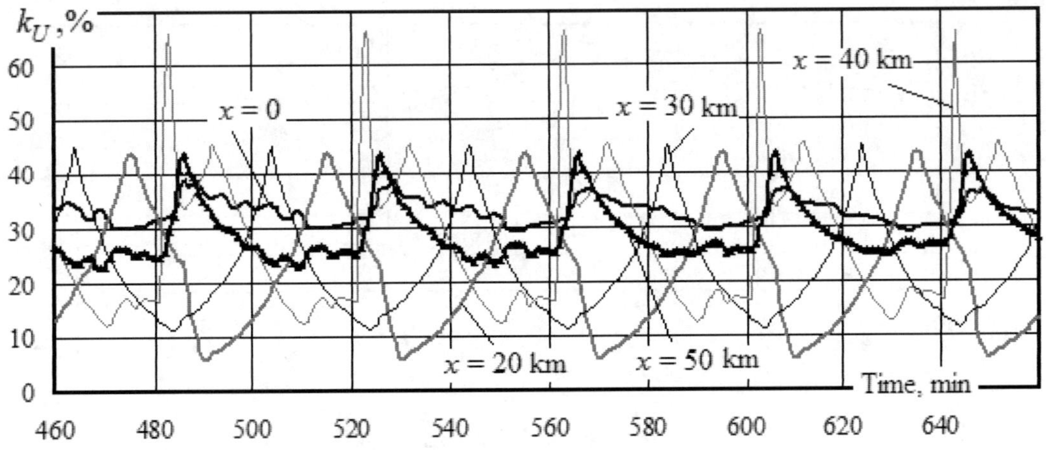

Figure 8. Time dependences of the total harmonic distortions with the approach width a = 100 m

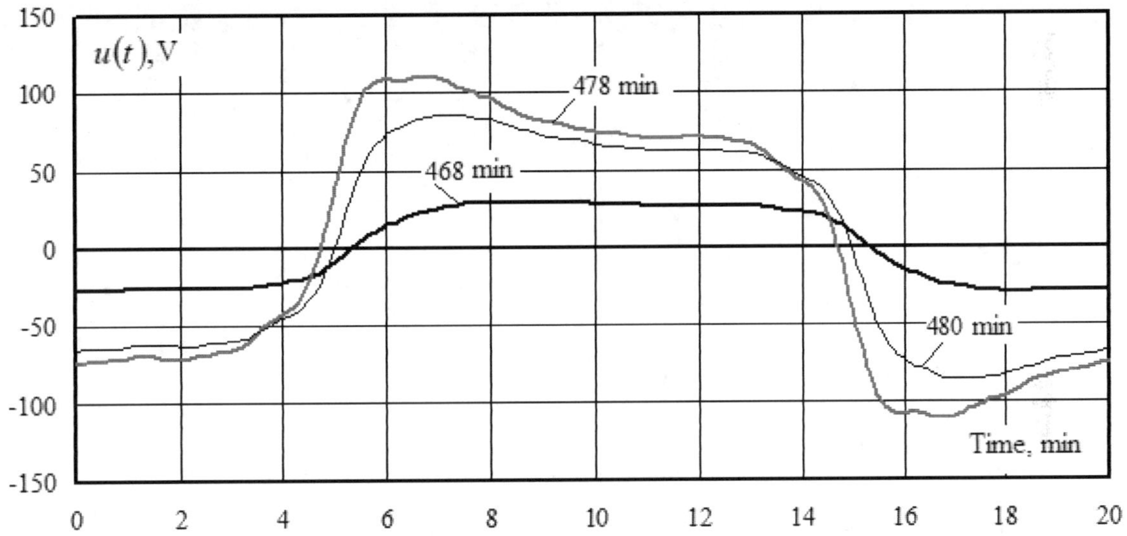

Figure 9. Shapes of induced voltage curves at the point corresponding to the coordinate x = 20 km

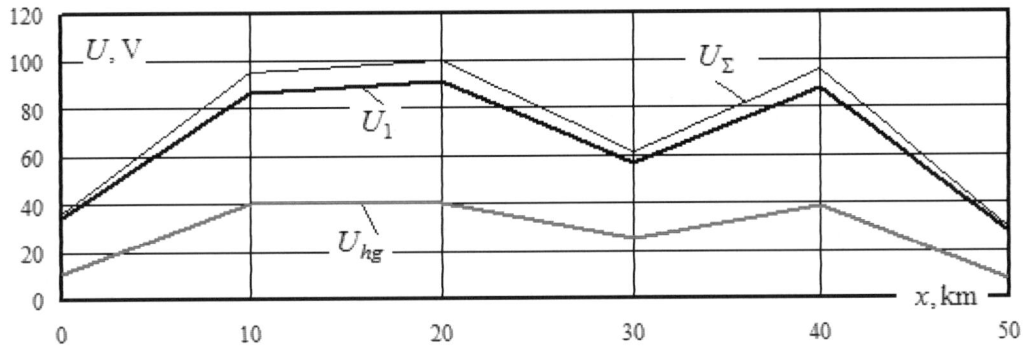

Figure 10. Maximum values of induced voltages

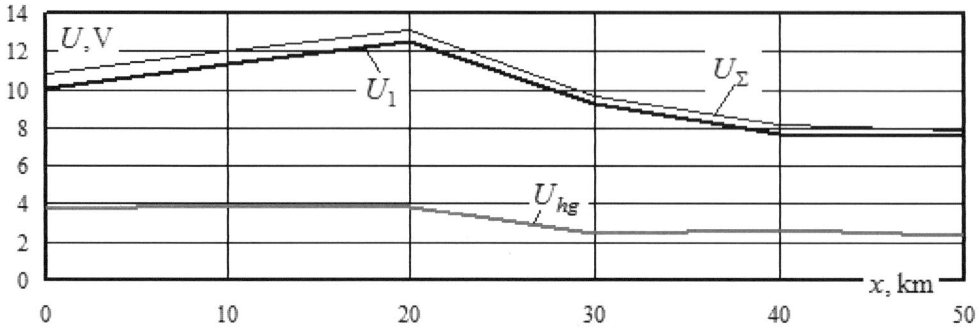

Figure 11. Average values of induced voltages

Table 1. Summary results

Indicator	Parameter	x, km					
		0	10	20	30	40	50
Maximum	U_1, V	34.50	86.40	91.60	56.50	88.10	27.60
	U_{hg}, V	11.05	40.03	39.83	25.49	38.26	8.51
	U_Σ, V	36.23	95.22	100	61.98	96.05	28.84
Average value	U_1, V	10.07	11.31	12.45	9.26	7.69	7.55
	U_{hg}, V	3.72	3.97	3.78	2.49	2.67	2.38
	U_Σ, V	10.77	12.02	13.06	9.64	8.17	7.95

The analysis of modeling results makes it possible to conclude the following.

1. For the considered electric traction network with a approach width of 100 meters, the maximum induced voltage reaches 100 V (Table 1), which is more than one and a half times higher than the permissible value [16]. To reduce it, it is necessary to provide for additional stationary grounding conductors or to increase the distance between the pipeline and the railway. This must be taken into account when designing advanced traction electric power supply systems in the areas where they are close to the main pipelines.

2. The magnitudes of the induced voltages change significantly over time, which is associated with the nonstationarity of the traction loads (Fig. 5-7).

3. There are significant harmonic distortions of induced voltages, the total harmonic distortion exceeds 60% (Fig. 8); therefore, the shapes of the curves differ significantly from the sinusoid.

4. During the operation of the facility, the magnitude of the "pipeline-to-ground" transition resistance may decrease [17], which will result in a decrease in induced voltages.

4. Conclusion

Computer technologies based on digital models of electric power systems in phase coordinates and allowing one to determine induced voltages on pipelines laid in electromagnetic influence zones of advanced overvoltage traction networks are proposed.

5. Acknowledgments

The research was carried out with the financial support on a grant of the Russian Ministry of Science themed as "Improving the quality of electric energy and electromagnetic safety in the electricity supply railway systems equipped with Smart Grid devices by applying methods and means of mathematical modeling based on phase coordinates".

References

[1] Gamzaev B A 2019 The state and peculiarities of the development of Russia's pipeline transport

at the present stage *Young scientist* **3 (241)** 155-159

[2] Ratner M P 1966 Inductive influence of electrified railways on electrical networks and pipelines (Moscow: Transport Publ.)

[3] Kotelnikov A V and Kosarev A B 1992 Electromagnetic effect of AC electrical traction networks on metal structures *Electricity* **9** 26-34

[4] Konnova E I and Kosarev A B 1990 Calculation of the electromagnetic effect of AC networks on metal supply lines *Bulletin of Research Institute of Railway Transport* **2** 17-19

[5] Yabluchansky A I 2008 Methods of assessing the dangerous effects of high-voltage electric power transmission lines on the gas pipeline currently being designed *Materials of the industry meeting on corrosion protection* pp. 110-123 (Moscow: IRTs Gazprom Publ.)

[6] Caulker D, Ahmad H, Mohamed Ali M S 2008 Effect of lightning induced voltages on gas pipelines using ATP-EMTP program *2008 IEEE 2nd International Power and Energy Conference* (IEEE)

[7] Ahmed S. Al Shahri, Nirmal-Kumar C. Nair 2016 Overhead shielding wire and induced potential on pipeline located close to high voltage transmission *Australasian Universities Power Engineering Conference (AUPEC)* (IEEE)

[8] Chuco Paucar B, Roel Ortiz J L, Pereira Pinto J O, Koltermann P I 2007 Induced Voltage on Gas Pipeline With Angle Between a Transmission Line *2007 IEEE Lausanne Power Tech* (IEEE)

[9] El-Sayed Mohamed El-Refaie, AbdEl-latif Sayed Ahmed, Sabry Mousa Mohamed, Hossam Mohamed Gaber 2017 Electromagnetic field interference between high voltage transmission lines and nearby metallic gas pipelines *2017 Nineteenth International Middle East Power Systems Conference (MEPCON)* (IEEE)

[10] Nagat M. K. Abdel-Gawad, Essam M. Shaalan, M. M. F. Darwish, Mahmoud A. M. Basuny 2019 Influence of Fault Locations on the Pipeline Induced Voltages Near to Power Transmission Lines *2019 21st International Middle East Power Systems Conference (MEPCON)* (IEEE)

[11] Zakaryukin V P and Kryukov A V 2015 The refined method of determining the mutual electromagnetic influences of adjacent electric power transmission lines *Proceedings of universities. Problems of power economy* **3-4** 29-35

[12] Zakaryukin V P and Kryukov A V 2005 *Complex asymmetrical modes in electrical systems* (Irkutsk: Irkutsk university Publ.)

[13] Vasilyansky A M, Mamoshin R R and Yakimov G B 2002 Improving the system of railway traction electricity supply electrified at a variable rate of 27.5 kV, 50 Hz *Railroads of the world* **8** 40-46

[14] Borodulin B M 2003 Balancing of currents and voltages on the existing AC traction substations *VNIIZHT Scientific Journal* **2** 14-17

[15] Carson J R 1926 Wave propagation in overhead wires with ground return *Bell Syst. Tech. J.* **5** 539–554

[16] Technische Richtlinien-71 (TRL-71). *EMR-Technic Kathodischer Korrosionsschutz fur Erdgasfernleitungen*

[17] Zinevich A M, Glazkov V I and Kotik V G 1975 Protection of pipelines and tanks against corrosion (Moscow: Nedra Publ.)

IOP Conf. Series: Materials Science and Engineering 1151 (2021) 012007 doi:10.1088/1757-899X/1151/1/012007

Tests to estimate the heat rate effect on the flexural durability of transport vehicle gear wheels

V V Skutelnik[1], S V Kolganov[1] and O L Malomyzhev[2]

[1] Irkutsk National Research Technical University, 83, Lermontov Street, Irkutsk, 664074, Russian Federation
[2] Irkutsk State Transport University, 15, Irkutsk, Chernyshevskogo Street, 664074, Russian Federation

E-mail: scutvv@gmail.com

Abstract. The article presents the test results concerning the effect of thermal loads on the flexural durability of gear wheels of transport vehicle high-loaded transmissions. The tests were performed using a universal hydraulic pulsation machine and a thermal loader at five fixed temperature levels from twenty to two hundred degrees Celsius. As a result of the performed tests, a significant dependence of the flexural durability of gear wheels on their thermal state was established. In the expected maximum temperature area, corresponding to the temperature of the low tempering of the gear wheel material, it decreased sharply more than ten times.

1. Introduction

The reliability and efficiency of multi-purpose tracked and wheeled transport vehicles are determined substantially by the transmission. The application of more powerful engines, in the case if overall dimensions of transmission units are not changed, is achieved by increasing the density of the layout of transmission elements that causes significant heat generation in them. The underestimation of the thermal calculation of transmissions was permissible in the condition of low heat generation in the unit up to 50 kW / m3 [1], being typical of low power vehicles. Currently, due to increase in the transmitted power and angular frequency of rotation of the teeth in the condition of limited dimensions of components and units, the issues of thermal calculation as well as strength are of paramount importance and require common solutions.

To provide the required performance and durability of transmission units of transport vehicles, it is necessary to carefully study their thermal conditions. It includes a complex of design and experimental works necessary to ensure the required level of temperature of parts and the unit as a whole for a trouble-free operation.

The first mention of the damages associated with the heating of gears was made in the middle of the last century. It was noted that the unit's heat exchange is usually based on the oil temperature and not on the coefficient of heat transfer from the oil to the ambient air, on the heat transfer coefficient from the crankcase surface. This can be a significant error, as the crankcase surface temperature differs from the oil temperature and the local temperatures on the housing surface can vary widely enough.

A quantitative measure of the thermal condition of a part, assembly or unit is the temperature, the spatial distribution of which over the entire volume forms the temperature field. Changing the

Content from this work may be used under the terms of the Creative Commons Attribution 3.0 licence. Any further distribution of this work must maintain attribution to the author(s) and the title of the work, journal citation and DOI.
Published under licence by IOP Publishing Ltd

temperature field over time is called the thermal mode and keeping the temperature within the limits from the minimum allowable to the maximum allowable during the entire period of the operation of transmission elements, providing a given thermal mode.

In the teeth contact zone, the temperature reaches high values of 800-900 °C [2]; the bulk temperature of the teeth is 25-30% higher than the oil temperature and can reach the low-temperature tempering of the steel the gears are made of [3], causing the premature wear of parts due to the reduction of their strength characteristics [4, 5].

12X2H4A, 12XH3A, 20X2H4A, 18X2BA grades of steel and other alloyed chromium-nickel steel are mainly used [6, 7] in units and drivetrain components of transport vehicles. These steel grades are characterized by high physics-mechanical properties, such as: high wear resistance and hardenability during hardening. However, the above-mentioned materials decrease their strength characteristics due to structural changes in the cemented layer [8] during a prolonged operation in the zone of elevated temperatures (if the oil temperature in the crankcase is above 140-150°C). According to [11], the permissible temperature should not exceed 170°C in the case of the mentioned above steel grades. The applied foreign steel (3310, 3310M, E9310) and that produced in Germany (14NiG10, 14NiG14, 14NiG18) have the compositions and characteristics that are similar to those of above-mentioned steel grades. These allow working temperature not higher than 150 °C [12].

The efficiency of gears, which ensures the guaranteed vehicle mileage during the entire operation period, is connected with the limit of endurance of the teeth while bending. Traditional tests of gear wheels on flexural durability, as a rule, are carried out without monitoring the thermal state and do not provide information how the thermal condition affect the being evaluated parameter.

At the same time, tests of material samples indicate a significant effect of the thermal condition the strength characteristics. The results of such tests in relation to various steel grades are given in reference books [13, 14].

Gear wheels of various drivetrain components and transmission units operate in a wide range of heat rate. Therefore, in order to estimate their durability properly and determine the maximum permissible temperatures, it is necessary to have information about the dependence of the strength characteristics of gears on the thermal state.

From the analysis of the aforesaid it follows that reliable and effective work of elements of transmissions essentially depends on a thermal mode. Exceeding the temperature of oil and parts in the units leads to rapid wear and tear and a reduction in the durability of transmission elements, which is one of the reasons limiting the performance of transmission mechanisms and may be the cause of premature failure of individual parts and units of the transmission as a whole.

Conducting experiments on the influence of thermal loads on the strength properties of parts, such as, for example, the dependence of the bending life of gear wheels, will reveal their dependence in the area of expected operating temperatures. In development of new and finishing of existing constructions, similar tests are extremely necessary for the most heat-stretched elements for the purpose of establishment of maximum permissible temperatures.

2. Gear wheel tests

Based on the foregoing, the objective of the study was formulated as follows: to test the gear wheels and to identify the thermal condition effect on their flexural durability. Full-size sun gears of planetary gear sets of transport vehicle transmission were used as test objects; they were manufactured according to standard technology, including an undercut tooth base, shot peening, cementation and grinding of the involute. Their parameters are presented in Table 1.

The objective was to modify the existing test facilities to estimate the flexural durability of gear wheels in order to determine the effect of the thermal conditions and to conduct tests using the example of a batch of gears of a planetary gearbox.

Table 1. Gear wheels test parameters

	Input values	Parameter values
1.	Steel grade	20X2H4A-III
2.	Module, m, mm	4
3.	Teeth number, z	32
4.	Original contour angle, α, degree.	20
5.	Working teeth width, b, mm	20
6.	Modification coefficient, x	0.64

The MUP-50 universal hydraulic pulsation machine and a device were used to perform the tests; the diagram is shown in Figure 1.

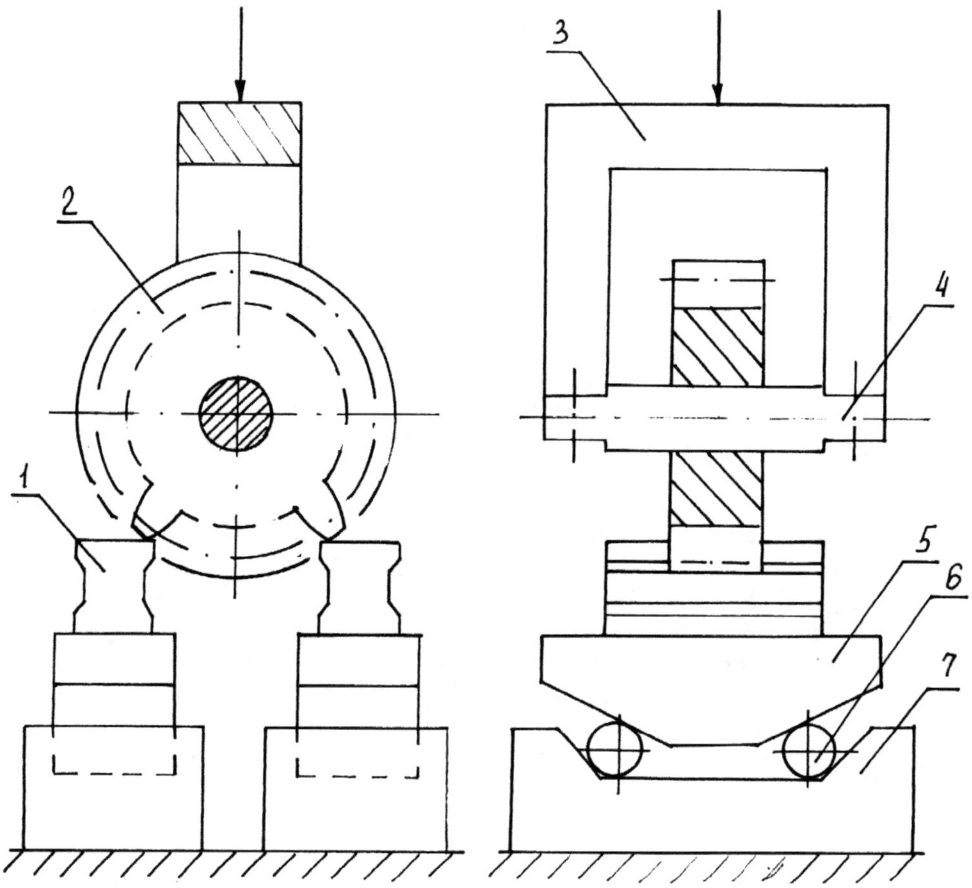

Figure 1. Load device circuit: 1 – punch, 2 – tested gear wheel, 3 – forking, 4 – axis, 5 – prism, 6 – roller, 7 – prism base.

The load was applied simultaneously to two teeth. The loaded teeth were selected according to the conditions of application of the load in a place corresponding approximately to the upper boundary point of one-pair gearing.

During dynamic tests, the maximum value of the initial load level per unit, the length of the contact line of the tooth were taken as WFmax≈0.6 Wv.sr. The minimum load value, taking into account the

IOP Conf. Series: Materials Science and Engineering 1151 (2021) 012007 doi:10.1088/1757-899X/1151/1/012007

geometric dimensions, was determined by the technical capabilities of MUP-50 machine (WFmin≤50 kgf / mm).

To ensure the pre-estimated thermal state of the gear wheels under test, the thermal loader was developed and manufactured to cover the loadable teeth from the ends and tops. To reduce heat loss to the environment, a heat shield was installed. The heating temperature was regulated by changing the voltage in the helix power supply system, the loader provided heating of gear wheels up to 200°C.

The load intensity was recorded by the pulsator dynamometer, the maximum load of the WF cycle amounted to 70000 N if the asymmetry coefficient was r-0.33 and the pulsation frequency - 690 cycles / min. All tests were performed at the same load level.

The temperature of the teeth under test was recorded with a potentiometer. A chromel-copel thermocouple installed in the cavity of the loaded tooth was used as a sensor. Asbestos insulation was used to protect the thermocouple from the flow of the heater.

The tests were performed at five fixed temperature levels of the tested wheels: 20, 100, 125, 150 and 200°C. The lower level of 20°C corresponded to the temperature in the laboratory room; the upper level slightly exceeded the temperature of the low tempering of the wheel material and corresponded to the actually possible values during the machine operation under extreme conditions.

To prevent additional thermal stresses in the loaded teeth, which may result from uneven heating of the teeth and punches, the gear wheel was heated to a predetermined temperature, until the teeth came into contact with the punches.

During the tests, the load and the thermal state of the test object were set and the number of loading cycles was determined until one of the teeth broke. At each temperature level, at least 7 tests were conducted. Each point is an arithmetic mean of 7 measurements.

3. Results

The test results are provided in Figure 2. Tests results confirmed the substantial dependence of gear wheels flexural durability on the thermal condition. At the same time, some interesting characteristics were identified. The flexural durability increased 3 times on average as the temperature increased from 20°C to 125°C. At temperatures of 100 ÷ 125°C, the maximum value of flexural durability was recorded. Due to further increase in temperature, within the range of its expected operational values, an intensive decrease in flexural durability of 15 times on average was noted.

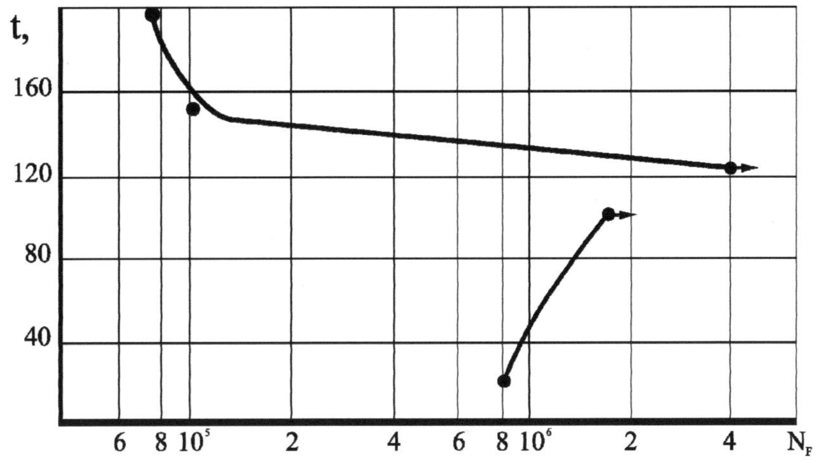

Figure 2. Dependence of Flexural Durability on Temperature

4. Conclusion

The explanation of the reasons for the observed dependencies requires additional tests on gears made of various materials and according to different technological processes. However, the following conclusions can be made as the results of the experiment:

1. Refinement of the existing gear flexural durability testing stands by installing thermal loaders allows one to expand significantly the range of conducted tests and to obtain information on the effect of the thermal condition on the estimated parameters.
2. The conducted tests identified a significant dependence of the flexural durability of gear wheels on the thermal state even at a moderate temperature range of 20 ... 200ºC.
3. The dependences of the flexural durability of the tested gear wheels on the thermal state are much greater than the similar dependences on the strength characteristics of the material the gears are made of and have a qualitatively different type.
4. In relation to a tested batch of gears, there is an extremum of flexural durability at the temperature range of 100 ... 200°C.
5. The flexural durability of the sun gear changed 11 times from 812,000 to 75,000 cycles at the temperature range of 20 ... 200ºC under constant load.
6. If the temperature changes in both directions from the extremum area, the durability drops, and when the temperature increases, it is 6 ÷ 7 times faster than when the temperature decreases.

References

[1] Terekhov A C 1973 *Thermal calculation of automobile gearboxes* Abstract of thesis, Dr. of technical science, Moscow
[2] Vulgakov E V (Ed) 1981 *Aircraft gears and gearboxes reference book* Moscow: Mechanical Engineering
[3] Sorokina V G (Ed) 1989 *Steel and alloys grades. Reference book* Moscow: Mashinostroenie
[4] Molyneux P 1975 *Oleodinamica, pneumatica* **16 (312)** 84-92
[5] Skutelnik V V, Malomizhev O L and Fadeev D S 2013 The effect of oil temperature conditions on the wear of transport vehicle transmission components *ISTU Vestnik* **11** 220-223
[6] Banas I P and Alekseeva G M 1985 Modern steel for high-loaded gear wheels *Mechanical engineering bulletin* **9** 12-15
[7] Dorfman V S 1978 *Modern materials in motor vehicle industry. Reference book* Moscow: Mashinostroenie
[8] Matveevsky R M 1971 *Temperature resistance of boundary lubricating layers and solid lubricating coatings during friction of metals and alloys.* Moscow: Nauka
[9] Matveevsky R M et al 1989 *Lubricants.* M. Mashinostroenie
[10] Pridantsev M V, Davidova L N and Tamarina I A 1990 *Construction steel. Reference book.* Mosocw: Mashinostroenie
[11] Nekrasov V I 1978 *Studies of the heat density of transmission units of heavy four-axle vehicles and ways to reduce it,* Diss. Ph.D, Kurgan
[12] Terekhov A S and Zaitsev A V 1998 On heat transfer in drive axles of automobiles *HUP Russian School on the Design of Heterogeneous Structures. Report thesis.* Miass, Chelyabinsk oblast.
[13] Nemzer G G and Aronov M A 1980 Study of steel thermo-physical properties *Forging and stamping production* **3** 26-30
[14] Neymark B E (Ed) 1967 *Physical properties of steel and alloys used in power engineering. Reference book* Moscow: Energiya.

IOP Conf. Series: Materials Science and Engineering 1151 (2021) 012008 doi:10.1088/1757-899X/1151/1/012008

Analysis of power losses in catenary wires in AC electric traction power supply systems when installing a device for shunt compensation of reactive power

A V Agunov, A D Komissarov, V V Seronosov, D A Sokolov and O A Stepanskaya

Emperor Alexander I St. Petersburg State Transport University, 9, Moskovsky ave., Saint Petersburg, 190031, Russia

E-mail: sokoloffffffff@gmail.com

Abstract. The research is aimed at studying the catenary wires power losses change by transverse capacitive compensation with considering the movement of trains on the intersubstation zone. Calculations instantaneous and average catenary wires power losses while passing through the intersubstation zone one train and two successive trains are presented in this article. Losses are determined for two cases: when installed compensating device and in its absence. The values of the differences of these losses are founded. The computer simulation in MATLAB-Simulink software was implemented. According to the results of calculations graphs were built, illustrating the change of catenary wires power losses while moving trains on the intersubstation zone when installed compensating device and without it, as well as the differences between the said losses. Average values of catenary wires power losses were determined by integrating these functions and their comparison is carried out. Calculations have shown that by selecting compensating device power by the average reactive power consumed by a train, its intersubstation zone passage with constant current causes the same average catenary wires power losses as in the absence of the compensating device. Increasing the number of trains, simultaneously located on the intersubstation zone at the same compensating device power, entails a reduction of catenary wires power losses with respect to a similar situation without the compensating device, but at the same time reactive power is compensated only partially.

1. Introduction

In the field of the AC railways, the energy efficiency of the traction power supply system when using shunt capacitive reactive power compensation is a very urgent problem. Due to the fact that the traction load has a sharply variable character, the choice of the optimal power of the compensating device is a certain difficulty.

When choosing the power of the compensating devices for the average reactive current consumed by trains, the most important features of the electric traction load are not taken into account - movement and change in the distances from the traction substations and the compensating device to the train.

Content from this work may be used under the terms of the Creative Commons Attribution 3.0 licence. Any further distribution of this work must maintain attribution to the author(s) and the title of the work, journal citation and DOI.
Published under licence by IOP Publishing Ltd

2. Calculations

As an example, let us consider an intersubstation zone with a length of L = 50 km. The voltage level in the catenary remains unchanged. The unit active resistance of the catenary is 0.1 Ohm/km. Let us assume that trains on this section consume a constant current I_{tr} = 100 A with a power factor cosφ = 0.8, then it is sinφ = 0.6. Based on the condition that the capacity of the compensating device is selected according to the average inductive current of the train I_{tr_ind}, the current of the compensating device I_{CD} will be 60 A. The scheme of the described power supply system is shown in Figure 1.

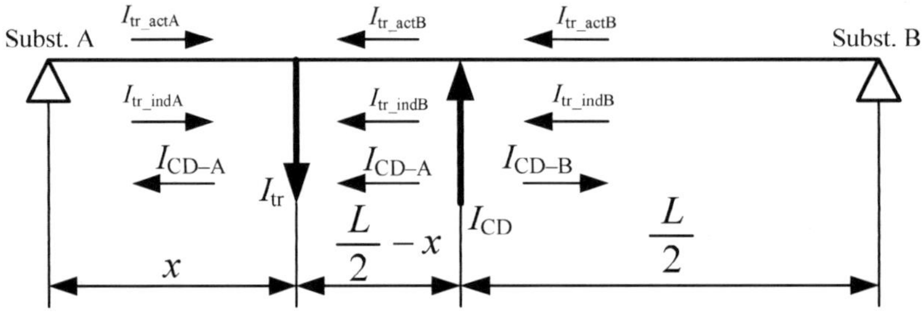

Figure 1. The scheme of the intersubstation area with one train indicating currents in the sections of the catenary.

Power losses in each section of the catenary are determined by the product of the square of the total current flowing in this section by its length and the unit active resistance of the catenary. It is worth paying attention to the section from the train to the compensating device: here, the reactive currents, unlike the other two sections, are summed up, and do not compensate each other. That is why the question arises about the effectiveness of installing a compensating device in terms of reducing power losses in the wires of the catenary, since in this area these losses exceed those in the absence of a compensating device. The total power losses in the catenary wires throughout the intersubstation area at each moment of time are determined by the sum of losses in the considered sections.

In the absence of a compensating device, the power losses in the catenary wires are the sum of losses in the sections from Substation A to the train and from the train to Substation B. Figure 2 shows the graphs of changes in the total power losses in the catenary wires in the presence (curve 1) and absence (curve 2) of a compensating device.

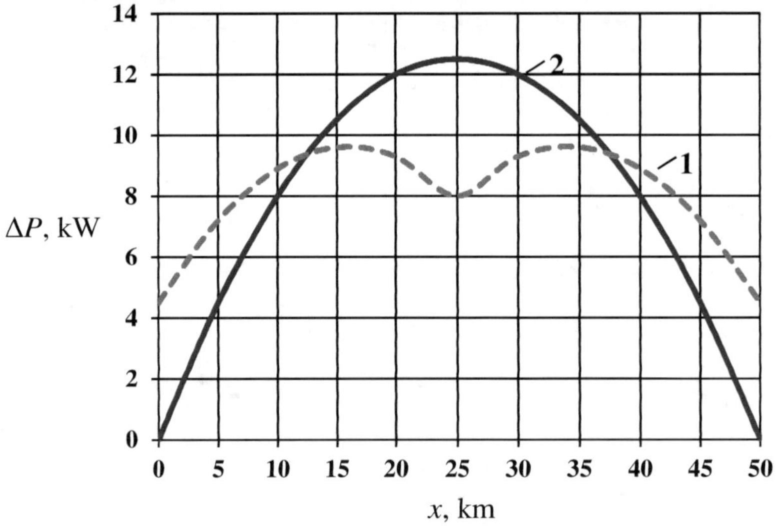

Figure 2. Distribution of power losses in the catenary wires when one train passes through the intersubstation area.

Integration of the functions of the considered losses allows determining their average values and making a comparison. The average value of power losses in the catenary wires with the passage of one train and the presence of a compensating device is 8.3 kW, in its absence it is 8.3 kW too.

That is, in this case, full compensation of the consumed reactive power occurs, and the average power losses in the catenary wires in the presence of the compensating device are equal to similar losses without it. However, when the train leaves the intersubstation area, the compensating device continues to cause a capacitive current to flow in the catenary, which entails power losses in it.

Let us consider the case when two trains appear on the same intersubstation area, following each other at a constant speed at a distance of 5 km, and each of them consumes an average current I_{tr} = 100 A (Figure 3). In this case, the capacity of the compensating device is still selected according to the average inductive current of one train. Graphs of changes in power losses in the wires of the contact network in the presence (curve 1) and absence (curve 2) of the compensating device are shown in Figure 4. In this case, x is the distance from Substation A to the first train.

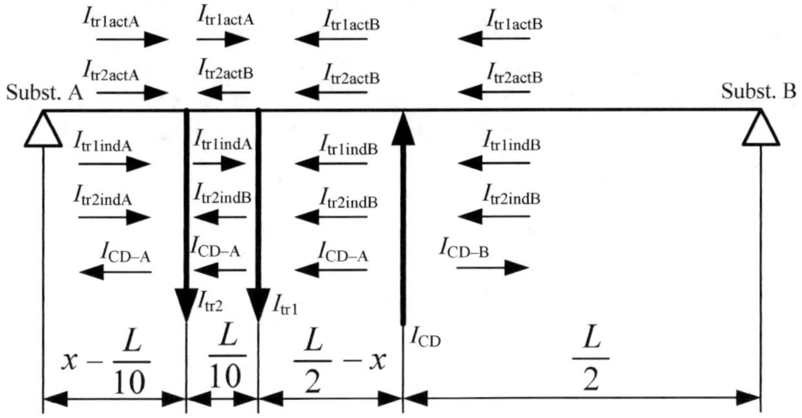

Figure 3. The scheme of an intersubstation zone with two trains showing currents in the sections of the catenary.

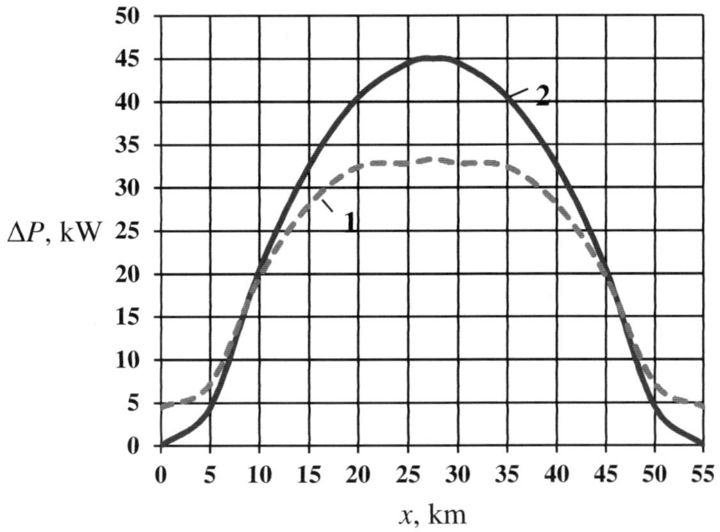

Figure 4. Distribution of power losses in the catenary wires when two trains passes through the intersubstation area.

The average values of power losses in the catenary wires obtained as a result of integration when two trains pass through the intersubstation area are as follows: in the presence of a compensating device $\Delta P_{CDav} = 21.1$ kW, and in its absence, $\Delta P_{0av} = 26.6$ kW.

To compare the power losses in the catenary wires in the absence of a compensating device and in its presence, the value of $\Delta\Delta P$ is introduced, determined by the difference of these parameters. If its value is greater than zero, then the effect of installing a compensating device in terms of reducing power losses in the catenary wires is positive, if it is less than zero, it is negative.

The average value of $\Delta\Delta P$ when two trains pass through the intersubstation area is determined by the difference in the average value of power losses in the wires of the contact network in the absence of a compensating device and in its presence: $\Delta\Delta P_{av} = \Delta P_{0av} - \Delta P_{CDav} = 5.5$ (kW).

That is, for the given parameters, the installation of the compensating device, the power of which is selected according to the average inductive current of one train, when passing through the intersubstation area of two successive trains leads to the decrease in the average power losses in the catenary wires by 5.5 kW. However, it should be remembered that the compensation of the reactive power consumed by trains in this case is carried out only partially.

In order to speed up the calculations, a model of the considered intersubstation zone was created in the MATLAB-Simulink, which also takes into account voltage losses in the catenary and resistance of traction substations. The graphs of the change in power losses in the catenary wires when one train passes in the presence (curve 1) and absence (curve 2) of the compensating device for the model in Simulink are shown in Figure 5.

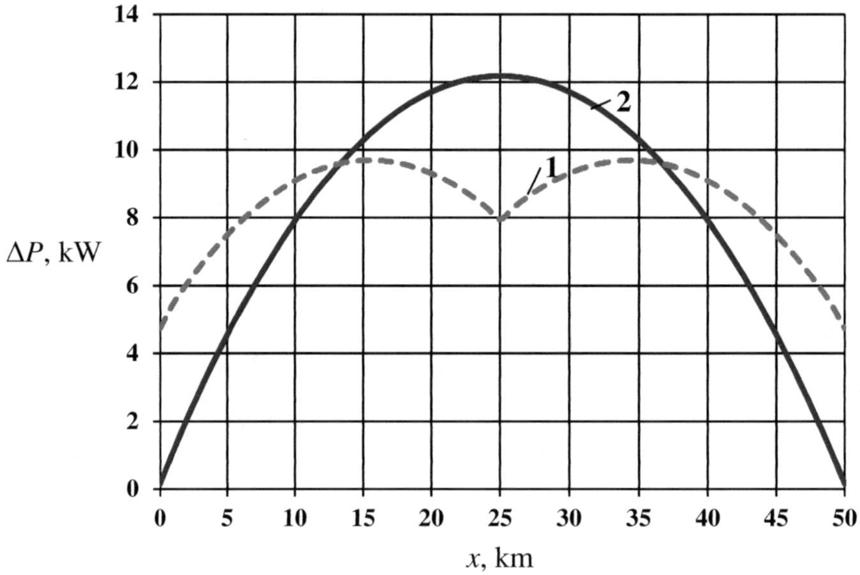

Figure 5. Distribution of power losses in the catenary wires when simulating the passage through the inter-substation zone of one train in Simulink.

The average value of power losses in the catenary wires when simulating in Simulink the passage of one train in the presence of a compensating device is $\Delta P_{CDav} = 8.4$ kW, in its absence – $\Delta P_{0av} = 8.2$ kW. The results of more accurate modeling are extremely close to the results of analytical calculations. In addition, the average power losses in the presence of the compensating device, taking into account additional influencing factors, turn out to be 2% more power losses in the catenary wires in the absence of the compensating device.

3. Conclusion

Ultimately, the choice of the capacity of compensating devices based on the average train current cannot be economically justified without analyzing the power losses in the catenary wires due to the fact that these losses increase with the installation of compensating devices in the intersubstation area, especially if trains run quite rarely, and it regularly remains without electric traction load.

At the same time, with intensive train traffic, such a choice of the compensating device power leads to a decrease in losses in the catenary, but the effect of reducing the consumption of reactive power in AC electric power supply systems decreases. An unreasonable increase in the power of compensating devices in order to increase the voltage also leads to an increase in power losses in the catenary wires. When choosing the power of compensating devices, taking into account the power losses in the catenary wires, it is necessary to consider the movement and the amount of current consumption of the train in the intersubstation area.

References

[1] German L A and Serebrykov A S 2015 *Regulated Installations of Capacitive Compensation in Traction Power Supply Systems of Railways* (Moscow: Training and Methodological Center for Education in Railway Transport)

[2] Lebedev A B 1937 *The Foundations of Electric Traction* (Leningrad – Moscow: Main Editorial Board of Energy Literature)

[3] Rozenfeld V E, Sidorov N N and Kuzin S E 1951 *Electric railways* (Moscow: Railway Transport Publishing House)

[4] Agunov A V, Marikin A N and Sokolov D A 2019 Computer modeling of the AC traction power supply system *Electronics and electrical equipment of transport* **5** 38-40

[5] Agunov A V, Varentsov V M, Marikin A N and Sokolov D A 2020 DC traction supply network power losses calculation with account of train traffic *Electrical engineering* **10** 27-30

The use of energy saving technologies in the service of a rolling stock

E A Milovanova[1] and M A Egorova[2]

[1] Railway Rolling Stock. Locomotives, Irkutsk state transport university, 15 Chernyshevskogo Street, Irkutsk, 664074, Russia
[2] Department of Foreign Languages, Baikal State University, 11 Lenin Street, Irkutsk, 664003, Russia

Abstract. The article deals with technical devices used at railway transportation for washing and cleaning of both rolling stock and its parts with a sufficiently large area of coverage of the product range which affect the treated objects or a jet of liquid with a detergent, or involve "bathing" of the object in the washing medium. At the same time the existing technology of cleaning and a design of washing machines do not allow to increase the efficiency of such devices, and the aggressive properties of liquids used in the process are not fully developed. The solution is seen in the intensification of technological processes of washing and cleaning of parts of the rolling stock by rational organization of the structure of the washing liquid flows to improve their performance. The article presents an assessment of the design features of traditional schemes of gas lift devices, the use of devices with forced gas supply for the implementation of technological processes in industrial production.

1. Introduction

Taking into account the fact that the development of transport communications is the basis of economic security of regions [1], Russia's transport strategy and development of competitive relations in transport [2] are associated with the use of new technologies, which is not limited to some specific fields (medicine, education, etc.). They can be used on railway transport for efficient management of logistics costs [3] as well as to minimize the negative impact on the environment. Furthermore, environmental and legal issues related to the use of environmental resources are already being considered [4] and environmental safety issues in the Russian Federation are being discussed [5].

JSC "Russian Railways" has adopted a Program to increase environmental responsibility. According to the adopted strategy, it was decided to reduce the burden on the environment from all types of activities 2 times by 2030. 2015 was declared the Year of waste management by the Company.

In recent years, systematic work on expansion of the landfill renewable energy project "smart station", a program for the elimination of coal-fired boilers, organized selective waste collection in office buildings and railway stations [6], development of approaches to creation of promising types of transport [7-11] have been also taken into account by today's demands for environmentally friendly projects.

In view of the above, the introduction of waste-free production and economical technological processes with the possibility of combining several technological operations attract much interest [12].

2. Discussion

Content from this work may be used under the terms of the Creative Commons Attribution 3.0 licence. Any further distribution of this work must maintain attribution to the author(s) and the title of the work, journal citation and DOI.
Published under licence by IOP Publishing Ltd

In various industries, various systems are used in technological operations, which mainly use the principle of interaction of gas and liquid [13]. Chemical and/or physical absorption is successfully used to separate gas mixtures and separate valuable components [14]. Chemical reactions such as hydroformylation, alkylation, oxidation, etc. are also widely used [15]. These technologies are implemented in various types of reactors (mixing and displacement reactors, reactor cascades, and multicellular apparatuses).

Gas lift devices provide high contact interaction between gas and liquid, are characterized by low energy consumption, while the design provides ease of retrofitting with heat exchange systems and the absence of moving parts, and also they are indispensable for cleaning and recycling of generated emissions and wastes [16-18]. That is why they are widely represented in various industries.

Due to the large amount of waste water and, as a result, high economic losses, JSC "Russian Railways" faces the urgent issue of creating and applying closed water circulation systems. In addition, the rational use of water resources is required. Thus, the annual water consumption of the East Siberian railway followed by the formation of wastewater is about 25 million m^3, and widely used methods such as sedimentation and filtration do not provide the proper level of wastewater treatment.

The solution of the problem in this area is seen in the improvement of existing and development of new methods of wastewater treatment, in changing the technology of designing and creating machines that increase the level of their safe operation, including environmental protection of service personnel.

A wide variety of designs of devices and machines for carrying out technological processes of cleaning [19-21] of parts and assemblies allows us to distinguish among them the most common ones in railway transport.

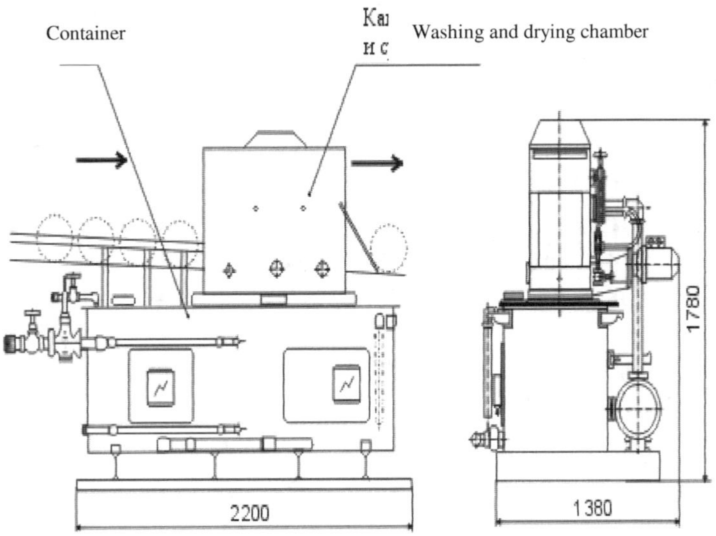

Figure 1. Rinsing apparatus MSP-1 for washing and drying roller bearings of box units of a rolling stock

The MSP-1 rinsing apparatus (Figure 1) is used for cleaning and drying car and locomotive bearings of box units.

The rinsing apparatus can be operated independently, as well as together with the bearing pressing position and the box rinsing machine, and all cycle operations are performed automatically. The feed of processing objects to the device is carried out along two lines, while the box bearings cross the washing chamber, in which they are rotated and treated with a

washing medium by means of a jet system at a temperature of up to 95 °C. Then the box bearings are sent to the drying chamber for processing with compressed air.
A washing chamber is designed for cleaning the parts of the rolling stock crew (Figure2). Cleaning of the surface of the wheel pairs is performed by means of metal brushes. The installation can be operated as an independent unit or as a part of an automated production line.

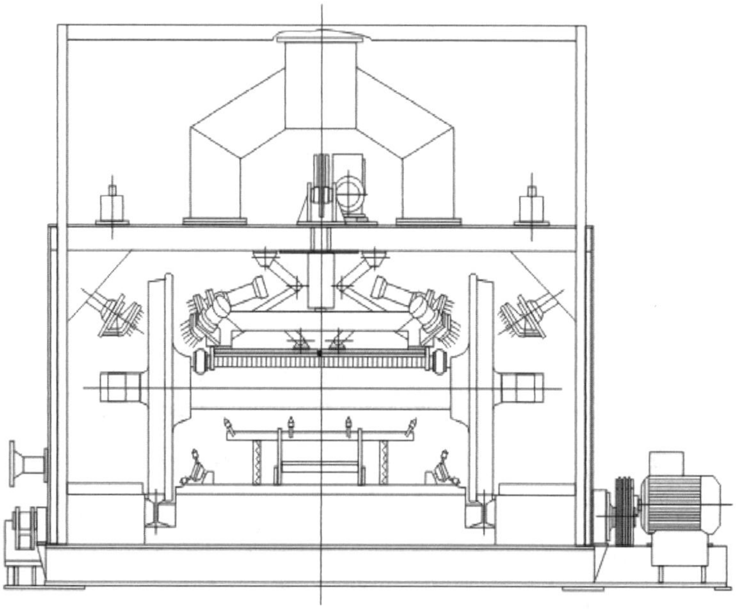

Figure 2. A cleaning chamber of wheel pairs of a rolling stock

Washing the surfaces of railway cars, wheel trolleys, components and parts of a rolling stock before their dismantling and disassembly is often carried out using a washing complex. The basis of it is a high-pressure hydraulic unit (30-150 barrels), which is an aggregate connected to the electrical network and the water supply network. The working parts of such installations are all kinds of spray heads. Water heating and a steam generator (heating boiler on diesel fuel) are provided, and there is some work with alkaline detergents for removal of especially heavy pollution.

The review of washing machines and devices for cleaning parts of a rolling stock given above shows that structurally they require an arrangement in a horizontal plane with the occupation of large production areas, which, in addition to the position directly for installing the washing machine, also require space for their maintenance and safe placement of maintenance personnel to prevent exposure to aggressive liquid media [22] during spray.

Such a technical design of washing machines limits the possibility of using the energy of the liquid medium flow in order to increase the efficiency of the technological process, since it mainly involves the impact on the treated objects either by a jet of liquid with a detergent, or by "bathing" the object in the washing medium. The existing cleaning technology and the design of washing machines do not allow increasing the efficiency of such devices, and the aggressive properties of liquids used in the technological process are not fully developed. Occasional attempts to introduce recycling cannot improve the current situation, which is complicated by the need for additional energy costs. The result is a large volume of toxic industrial effluents in the form of aggressive fluids.

At the same time, washing machines used for cleaning units and parts of a railway rolling stock often have a low degree of unification, which leads to an increase in the cost of equipment, often represented by piece production.

3. Results

The solution of the problem is seen in the intensification of technological processes for washing and cleaning parts of a rolling stock by organizing rationally the structure of the washing fluid flows to increase their productivity.

The achievement of this goal is based on the study of the influence of hydrodynamic factors and design parameters of gas-lift devices on the performance of the cleaning and rinsing process.

The classical design schemes of gas-lift reactors [19, 23] differ in generalized design features, such as: a vertically located container (reactor), usually cylindrical in shape with a partition that creates zones of ascending and descending gas-liquid flows (upstream zone (USZ) and downstream zone (DSZ)) with the possibility of communicating ascending and descending zones in the upper (above the partition) and lower (under the partition) parts of the reactor. Without a structural partition (material boundary) in the reactor the interaction of ascending and descending flows occurs throughout the entire volume of the reactor. A bubbler located at the bottom of the reactor in the downflow zone is supplied with gas (steam), which creates a moving circulation circuit of the gas-liquid mixture.

Schematic diagrams of gas-lift reactors are shown in Figure 3.

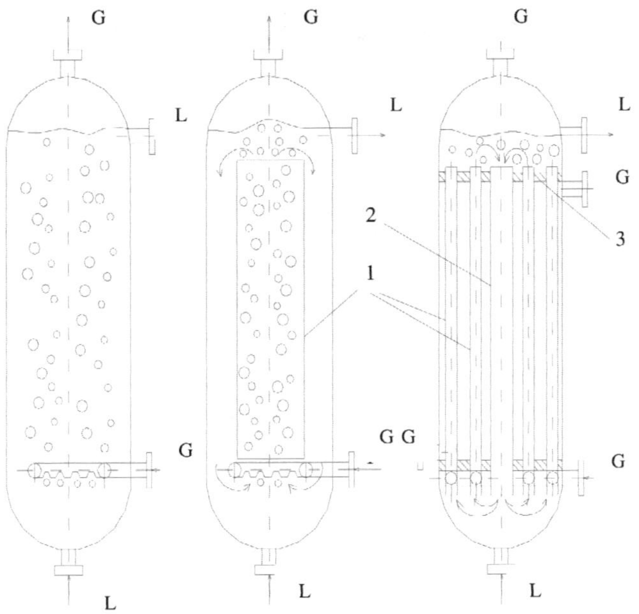

Figure 3. Schematic diagrams of gas-lift reactors: 1 - upstream zone; 2 - downstream zone; 3 – partition; L – liquid; G – gas

The best technical solutions, which originality is confirmed by patents for the invention and issued author's certificates [24-26], are presented collectively in the schematic diagrams of classical structures, and the increase in the intensity of the liquid flow is traditionally carried out by using various types of flow elements and heads with technological holes in them and other modifications.

The recognized advantages of reactors made according to traditional schemes are: technical simplicity of the design; ease of its modification by adding additional elements to the reactor design that allow the device to be used for a wide range of needs in various industries; the ability to regulate smoothly the gas content by changing the intensity of gas (steam) supply; immobility of the basic design of elements; the ability to develop aggressive properties of the working fluid due to the optimal choice of its number of circulation cycles in the gearbox design.

The offered improvement of traditional schemes of gas-lift reactors is based on the experience of inefficient use of energy of descending liquid flows in the DSZ, which is known from experience, to ensure a high-quality impact on processing objects, and, thus, increase the productivity of this process [27, 28].

In order to eliminate the considered shortcomings, constructive measures are suggested to introduce "passive" elements into the reactor device, i.e., elements that do not require additional energy consumption for their operation, which contribute to the intensification of mass exchange processes by changing the structure of gas-liquid flows.

Increase in the intensity of mass transfer processes in the downstream zone is provided, in addition to the open jet stream, by additional design changes, such as:

1. an additional low-flow bubbler is added to the downstream zone (at the bottom of the device), the capacity of which is about 15% of the total gas capacity (taking into account the bubbler located in the upstream zone).

2. a "passive" element is introduced into the design of the device, which is a horizontal partition with through holes in the form of jet nozzles for creation of an open jet flow of a gas-liquid mixture in the downstream zone. The passage of gas in the DSZ is carried out due to the valves made in the horizontal partition. The offered changes are reflected in the diagram of the reactor with an open jet flow in Figure 4.

Laboratory tests performed with the reactor corresponding to the diagram in Figure4 showed an increase in the level of gas saturating, compared with traditional schemes, in the area of downstream at about 15 % to equalize the level of the gas-liquid mixture in the zones of upward and downward flows, taking into account the gas content of φ_g =0.4 in DSZ.

When testing, the intensity changes of the adsorption rate of oxygen with a solution of sodium sulfate are in dependence on the geometric parameters of flow holes F_c in the horizontal partition and the area F_{DSZ} sectional area downstream. And depending on the level of placement of horizontal partitions in the zone, the downward flow was determined.

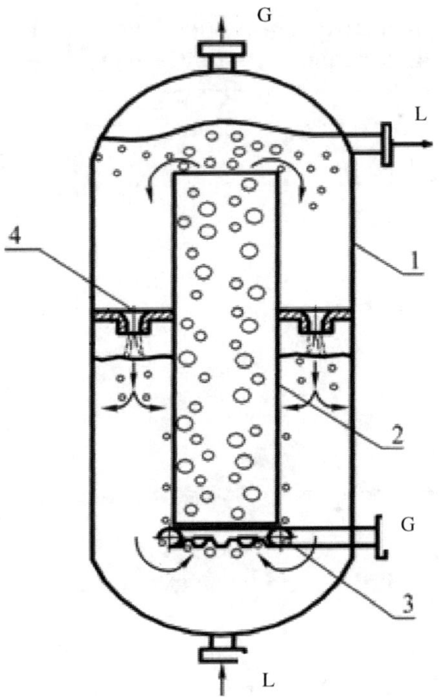

Figure 4. Schematic diagram of a reactor with an open jet flow in the downstream zone: 1 - chamber; 2 - circulation pipe; 3 - bubbler; 4 - horizontal partition

The obtained data on changes in the oxygen sorption rate are shown in tables 1 and 2, where the number of rows is the number of measurements, and the columns reflect structural changes in the classical scheme.

Table 1. Dependence of the change in the oxygen sorption rate on the diameter of the flow hole in the horizontal partition in the downstream zone

No. n / a	F_c / F_{DSZ}						Without a partition in DSZ
	0.01	0.02	0.03	0.04	0.0 5	0.06	
1	2.8	4.5	4.6	4.6	4.2	4.1	3.6
2	2.9	5.0	4.9	4.8	4.2	4.2	3.8
3	2.7	4.5	4.6	4.7	4.4	3.9	3.8
4	3.2	5.2	5.1	4.9	4.6	4.5	3.7
5	3.1	5.2	5.0	5.0	4.8	4.5	4.1
6	3.2	5.1	5.3	4.9	4.5	4.6	4.3
7	3.1	5.2	5.1	4.9	4.7	4.5	4.1
8	3.3	4.5	4.5	4.4	4.3	3.9	4.2
9	2.9	4.4	4.6	4.2	4.3	3.8	3.8
10	2.7	4.4	4.3	4.2	4.0	3.8	3.5
in average	3.0	4.8	4.8	4.6	4.4	4.2	3.9
Δ%	-23	+23	+23	+18	+13	+8	–

Table 2. Dependence of changes in the oxygen sorption rate on the level of placement of the horizontal partition in the downstream zone

No. n /a	h_p/ h_{vp}						
	0.2	0.3	0.4	0.5	0.6	0.7	0.8
1	2.1	3.9	4.1	4.7	4.8	3.3	2.6
2	2.3	4.0	4.2	5.0	5.1	3.5	2.8
3	2.3	3.7	3.9	4.7	4.8	3.5	2.8
4	2.2	4.3	4.5	5.2	5.3	3.4	2.7
5	2.6	4.3	4.5	5.1	5.2	3.8	3.1
6	2.8	4.4	4.6	5.4	5.5	4.0	3.3
7	2.6	4.3	4.5	5.2	5.3	3.8	3.1
8	2.7	3.7	3.9	4.6	4.7	3.9	3.2
9	2.3	3.6	3.8	4.7	4.8	3.5	2.8
10	2.0	3.7	3.9	4.4	4.5	3.2	2.5
in average	2.4	4.0	4.2	4.9	5.0	3.6	2.9
Δ%	-38	+23	+6.9	+25	+27	-8.4	-27

The results of the experiment are shown in Figure 5. The absence of a partition in the downstream zone (the classical scheme) shows an average oxygen sorption rate which is close to the recommended value [29].

The results of the measurements of sorption of oxygen in Table 1 correspond to a fixed position of the horizontal partition in the DSZ in height (h_p/ h_{vp}=0.6) in Table 2 corresponds to a constant cross-sectional area of the flow hole in the horizontal partition (F_c/ F_{DSZ}=0.02). The last line of Tables 1 and 2 informs about the percentage change in the oxygen sorption rate compared to the device operating according to the traditional scheme (the "+" sign corresponds to an increase in the rate of oxygen dissolution; "–" – to a decrease).

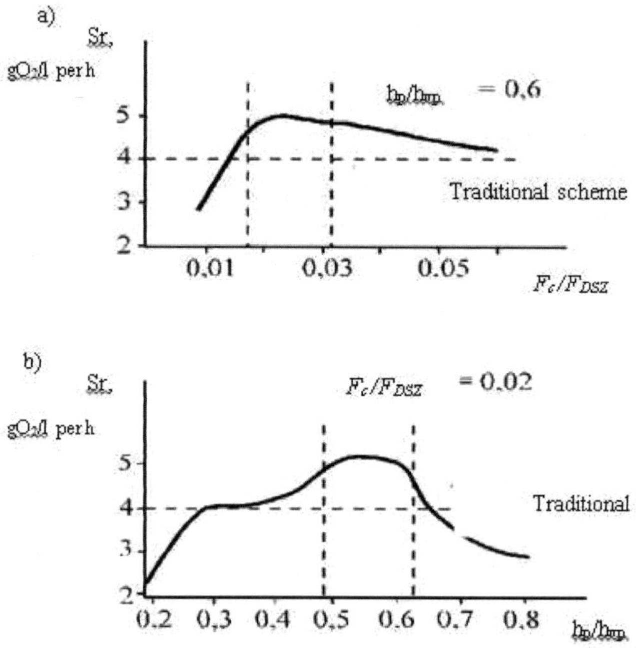

Figure 5. Change in the rate of oxygen sorption rate (Sr) a) depending on the diameter of the flow hole in the horizontal partition in the DSZ; b) depending on the level of placement of the horizontal partition in the DSZ

4. Conclusion

The results of comparative mass transfer studies allowed us to formulate recommendations for choosing the relative area of the flow hole in the horizontal partition of the F_c/F_{DSZ}=0.025 and the level of its placement h_p/h_{vp}=0.55. Gas-liquid reactors with a modified flow structure of the gas-liquid mixture, made by introducing additional structural elements into the design, allowed increasing the productivity of the technological process by up to 30 % compared to the traditional scheme.

References

[1] Shalak A V 2017 Transport communications as basis of economic security of the Siberian region: geopolitical aspect *Proc. Int. Conf. on Evroaziatskoe sotrudnichestvo, Materialy mezhdunarodnoj nauchno-prakticheskoj konferencii* (Irkutsk: Baikal state university) pp 219-225

[2] Trofimov S V 2018 The Russian transport strategy and the development of competition relationships in transport *Competition law* **3** 14-16

[3] Astrahanceva A S 2018 Aproaches to the use of innovation technologies in the railway transport as a means of effective management of logistics costs **4** 147-151 (Moscow: "Jekonomika")

[4] Rybinskaja E T 2017 Ecological and leagal problems of the use of energy resources *Sovremennye tendencii v socialno-jekonomicheskih i gumanitarnyh naukah: teorija i praktika sbornik nauchnyh trudov* (Irkutsk) pp 238-243

[5] Shobodoeva A V 2018 Environmental safety of the Russian federation: theoretical and methodological aspects *Izvestija Irkutskogo gosudarstvennogo universiteta Serija: Politologija. Religiovedenie* **25** 35-42

[6] Gapanovich V A 2015 Environmental strategy in action *Lokomotiv - The journal is a specialized monthly* **12 (708)** 2 – 4

[7] Milovanova E A, Milovanov A A and Milovanov A I 2014 Search for new ways of development of the transport system in the Eastern TRANS-Urals of Russia *Magnitolevitacionnye transportnye sistemy i tekhnologii- Peer-reviewed research journal* 113-125

[8] Gozbenko V E, Kargapoltsev S K, Minaev N V, Karlina A I 2016 Simulation of the vibration of the carriage asymmetric parameters in mathcad *International Journal of Applied Engineering Research* **23** 11132-11136

[9] Ahmadeeva A A, Gozbenko V E 2010 Dynamic properties of the car with two-stage spring suspension *Sistemnyj analiz. Modelirovanie - The journal of Irkutsk state transport university* **3 (27)** 60-69

[10] Ahmadeeva A A, Gozbenko V E 2011 Determination of the main coordinates of the car with a two-stage spring suspension *Sovremennye tekhnologii. Sistemnyj analiz. Modelirovanie - The journal of Irkutsk state transport university* **4 (32)** 71-76

[11] Gozbenko V E, Kargapolcev S K, Karlina A I 2016 Bringing a dynamic system with three degrees of freedom to the main coordinates *Sovremennye tekhnologii. Sistemnyj analiz. Modelirovanie - The journal of Irkutsk state transport university* **3 (51)** 35-38

[12] Timofeev V S, Serafimov L A 2003 *Principles of basic organic and petrochemical synthesis technology* (Moscow: Vysshaya shkola Publ.)

[13] Dankverts P V 1973 *Gas-liquid reactions* (Moscow: Himiya Publ.)

[14] Frolov V F 2003 *Lectures on the course "Processes and devices of chemical technology"* (Pb.: Himizdat Publ.)

[15] Potekhin V M, Potekhin V V 2005 *Fundamentals of the theory of chemical processes technology of organic substances and oil refining* (SPb.: Himizdat Publ.)

[16] Kutepov A M, Bondareva T I, Berengartep M G 2004 *General chemical technology* (Moscow: IKC «Akademkniga» Publ.)

[17] Krivoshein D A, Kukin P P 2003 Engineering protection of surface water from industrial effluents (Moscow: Vysshaya shkola Publ.) 344 p
[18] Haliullin A K, Salaurov V N 2002 Basics of industrial ecology (Irkutsk: Ottisk Publ.) 268 p
[19] Sokolov V N, Domanskij I V 1976 Gas-liquid reactors (Leningrad: Mashinostroenie Publ.) 216 p
[20] Bulatov S N, Kernerman VA 1982 *Avtorskoe svidetelstvo SSSR No.939060*, 30.06.
[21] Shishkin A V, Barabash I I 1986 *Avtorskoe svidetelstvo SSSR No.1214191*, 28.02.
[22] Kargapolcev S K, Volkov A N, Milovanov A I, Milovanova E A 2003 Detergents Deponirovannaya rukopis No. 629.488.2 14.11.
[23] Kasatkin A G 1973 Basic processes and apparatus of chemical technology (Moscow: Himiya
[24] Erdyakov YU V, Ermakov S B 1980 *Avtorskoe svidetelstvo SSSR No.766629*, 30.09.
[25] *Patent No.1525930, 1978.*
[26] Patent USA No.3630848
[27] Kardashev G A 1990 Physical methods of intensification of chemical technology processes (Moscow: Himiya Publ.) 206 p
[28] Viestur U E, Kuznecov A M, Savenkov V V 1988 *Sistemy fermentacii* (Fermentation systems) (Riga: Zinatne Publ.) 368 p
[29] Method of determining the sulfite number using a dissolved oxygen sensor 1976 (Moscow, VNIIENERGETIKA) 54 p

Modeling a boiler unit in the automation systems of transport infrastructure

N N Kutsy and A A Podkorytov

Irkutsk National research technical University, 83, Lermontova st., Irkutsk, 664074, Russia

E-mail: podkorytovleha@mail.ru

Abstract. The goal is to develop a model that takes into account the dynamics of the operation of the combustion device of the boiler unit. The article discusses the application of the technique of D.Zh. Box and G. Jenkins to identify the steam production process. A boiler unit used in transport infrastructure automation systems was chosen as the object of research. The boiler unit is a dynamic stochastic object with uncontrolled disturbing influences. Using statistical methods, a mathematical model has been obtained that can be used to predict and regulate the operating modes of the boiler unit.

1. Introduction

It is hard to imagine today's life without equipment that uses the energy and properties of water vapour. This includes various types of boiler units, turbine generators, etc. A boiler unit as a control object is widely used in almost all spheres of life: agriculture, light and food industries, and the chemical industry. It is also worth noting the widespread use of boiler units in transport infrastructure systems for heating railway stations, heating fuel, etc.

The efficient operation of the boiler unit is impossible without the organization of a high-quality regulation process at each of its stages, which starts from water treatment, water supply to the boiler drum and ends with ensuring the correct operation of the combustion chamber. To provide a given combustion process, it is necessary to constantly supply a certain amount of fuel and the corresponding amount of secondary air to the furnace. This function is performed by the fuel-air regulator, which maintains the specified ratio between the amount of fuel and air throughout the entire range of fuel delivery. The amount of supplied fuel varies depending on the steam load of the boiler unit. However, this regulator has its drawbacks, since it does not take into account the quality indicators of the fuel: composition, temperature, humidity, etc.

An alternative to this regulator is a regulator that takes into account the oxygen content in the flue gases, keeping it within the specified limits (usually 1-1.5%). The oxygen content in the flue gases is a qualitative indicator indicating the optimal ratio of fuel and air supplied to the furnace.

2. Identification

This article deals with the development of a dynamic stochastic model that takes into account the relationship between the secondary air flow rate and the oxygen content in the exhaust gases.
The following technological parameters were selected as the studied data:

Content from this work may be used under the terms of the Creative Commons Attribution 3.0 licence. Any further distribution of this work must maintain attribution to the author(s) and the title of the work, journal citation and DOI.
Published under licence by IOP Publishing Ltd

- secondary air pressure, X
- oxygen content in flue gases, Y.

To obtain data, the method of passive experiment was used. The data were obtained by observing the normal course of the steam generation process on a steam boiler of the BKZ-420-140 brand.

The volume of data for the study is 1800 measurements with a sampling rate of 10 seconds.

Since the measured values of these parameters are in different ranges (percentage of oxygen in flue gases and secondary air pressure in kPa), for further analysis it is necessary to bring these time series to a single form. To do this, we will perform the alignment procedure for each row.

As an example, Figure 1 shows the time series: secondary air pressure and oxygen content in exhaust gases in a standardized form [1].

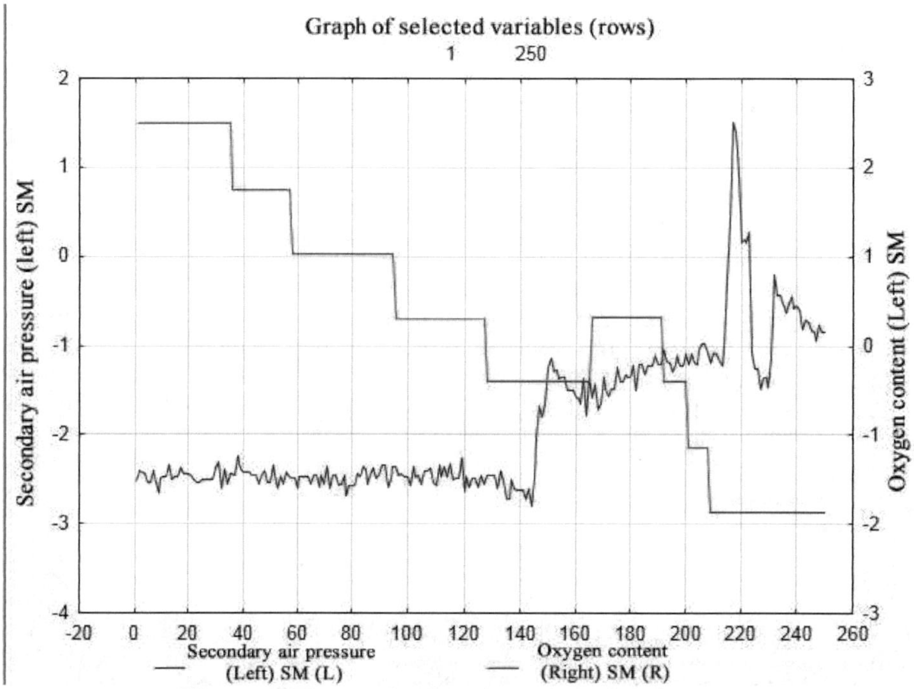

Figure 1. The graph of the investigated time series

In the course of studying the time series for stationarity, it turned out that these time series are non-stationary. Since the method of Box and Jenkins can be used only for stationary series, it is necessary to bring the studied series to a stationary form by taking successive differences. In this regard, for the above time series, difference time series were obtained using the operator for taking differences:

$$x_t = \Delta^d X_t;\ y_t = \Delta^d Y_t \ with\ d > 0, \quad (1)$$

where d is the order of the difference; xt, yt - normalized values of time series:

$$x_t = (X_t - \overline{X}_t)/\sigma_x,\ y_t = (Y_t - \overline{Y}_t)/\sigma_y, \quad (2)$$

in which

$\overline{X}_t$, $\overline{Y}_t$ – mean values of the series, σ_x, σ_y – standard deviation.

Bringing the series to a stationary form in the future will make it possible to use the method of cross correlation functions to determine in the structure of the model those lag times for which the coupling coefficients between the secondary air pressure and the oxygen content in the exhaust gases are most significant [2].

As an example, Figure 2 shows a graph of the cross-correlation function Rxy (k) between the observed series, obtained as a result of processing statistical material.

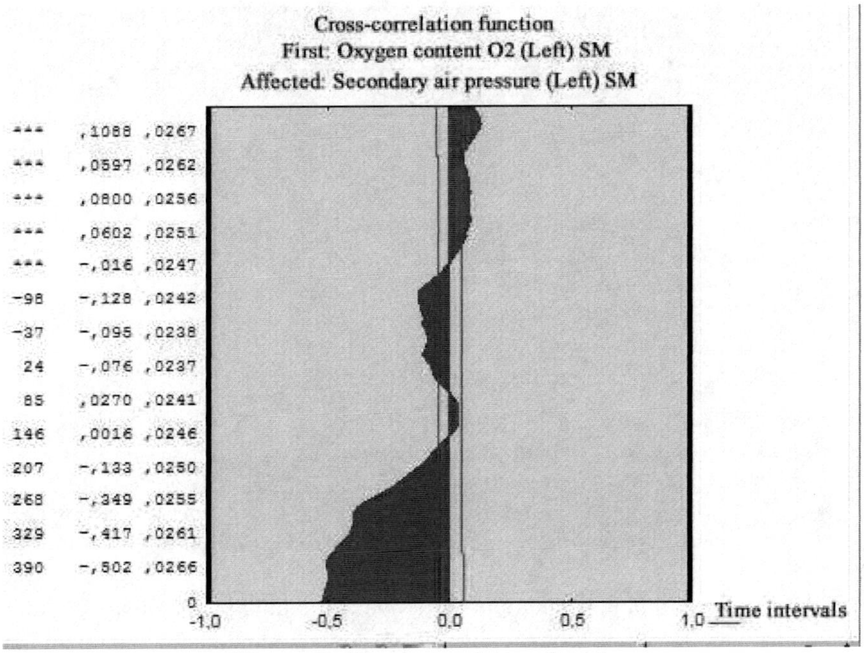

Figure 2. Cross-correlation function

A visual analysis of this graph does not allow making an unambiguous conclusion about those time intervals at which the secondary air pressure on the left side significantly affects the oxygen content in the flue gases on the left, since the mechanism of interaction is veiled by the correlation of the values of the input series, but it helps to determine the significance of the coefficients of cross-correlation [3].

Further, in accordance with the methodology, we will carry out a preliminary identification of the studied time series to establish the correspondence of the selected class of the model with the available experimental data. The main criterion for identification is the behavior of the autocorrelation and partial autocorrelation functions.

Let us construct for the investigated time series models of autoregressive and moving average (ARSS) [4]:

$$\alpha_t = x_t - \sum_{i=1}^{p} \Phi_i x_{t-i} + \sum_{j=1}^{q} \Theta_{t-j} \alpha_{t-j};$$

$$\beta_t = y_t - \sum_{i=1}^{p} \Phi_i y_{t-i} + \sum_{j=1}^{q} \Theta_{t-j} \beta_{t-j}. \quad (3)$$

where α_t, β_t – aligned series, respectively, for input and output difference series; Φ_i – parameter values for auto regression model; Θ_{t-j} – parameter values for the moving average model; p – autoregressive model order; q – order of the moving average model.

The time series of the secondary air flow rate is described by the ARIMS autoregressive model (2 1 0) p = 2; q = 0. The time series of the oxygen content in the exhaust gases is described by the model (1 1 0) p = 1; q = 0.

Next, for the obtained time series, we construct a cross-correlation function (Figure 3).

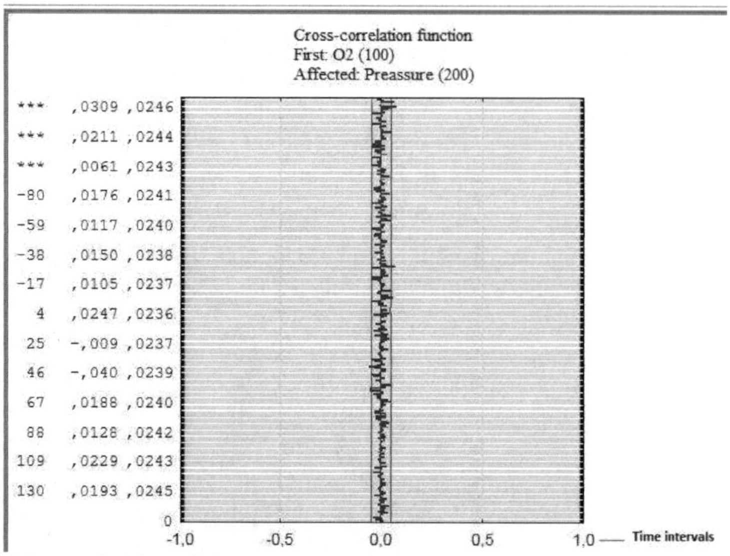

Figure 3. Plot of the cross-correlation function of aligned series

To obtain estimates of p, q, ФI, a nonlinear least squares algorithm was applied [5].

3. Estimation

Dynamic stochastic models of the influence of secondary air on the left side on the oxygen content in the exhaust gases were obtained using the Box-Jenkins method in the class of models yt = δ-1 (B) ω (B) xt-b + nt,
where B – operator of shift back one step, b - delay parameter [6]:

$$\delta(B) = 1 - \delta_1 B - \delta_2 B^2 - \ldots - \delta_r B^r;$$

$$\omega(B) = \omega_0 - \omega_1 B - \omega_2 B^2 - \ldots - \omega_s B^s. \tag{4}$$

The developed dynamic stochastic models are presented below.

Dependence of the oxygen content in the exhaust gases on the secondary air flow rate:

$$(1 - \underset{\pm 0.023}{0.307} B - \underset{\pm 0.023}{0.049} B^2) \Delta O2_{left} (t) =$$

$$= -\underset{\pm 0.065}{0.035} \Delta f(t-34) - \underset{\pm 0.065}{0.05} \Delta f(t-39) - \underset{\pm 0.065}{0.047} \Delta f(t-55) - \underset{\pm 0.065}{0.044} \Delta f(t-56) - \underset{\pm 0.065}{0.031} \Delta f(t-64) + \underset{\pm 0.065}{0.05} \Delta f(t-239) + \underset{\pm 0.065}{0.043} \Delta f(t-254) - \underset{\pm 0.065}{0.049} \Delta f(t-364) - \underset{\pm 0.065}{0.048} \Delta f(t-371) + n_t \tag{5}$$

where Δ are the first differences, and the values under the coefficients are their standard errors.

4. Diagnostic check

The obtained models are analysed for the adequacy of the real process of influence of the secondary air flow rate on the oxygen content in the exhaust gases in the furnace. The analysis is performed using a diagnostic check [7]. This check consists of two stages. Initially, the coefficient χ^2 is calculated - statistics for the values of the autocorrelation function of residual errors r_{aa} (k) as Q = (N-s-b-r); $\sum_{k=1}^{k} r$ αα^2(k). Here N is the number of observations, k is the maximum delay of autocorrelations and cross-correlations, S is the number of "right-handed" parameters of the dynamic stochastic model, r is the number of "left-handed" parameters. Then χ^2 statistics are calculated using the cross correlation functions r aa (k) between the equalized input series t and a series of residual errors, and now $H = (N\text{-}s\text{-}b\text{-}r) \sum_{k=1}^{k} r\ (k)$[8,9].

In the first case, Q is compared with χ^2 – distribution with K-p-q degrees of freedom, and in the second - H is compared with χ^2 – distribution with K-r-S degrees of freedom [10]. The calculated value of H should not exceed the tabular value for the investigated number of degrees of freedom with a significance factor of 0.05.

As the results of the diagnostic check in table 1 show the values of the coefficients χ^2 – statistics on the autocorrelation function.

Table 1. Coefficient values χ^2 statistics.

Input	output	Number degrees of freedom	*H*	Number degrees of freedom	*Q*
air pressure	O^2	30	38.45	30	43.8

Diagnostic testing for autocorrelation and cross-correlation functions using the values of χ^2 statistics does not give grounds for doubting the adequacy of the model [11,12].

5. Conclusion

As a result of the research carried out, a model has been obtained that makes it possible to evaluate the effect of the secondary air flow rate on the oxygen content in the exhaust gases.

The model can be used to predict and control the boiler units used in the transport infrastructure automation system.

6. Acknowledgments

The reported study was funded by RFBR, project number 19-31-90083\19

References

[1] Pletnev G P 2007 *Automation of technological processes and production in heat power engineering. 4th ed.* Moscow: Publishing House. 115-156

[2] Rotach V Y 2005 *Automatic control theory* Moscow: Ed. House of MEI

[3] Voloshenko A V 2011 *Schematic diagrams of steam boilers and fuel supplies.* Tomsk: TPU Publishing House.

[4] Serov E P and Korolkov B P 1981 *Steam generator dynamics.* Moscow: Energoizdat Publishing House.

[5] Khapusov V G and Baev A V 2014 Mixed autoregressive models and forecasting of the steam generation process *Bulletin of the Irkutsk State Technical University* 29–34

[6] Khapusov V G 2007 *System modeling.* Irkutsk: Publishing house of IrNITU

[7] Shorokhov V A and Smolnikov A P 2009 Development of a dynamic model of a multiply connected ACS of a pulverized coal block with direct injection of dust *Teploenergetika* 56–61

[8] Arakelyan E K and Pikina G A 2007 *Optimization and optimal control* Moscow: MPEI Publishing House.

[9] Box G and Jenkins G M 1976 *Time Series Analysis, Forecasting and Control, San Francisco: Holden-Day*, pp. 208-214

[10] Kuprienko N V, Ponomareva O A and Tikhonov D V 2015 *Statistics, time series, trend analysis and forecasting, study guide*

[11] Afanasiev V N and Yuzbashev M M 2010 *Time series analysis and forecasting, finance and statistics,* pp. 143-158.

[12] Sadovnikova N A and Shmoilova R A 2016 *Time series analysis and forecasting. Textbook,* Sinegeria University Publishing House, pp. 212-243.

IOP Conf. Series: Materials Science and Engineering 1151 (2021) 012011 doi:10.1088/1757-899X/1151/1/012011

Features and prospects of development of transit container transportation on the Northern sea route

E S Yudnikova[1] and N M Bedashov[1]

Eemperor Alexander I St. Petersburg State Transport University, 9 Moskovsky Ave., St. Petersburg, 190031, Russia

E-mail: ya.kolia60@yandex.ru

Abstract Due to the congestion of the main transit container corridors, there is an urgent need for the development of alternative container lines. This trend prompted us to conduct a study of an alternative sea corridor to the Suez Canal. In this paper, we have considered the features and prospects for the development of transit container transportation along the Northern Sea Route. We analyzed many factors that influenced the complexity of transportation by the Northern Sea route, such as severe weather conditions, undeveloped infrastructure, and others. We studied the geography of the Northern Sea Route, the natural conditions along its entire length, and identified the main ports along the route. We conducted a comparative analysis between the Northern Sea Route and its main competitor, the Suez Canal. We found out that there are prospects for the development of the Northern Sea Route and its economic benefits over competitors. We were convinced that the Northern Sea Route is a strategically important project for Russia. We reviewed the transport strategy for the development of the Northern Sea Route until 2030. They presented forecasts for the development of the Northern Sea Route. They gave an example of an experimental flight of a Maersk vessel passing along the Northern Sea Route. Based on our research, we were able to make forecasts for the development of transit container traffic along the Northern Sea Route.

1. Introduction

Prior to the coronavirus (COVID-19) pandemic, maritime transport was considered one of the most cost-effective and mass-market modes of transport with more than 80% of all global cargo flows accounted for by sea. For the further development of sea container transportation, it is necessary to explore new routes for cargo delivery. This route is the Northern Sea Route, perhaps the most discussed transport corridor in recent times.

The Northern Sea Route runs along the Arctic Ocean, along the Siberian coast of Russia, connecting all the northern ports, river mouths, into one single shipping hub. From the west, the NSR is bounded by Cape Desire, and from the east by the Coastal Strait.

It is important to note that the length of the NSR is 5.600 kilometres

Many experts predict an increase in the volume of cargo transportation along the Northern Sea Route, but for the development of this direction, it is necessary to create a modern infrastructure. For Russia, the development of the Northern Sea Route is a strategically important task, as the Northern Sea Route unites about 70% of the Russian space.

Content from this work may be used under the terms of the Creative Commons Attribution 3.0 licence. Any further distribution of this work must maintain attribution to the author(s) and the title of the work, journal citation and DOI.
Published under licence by IOP Publishing Ltd

2. Features of transit traffic by the Northern Sea Route

One of the main features of cargo transportation by the northern sea route is its climate, all the seas that are part of the NSR are characterized by extremely severe weather conditions. The average temperature in summer usually does not rise above +7 degrees Celsius, and the temperature in winter sometimes drops to -35 degrees Celsius.

In addition to the very cold temperature, the NSR has another feature, it is an ice layer throughout its entire length, and although in summer in some places the ice cover goes away for 2-3 months, but there are places where the ice will remain all year round. But every year, due to global warming, the summer period increases, which is the prospect of the development of the NSR [1].

In recent years, Russia has taken the development of the Arctic quite seriously. This development will be impossible without modern port infrastructure, now the main strategic ports along the length of the NSR are:

- Pevek;
- Tiksi;
- Providence Bay;
- Igarka;
- Dudinka;
- Dixon.

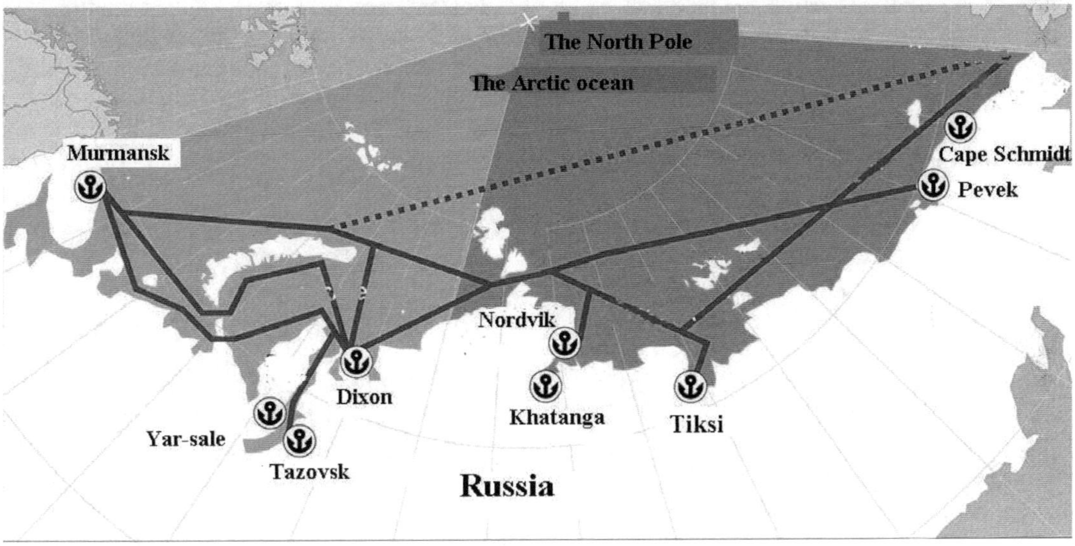

Figure 1. Cargo transportation along the Northern Sea Route

In the last few years, the Government of the Russian Federation has been actively implementing an economic project to convert the port of Petropavlovsk-Kamchatsky into a major hub port of the Northern Sea Route. The port of Petropavlovsk-Kamchatsky occupies a very favorable position in its region, with the presence of modern infrastructure, it can become an important transport hub between northern Europe and the east coast of the Pacific Ocean. In addition, the port of Petropavlovsk-Kamchatsky has a number of advantages, such as:

- year-round navigation;
- non-freezing bay;
- ability to sort and accumulate cargo [2].

Figure 1 shows the main transport corridors along the Northern Sea Route.

3. Competitors of the Northern Sea Route

The main competitor in the global shipping market for the Northern Sea Route is the Suez Canal. Its length is 193 km, and it is one of the busiest arteries in the world. Due to the increase in the size of ships, the Suez Canal ceases to cope with the load and its main problem today is the limitations on the size of ships that it can accept, which leads to the need to reload cargo from large vessels to the vessels of the Suez Canal Administration. In turn, this affects the time and financial costs. Therefore, the main problem of the Suez Canal is low capacity with growing demand for its services. This opens up growth opportunities for the Northern Sea Route.

The main role of the Northern Sea Route, like the Suez Canal, is to optimize transport routes between the Asia-Pacific region and Europe, but compared to the Southern Suez Canal, the Northern Sea Route has a number of advantages, and also because of the fairly recent start of its development, it has much more prospects in the future. Let's take a closer look at them.

In 2019, only 660 vessels passed through the waters of the NSR. For example, this is the average amount of time that passes through the Suez Canal in 10 days. In total, about 24,000 ships pass through the Suez Canal every year. Such a huge number of cargo ships often entails downtime during the passage of the Suez Canal, sometimes ships stand from 3 to 7 days waiting for their turn to pass.

Compared to the Suez Canal, the volume of traffic along the Northern Sea Route is not significant and amounted to 27 million 593 thousand 76 thousand tons as of November 29, 2019, which is almost 70% higher than in the same period last year.

At the same time, the volume of transit cargo traffic showed an increase and amounted to about 40% (697.277 thousand tons compared to 491,342 thousand tons in the comparable period of 2018) .

The total tonnage of ships passing through the Suez Canal in 2019 was 1.17 billion tons.

The Suez Canal can be traversed in an average of 24 hours. Approximately 48 vessels pass through the NSR in a day, and the NSR can be crossed in 7-15 days under favorable external conditions and must be accompanied by a Russian icebreaker [3].

Through the NSR, navigation is possible only 5-7 months a year, unlike the Suez Canal, where navigation is possible all year round.

For the transition from China to Europe via the Suez Canal, an average of 35 days and 875 tons of fuel oil are needed, compared to the Northern Sea Route, where the figures will be 25 days and 625 tons of fuel oil [3].

A passage through the Suez Canal costs $ 250.000. The NSR does not have direct fees for the passage itself. However, the fee for icebreaking support costs about $ 300.000. It is also worth including such a line of expenses as an additional insurance premium due to piracy in the Gulf of Aden and the Red Sea. That's another $ 120.000. Insurance for passage on the NSR costs $ 70.000 [3].

For comparison, let's give an example of the cost of sending a ship from the port of Yokohama to the port of Kirkines, through the Northern Sea Route, the duration of the journey will be 26 days and $ 500.000 for maintenance, when through the Suez Canal the journey will take 44 days and cost $ 1.000.000, respectively [3].

We also need to compare the distance and time of delivery of various goods through the Suez Canal and the NSR.

The way of the container ship from Murmansk to Japan will be: through the Suez Canal – 12291 miles and 37 days, and through the NSR-6010 miles and 18 days, total savings – 19 days and 6281 miles.

Cargo delivery from Rotterdam to Asian ports (Japan, Republic of Korea, China): through the Suez Canal-an average of 10.600 miles and 33 days, through the NSR 7610 miles and 23 days, total savings – 10 days [3].

The above figures show us that, in theory, the passage through the NSR is cheaper and faster than that through the Suez Canal. However, all these advantages are offset by a number of factors:

- high risks due to outdated infrastructure, poor quality of services;
- lack of icebreaker fleet to escort in the right place and at the right time to avoid the threat of ice compression;

- NSR can only be used by vessels with enhanced protection measures against ice and hummocks [4].

The conclusions of section 2 are presented in table 1.

Table 1. Comparison of routes through the NSR and through the Suez Canal.

	The Suez canal	Northern Sea Route
Distance	10.600 miles	7.610 miles
Travel time	44 days	33 days
The cost of the route	1 000 000 $	500 000 $
Pirates	Yes	No
Navigation	All year round	6-7 months
Climatic conditions	Suitable for all loads	Not suitable for many loads

4. Prospects for the development of the Northern Sea Route

Russia invests huge efforts and funds in the development of the NSR, this is clearly demonstrated by the transport strategy for the period up to 2030 on the development of the Northern Sea Route. The following are the main provisions of measures for the goals, objectives and stages of implementation of the Transport Strategy of the Russian Federation for the period up to 2030:

1. Development of the Northern Sea Route:

- construction of 3 universal nuclear icebreakers of the Russian Maritime Register of Shipping class Icebreaker 9 A2 with a shaft capacity of about 60 MW;

- construction of 3 diesel-electric icebreakers of the Russian Maritime Register of Shipping Icebreaker 8 AUT2 class with a capacity of about 25 MW on propeller shafts;

- construction of 2 linear icebreakers of the LK-16 class of the Russian Maritime Register of Shipping Icebreaker 6 AUT1-ICS with two full-turn steering columns with a capacity of 2 x 8 MW [5];

- construction of 3 hydrographic pilot vessels of the Russian Maritime Register of Shipping KM* Arc 7 class;

- construction of three shallow-draft pilotage vessels of the Russian Maritime Register of Shipping Ice 3 class.

2. Development of the Northern Sea Route:

- modernization of berths and port facilities in Dixon, Tiksi, Pevek for the purpose of basing emergency rescue and hydrographic vessels, storage of property of emergency rescue groups(ASG), oil spill response (OSR), bunkering of vessels with fuel, water, replenishment of ship stocks and repair [12];

- construction and modernization of coastal navigation equipment (SNO);

- implementation of area survey of the bottom relief of the Arctic seas along the routes of transportation of hydrocarbon raw materials;

- development of the basic production infrastructure for the creation of publishing originals of navigational charts, manuals and manuals for navigation;

- creation of onshore production and technological complexes for the repair of navigation equipment in Arkhangelsk, Tiksi, Zeleny Mys, Pevek.

3. Construction of new ports and transshipment complexes for the integrated development of new territories and deposits:

- development of the oil handling complex in Kozmino Bay (Vostochny port) to increase capacity by 15 million tons;

- development of production capacities up to 8 million tons of coal of JSC "Daltransugol" (Vanino port);

- construction of the seaport of Sabetta (Yamal Peninsula) with a capacity of 30.7 million tons, including LNG-25 million tons, GC-2.2 million tons, oil-3.5 million tons;

- construction of a seaport for transshipment of LNG and GC in the area of Teriberskaya Bay (Murmansk region);

- construction of a specialized coal transshipment complex in Muchke Bay (Vanino Port) with a capacity of 45 million tons;

- construction of a coal port in the area of Cape Izylmetyev (Shakhtersk port, Sakhalin Region) with a capacity of 10 million tons [4].

Thus from the text of the Transport strategy there is a clear understanding that SMEs will play an important role in the development of navigation in Russia that will improve as the quality of services in the field of freight transport , and in the future on the basis of the Northern sea route an international transport corridor.

Two groups of factors are important as prospects for the development of the Northern Sea Route. The most important thing is the development of the infrastructure of the Arctic territories, which in turn will lead to an increase in oil and gas production on the Arctic shelf. It is oil and gas that should become the main cargo transported along the Northern Sea Route [6].

An equally important factor may be the transformation of the NSR into an international transit corridor, which will open a new route of cargo transportation between the countries of Asia and Europe [13].

However, the prospects for the development of the NSR face a number of circumstances:

- compliance with the delivery schedule is more important than speed;
- cost savings due to traffic volumes;
- transportation via the NSR is possible only on the route from the starting point to the final point, unlike transportation through the Strait of Malacca, which allows calling at dozens of intermediate ports;
- established supply chain of most of the world's largest shipping companies and a network of counterparties along the route through the Suez Canal [14].

The analysis of the situation allowed us to identify a number of reasons that hinder the development of the NSR:

1) "narrowness" of the niche for transportation;

2) established traditions and principles of maritime transport organization;

3) the absence of large shipping companies that act as transit organizers;

4) lack of modern port terminals [7].

To strengthen Russia's position in the maritime transport market, it will be necessary to provide the NSR with the status of a world transport corridor that meets all the standards of world shipping. In order for the NSR to be perceived not as exotic, it is necessary to increase the cargo flow to 1% of the global sea traffic, which will amount to more than 90 million tons.

Major world powers are not yet ready to consider the NSR as a cost-effective route. However, a well-designed policy for the development of competitive advantages, as well as the attraction of foreign capital for the joint construction of infrastructure, and further reduction of transportation tariffs can raise the importance of the NSR as the most profitable world sea route [8].

5. Forecasts for the development of transit container traffic along the Northern Sea Route

One of the most important segments of the market, for the development of which it is necessary to make efforts, is the transportation of transit containers along the NSR. According to experts, by 2030, if the necessary conditions are created, it is possible to redistribute the transit of container cargo between such countries as Japan, China and South Korea with European countries up to 6 million tons.

The developed model involves the creation of an operator company with state participation and a regular (operating for 9 months a year) Arctic container line between the ports of Murmansk and Petropavlovsk-Kamchatsky. The rationale for the creation and development of this line is given in the works of famous Russian scientist V. S. Zbarashchenko [5].

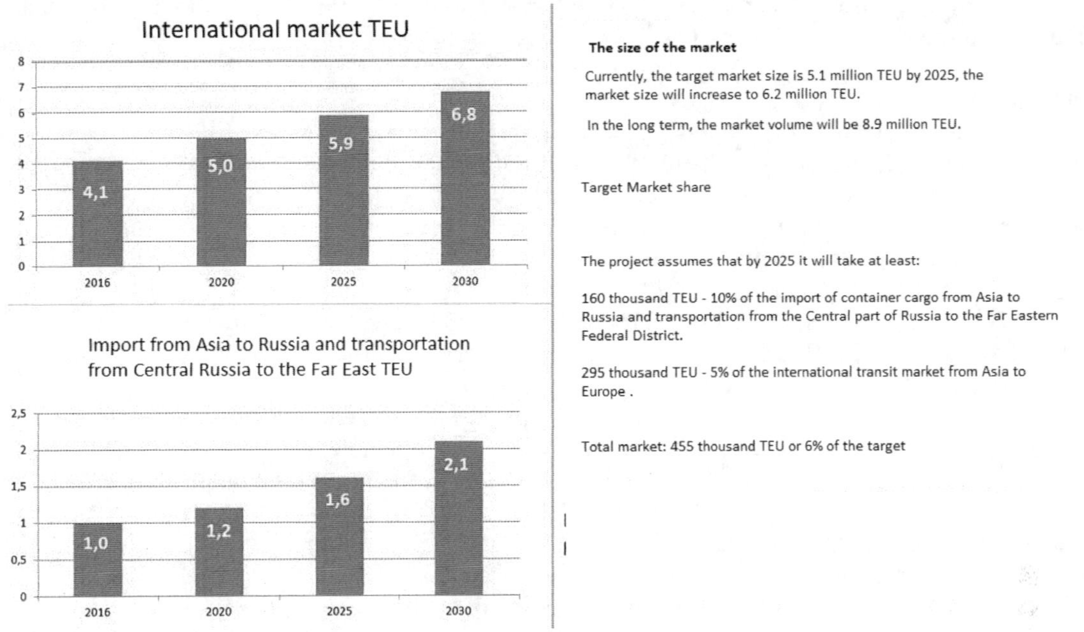

Figure 2. Dynamics of target markets and target market share [5]

The calculations showed that it is possible to take 10% of the market of import of container cargo from Asia to the Russian Federation and 5% of the market of international transit from Asia to Europe. The total volume of container traffic on the NSR can reach 455 thousand containers (Figure 1). Thus, the NSR will remain a niche route with its competitive advantages [9].

A year ago, the test transportation of a Maersk container ship along the Northern Sea Route along the Busan-Vladivostok-Nakhodka-Bering Strait-Bremerhaven-St. Petersburg route was successfully completed. Almost the entire route passes through Russian territorial waters.

The press service of Maersk stated that this was done to study the potential of container transportation in northern waters [10]. It is known that the Danish ship spent 26 days on the entire journey instead of the standard 34, without taking into account calls to the ports of Vostochny and Busan [11]. This is not surprising, since the Northern Route is shorter than the Southern One by seven thousand nautical miles. Experts at Maersk assure that they are ready to consider the NSR as a commercial alternative to the existing logistics schemes, domestic experts do not doubt that the Danes have estimated the economic benefits of the new project [9].

6. Conclusion

The development of the Arctic region and the Northern Sea Route is a strategically important task for Russia. Although sea container transportation takes a small share in the total cargo flow, but do not forget that the cost of transportation of container cargo per ton is much higher than liquid or bulk cargo. In addition, the creation of an international container corridor along the Northern Sea Route will reduce transport costs and change the port of Petropavlovsk-Kamchatsky into a major international logistics center. In turn, this will allow one to unload the Trans-Siberian Railway, as well as create the necessary conditions for the formation of infrastructure along the coast of the NSR.

Thus, the study showed the prospects of sea container transportation through the Northern Sea Route, but it is necessary to take into account that the full realization of the competitive advantages of the NSR is impossible without creating a product — international container transportation by the Arctic container line. The income received from the sale of this product will serve to return on investment, develop the infrastructure of the NSR and ensure the profitability of an important and large-scale project.

References

[1] Likhachev A 2018 *PortNews*, available at: http://portnews.ru/ne-ws/288164/
[2] Gordeev M 2019 *Infranews*, available at: http://infranews.ru/logisti-ka/54616-doxody-egipta-ot-tranzita-sudov-cherez-suec-vyrosli-na-5-do-59-mlrd/
[3] Polyakova I 2019 Northern Sea Route: a vector of development *Transport of Russia*, available at: http://www.transportrussia.ru/.
[4] Yakovlev M 2019*Arctic road of life*, available at: http://yourarctic.com/
[5] D.Ivanov D 2020 The ice is broken *Rossiyskaya Gazeta*, available at: http://rg.ru/.
[6] Peresypkin V 2020 Development of the Northern Sea Route *Morskie Vesti*, available at: http://morvesti.ru/.
[7] Semushin D 2019 *The first Chinese container ship set off to Europe through the Arctic: prospects for the Northern Sea Route*, available at: http://www.regnum.ru/.
[8] Bakylin N 2019 *Russia and the Northern Sea Route*, available at: http://voprosik.net/
[9] 2008 Transport strategy of the Russian Federation for the period up to 2030 No. 1734-r, available at: https://mintrans.gov.ru/documents/3/1009
[10] 2008 Changes to be made to the order of the Government of the Russian Federation No. 1734-r.
[11] Gordeev I 2020 Northern Sea Route and the Suez Canal *DeloNovosti*, available at: https://delonovosti.ru/analitika/3921-severnyy-morsk-oy-put-i-sueckiy-kanal.html0)
[12] Petrakov V and Roy O 2020 *Prospects for the use of the Northern Sea Route for transit container transport*, available at: https://cybe-rleninka.ru/article/n/perspektivy-ispolzovaniya-severnogo-morskogo-putid-lya-tranzitnyh-konteynernyh-perevozok
[13] Zbarashchenko V S 2020 *Strategic transport corridors*. "New Silk Road: Business Cooperation and Prospective of Economic Development" available at: *https://www.researchgate.net/publication/340553452_International_Transport_Corridors_Impact_on_Territorial_Development*
[14] Nikulin V P 2013 The state of Russian maritime transport, *Sea news of Russia* **5**

IOP Conf. Series: Materials Science and Engineering 1151 (2021) 012012 doi:10.1088/1757-899X/1151/1/012012

Construction schedule planning for railway station facilities

V Z Velichkin, M V Petrochenko, E B Zavodnova, A Yu Gorodishenina

Peter the Great St. Petersburg Polytechnic University, 29, Politechnicheskaya st., St. Petersburg, 195251, Russia

E-mail: gorodish_au@spbstu.ru

Abstract. The article discusses the improvement of the network modeling technique for solving problems in the design of the organization of work for the construction of station buildings and structures necessary for servicing passengers intended for operation on railway transport in the presence of various technological processes. The network model in the design of the organization of work on the construction of rather complex structures and in the presence of various technological processes becomes very cumbersome and poorly visible in terms of taking into account the interconnection of processes and the coordination of their movement along the work fronts. In addition, on such a network model it is rather difficult to consider the options for applying various methods and technologies of work organization. An example of the formation and calculation of a matrix-network model for the construction of a station complex is given, including 14 technological processes, showing the relevance of the chosen method.

1. Introduction

When designing the organization of work on the construction of rather complex structures and in the presence of various technological processes, the most convenient form of reflection of the entire interconnection and sequence of construction and installation works is the network model [1-6]. The network model allows you to trace the sequence of execution of construction processes and their interrelationship during implementation.

In such structures as station buildings and structures intended for operation on railway transport, the sequence and timing of work are significantly influenced by taking into account their interconnection in individual parts of structures and organized fronts of ongoing construction processes.

In these cases, the network model becomes very cumbersome and poorly visible in terms of taking into account the interconnection of processes and coordinating their movement along the work fronts. In addition, on such a network model it is rather difficult to consider the options for applying various methods and technologies of work organization.

These methods, first of all, include the method of the critical path, the method of continuous use of technical and labor resources, the method of continuous development of work fronts and the method of the fastest deployment of work (the method of rank connections).

2. Materials and methods

At first glance, in the case of interconnection of the complex of works, it is desirable to apply the matrix method of reflecting the technology of performing works on various fronts. However, the

Content from this work may be used under the terms of the Creative Commons Attribution 3.0 licence. Any further distribution of this work must maintain attribution to the author(s) and the title of the work, journal citation and DOI.
Published under licence by IOP Publishing Ltd

matrix provides for a strict sequence (technology) of work.

It seems impossible to reflect the complex spatial interrelation of works (network topology) on such a matrix without special algorithms and schemes. One of the possible approaches to solving such problems may be the use of matrix-network planning.

In our opinion, it will be more convenient for practical calculations to build a matrix model based on a simplified and compact network model of the entire set of works.

The proposed matrix model in such a complicated version reflects the breakdown of the structure into separate sections (work fronts) in relation to each technological process and the entire complex of works performed. In the published works, such a matrix is considered as a matrix with different-sized fronts of works [7-10].

The complex of construction processes performed is reflected in such a matrix in a sequence corresponding to the topology of the developed network model for the construction of the structure under consideration.

The sequence of work execution in the matrix is fixed by layers corresponding to the main paths of the graph (network model). Jobs connecting the main paths of the graph are reflected by the presence of several layers in the matrix for these jobs. The calculation of the timing of work at each seizure is determined based on the readiness of the work front and in the corresponding topological sequence.

The start and end dates of the «j work» on the «i front» can be described by the following expressions:

$$T_{st\,ij} = max\begin{cases} t_{fin\,i(j-1)} \\ t_{fin\,(i-1)j} \end{cases} \qquad T_{st\,ij} = T_{fin\,ij} + t_{ij} \tag{1}$$

where:

$T_{st\,ij}$ – start date of the «j work» on the «i front»,

$T_{fin\,ij}$ – end date of the «j work» on the «i front»,

t_{ij} – duration of «j work» on the «i front».

3. Results and discussion

As an example, let us consider a model for the construction of a station complex. The main technological processes for the construction of the station complex necessary for serving passengers are summarized in Table 1.

Table 1. Complex of technological processes indicating the duration of work

№	Name of the type of work	Duration of work, days
1	Installation of reinforced concrete frames	16
2	Installation of cover plates	18
3	Installation of wall panels	17
4	Roofing device	22
5	Rough floor covering	13
6	Construction of aerated concrete partitions	14
7	Installation of door blocks	13
8	Preparatory work for the installation of window blocks	4
9	Installation of window blocks	19
10	Installation of ventilation equipment	14
11	Installation of demonstration equipment	7
12	Installation of power supply system equipment	14
13	Rough finishing of premises	23
14	Installation of roof equipment	10

This complex of technological processes was linked by the sequence of their execution using the network model shown in Figure 1, the work numbers are highlighted, for each work the start and end dates of the process are given.

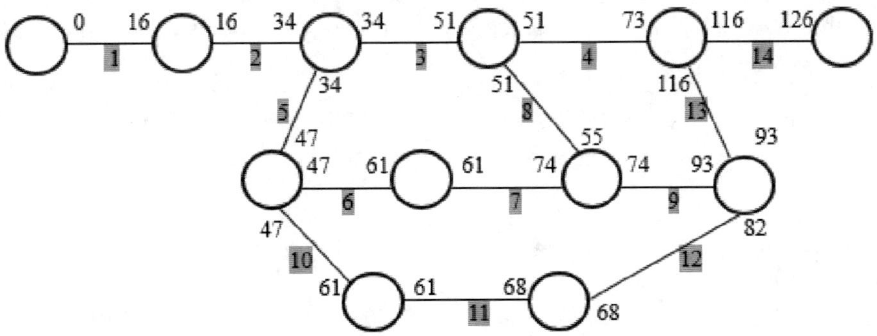

Figure 1. Network model of the construction of the station complex.

This network model does not take into account the combination of work performed in different sections (captures) of the structure. It takes into account only the timing of completion of work in accordance with the accepted technology (topology) of the construction of the structure. On the basis of this network model, a matrix model of the construction of a structure has been developed, taking into account the division of the structure into separate processes into a number of work fronts (seizures). This is a model of streamlined organization of work, which reduces construction time by combining work.

Such a model is shown in Table 2. Lines show the sequence of technological processes. In each cell, the timing of work on the corresponding grip is calculated. The critical path is shaded.

Table 2. Matrix model of erection of the station complex

graph paths / fronts	1	2	3	4	5	6	7	8	9	10	11	12	13	14
1	4 0-4	5 4-9	8 13-21	6 21-27	5 32-37	4 37-41	3 41-44	1 44-45	5 45-50	3 50-53	4 71-75	3 75-78	5 78-83	4 89-93
2	4 4-8	4 9-13		5 27-32		3 41-44	7 47-54	1 54-55	4 55-59	7 64-71		4 78-82	6 83-89	
3	4 8-12	4 13-17	9 22-31	5 32-37	4 37-41	3 44-47		1 55-56	5 59-64		3 75-78	4 82-86	6 89-95	6 101-107
4	4 12-16	5 17-22		6 33-43	4 43-47	4 47-51	3 54-57	1 57-58	5 64-69	4 71-75		3 86-89	6 95-101	
Σ	16	18	17	22	13	14	13	4	19	14	7	14	23	10

The network model does not take into account the breakdown of processes into separate seizures (work fronts), but when it was drawn up, the possibility of simultaneous execution of some work on certain parts of the structure was taken into account. To reduce the construction time, it is advisable to determine the work fronts (seizure) for each process and organize the flow technology of the construction.

For this example, we will build a matrix indicating the layers according to the sequence of work execution with the selection of grips for each work (Table 2). On the matrix, the paths of the graph are shown by layers (lines).

Here are the following paths from the beginning of the fork to the final work (graph arcs):
-1-2-3-4-14; -2-5-6-7-9-13-14; -3-8-9-13-14; -5-10-11-12-13-14.

As can be seen from the example (Table 2), taking into account the sequence of work and observing the possibility of working on one front of only one brigade, the construction period was reduced from 126 days to 107 days.

When calculating on the matrix, the critical path method was adopted in the same way as on the network diagram, but a reduction was achieved due to the stream organization of work on the selected work fronts. On the network model (Figure 1), the possibility of simultaneous execution of several jobs on one front is accepted. Then, when constructing and calculating the matrix model in full accordance with the network model with the flow organization of work, we get a reduction in construction time from 126 days to 65 days (Table 3). Lines show the sequence of technological processes. In each cell, the timing of work on the corresponding grip is calculated. The timing of the work is calculated in accordance with the established technology and when working on the capture of only one specialized brigade. The critical path is shaded. The reduction was achieved due to the ability of several brigades to work on one front.

Table 3. Matrix model of erection of the station complex

graph paths / fronts	1	2	3	4	5	6	7	8	9	10	11	12	13	14
1	4 0-4	5 4-9	8 13-21	6 21-27	5 13-18	4 18-22	3 22-25	1 21-22	5 25-30	3 12-21	4 19-33	3 33-36	5 36-41	4 47-54
2	4 4-8	4 9-13		5 27-32		3 22-25	7 28-35	1 22-23	4 35-39	7 22-29		4 36-40	6 41-47	
3	4 8-12	4 13-17	9 22-31	5 32-37	4 18-22	3 25-28		1 31-32	5 39-44		3 33-36	4 40-44	6 47-53	6 59-65
4	4 12-16	5 17-22		6 37-43	4 22-26	4 28-32	3 35-38	1 32-33	5 44-49	4 29-33		3 44-47	6 53-59	
Σ	16	18	17	22	13	14	13	4	19	14	7	14	23	10

Based on this matrix model (Table 3), a normalized matrix was built based on the works of prof. V.A. Afanasyev [1] (Table 4). The timing of work on the normalized matrix fully corresponds to the timing of the matrix in Table 3.

The normalized matrix shown in Table 4 is calculated by the method of continuous development of work fronts. The term of the work is 107 days. Calculation according to the normalized matrix execution of works according to the method of rank relationships leads to a duration of 122 days.

Table 4. Normalized matrix model for the construction of the station complex

graph paths / fronts	1	2	3	4	5	6	7	8	9	10	11	12	13	14
1	4 0-4	5 4-9	-	-	-	-	-	-	-	-	-	-	-	-
2	4 4-8	4 9-13	8 13-21	6 21-27	-	-	-	-	-	-	-	-	-	-
3	4 9-13	4 13-17	-	-	-	-	-	-	-	-	-	-	-	-
4	-	-	-	5 27-32	5 32-37	4 37-41	3 41-44	1 44-45	5 45-50	3 50-53	-	-	-	-
5	-	-	-	-	-	3 41-44	-	-	-	-	-	-	-	-
6	4 17-21	5 21-26	9 26-35	5 35-40	4 10-44	3 44-47	7 47-54	1 54-55	4 55-59	-	-	-	-	-
7	-	-	-	-	-	-	-	1 58-59	5 59-64	7 64-71	4 71-75	3 75-78	5 78-83	-
8	-	-	-	-	-	-	-	-	-	-	-	4 79-83	6 83-89	4 89-93
9	-	-	-	6 55-61	4 61-65	4 65-69	3 69-72	1 72-73	5 73-78	4 78-82	3 82-85	4 85-89	6 89-95	-
10	-	-	-	-	-	-	-	-	-	-	-	3 92-95	6 95-101	6 101-107

4. Conclusion

1. The proposed method for constructing a matrix model based on a network model is simple and allows us to consider several options for organizing work and choose the best one.

2. The criterion for choosing a method of organizing construction can be the construction period, the cost of performing the work under consideration, the operating time of the leading machines and others.

3. The construction of a matrix model can be carried out by observing the accepted topology of the work performed without building a network model.

References

[1] Afanasyev V A 1990 *Stream organization of construction* (L: Stroyizdat)
[2] Goldratt E 2006 *Critical chain* (Moscow: TOS Center)
[3] Czarnecki A and Nowak A 2008 Time-variant reliability profiles for steel giarder bridges *Structural Safety* 30(1) 49-64
[4] Au S, Ching J and Beck J 2007 Application of subset simulation method to reliability benchmark problems *Structural Safety* 29(3) 183-93
[5] Butsenko E 2016 The practice of using network economic and mathematical modeling of the

investment design process *Bulletin of Tomsk State University. Economy* 1(33) 147-58
[6] Emelyanov D, Ponyavina N and Chesnokova E 2017 Solving the problem of planning complex production processes in an enterprise based on network planning methods *Proceedings of higher educational institutions. Textile industry technology* (Publisher: Ivanovo State Polytechnic University) 5 (371) 28-32
[7] Ivanov M 2013 *Automation of network planning and management. Systems. Methods. Technology.* (Bratsk: Bratsk State. un-t) 2(18) 63-9
[8] Rowinski L 1979 Organizacia procesow budowlanych (Warszawa: PWN)
[9] Velichkin V, Petrochenko M and Ptukhina I 2019 Matrix-network model of planning complex work complexes *Proceedings of the A.F. Mozhaisky* 669 13-7
[10] Vladimirov I 2018 Application of a matrix algorithm for calculating the duration of construction of monolithic buildings using the flow organization of construction production. *Science and business: ways of development* 4(82) 89-96

IOP Conf. Series: Materials Science and Engineering 1151 (2021) 012013 doi:10.1088/1757-899X/1151/1/012013

Using train-passing techniques on the rail network to prevent flooding on rivers

L Haltar[1], A V Dmitrenko[2], E V Lesnykh[2], S A Becher[2], N S Buryanina[3]

[1] Ministry Transport of Mongolia, Government House No. 13, Chingis Avenue, Ulan Bator, 14251, Mongolia
[2] Siberian transport University, 191, Dusi Kovalchuk st., Novosibirsk, 630049, Russia
[3]Chukotka branch of the M. K. Ammosov North-Eastern Federal University, 3Studencheskaya st., Anadyr, 698000, Russia,

E-mail: abbiel@mail.ru

Abstract. In order to prevent floods on rivers in the presence of a cascade of hydroelectric power plants, it is possible to use the methods of railway transport in the management of their operation, which ensure the prevention of prolonged congestion in the movement of trains due to the accumulation of an increased fleet of wagons on certain landfills. In railway transport, the most dangerous in operational activities are unloading points located mainly on the outskirts of the country or the continent of Eurasia. For railway transport, the most important point in management is the reception of trains during the overhaul of tracks on main lines, mainly located in the central geographic center of the country, located in Kuzbass in Russia. On rivers, in the event of electricity generation, the most dangerous place is the hydroelectric power station located in the middle of an extended river basin. However, the most important and effective points in managing their operation are the upper and lower hydroelectric power plants. The lower hydroelectric power plants provide a quick discharge of excess water, which helps prevent floods. Upper hydroelectric power plants allow one to raise the water level of the cascade of the entire extended river basin at a lower cost.

1. Introduction

In order to prevent floods on rivers in the presence of a cascade of hydroelectric power plants, it is possible to use the methods of railway transport in the management of their operation, which ensure the prevention of prolonged congestion in the movement of trains due to the accumulation of an increased fleet of wagons on certain landfills. In railway transport, the most dangerous in operational activities are unloading points located mainly on the outskirts of the country or the continent of Eurasia. For railway transport, the most important point in management is the reception of trains during the overhaul of tracks on main lines, mainly located in the central geographic center of the country, in Kuzbass in Russia.

On rivers, in case of electricity generation, the most dangerous places are hydroelectric power plants located in the middle of an extended river basin. However, the most important and effective points in managing their operation are the upper and lower hydroelectric power plants. The lower hydroelectric power plants provide a quick discharge of excess water, which helps prevent floods. Upper hydroelectric power plants allow to increase the water level of the cascade of the entire

Content from this work may be used under the terms of the Creative Commons Attribution 3.0 licence. Any further distribution of this work must maintain attribution to the author(s) and the title of the work, journal citation and DOI.
Published under licence by IOP Publishing Ltd

extended river basin at lower costs, which makes it possible to increase the generation of electricity.

In spite of the presence of hydroelectric power plants, long-term destructive floods occur in long river basins, leading to huge losses. Long-term traffic disruptions occur in railway transport, leading to a deterioration in the operational performance of long-distance trunk lines. The study showed that huge losses in river basins on the territory of a large country are caused by insufficient use in practice of rational methods of managing production processes in these sectors of the country's economic and economic activity [1,2]. The article describes the peculiarity of the application in practice of rational methods that ensure the improvement of economic indicators of the work of these sectors of economic and economic activity of the whole country.

Technological progress in various sectors of economic and economic activity made it possible to achieve high economic indicators both by increasing labor productivity and by reducing costs associated with an increase in production and the growth of created material assets. Thus, the construction of a hydroelectric power station made it possible to obtain a huge effect through the use of renewable energy sources. Electricity has made it possible to improve the production process, ensuring the development of lighting and increased comfort in people's lives.

Regulation of the order of water accumulation in reservoirs in spring allows avoiding large floods for a long period. However, despite the management measures taken at the constructed hydroelectric power plants, it was not possible to avoid large floods with catastrophic consequences, as it was in the summer of 2019 in Tulun.

2. Materials and methods

The occurrence of floods in recent years on extended river basins of a large area is explained by two main reasons.

- The presence of a large volume of water in the upper reservoirs, which are accumulated to increase the production of cheap electricity for the purpose of its use in production processes.
- Lack of rational methods of managing the order of electricity generation in the whole cascade of hydroelectric power plants, providing the opportunity to simultaneously: both prevent floods on rivers, and also achieve an increased volume of electricity generation throughout the year.

The analysis performed showed that the water management and power generation system mainly takes into account the management procedure for individual isolated hydroelectric power plants. The proposed measures made it possible to take into account the uneven inflow of water by regulating only individual hydroelectric power plants, preventing floods [3].

However, the scientific studies carried out have shown that regulating the flow of water only for individual hydroelectric power plants does not provide flood prevention in large river basins, as was the case repeatedly on the Amur River. It will be possible to achieve great success due to the integrated regulation of the operational work of several hydroelectric power plants of the entire extended river basin at once, taking into account their capabilities, due to being at different heights above sea level.

Taking into account the peculiarities of complex regulation, it will be possible to achieve great success in the operation of all hydroelectric power plants in the basin by using the possibilities of controlling the movement of trains on railway transport, which make it possible to achieve success in ensuring the transportation process on a national scale.

To assess ways to improve the efficiency of controls in the technical and economic calculations, the presence of the main factors encountered both in railway transport and in the operation of a cascade of a hydroelectric power plant in an extended river basin was taken into account.

- The most dangerous points in which the greatest losses from emergency situations will occur, both in railway transport and in the operation of hydroelectric power plants of an extended river basin.
- The most important control points from which it will be possible to achieve the greatest

possible and cost-effective improvement in the overall operational environment.

- The procedure for complex regulation as a whole for the entire production process throughout the country. Here the question becomes about the advisability of different ways of owning large expensive technical objects: private or public way.

3. For railway transport of the continent of Eurasia

Long-term disruptions in train traffic have occurred and continue to occur in rail transport at long-range landfills. They arise not only due to insufficient bandwidth of the individual limiting elements. They also occur mainly due to the appearance of an increased fleet of wagons on individual landfills and adjacent main railway lines.

The analysis showed that in recent decades, the most prolonged traffic jams and downtime of trains en route have arisen due to untimely unloading of cars, occurring mainly on the outskirts of a large country [4]. Thus, throughout the continent, the initial long delays of freight trains en route occurred due to untimely unloading of cars at sea and river ports within the country and in the sea ports of the continent of Eurasia. The resulting long delays in freight trains caused the appearance for a long period of an increased fleet of cars, which subsequently paralyzed the operational work of long main railway lines. Therefore, the outskirts of the continent and border crossings between states are the most dangerous points at which long-term complications begin in the transportation process on the railway transport of the continent of Eurasia.

The improvement of the train traffic control system will be possible in two main ways. It will be necessary to choose a method that allows you to achieve the greatest success in managing the movement of freight trains with lower costs.

- At present, if there are separate roads and in the presence of private property, the control of the passage of trains is carried out at these specific enterprises or sections in case of difficulty in the movement of trains. Planning is carried out due to longer stops of freight trains in front of the point of main difficulties in the movement of trains on main railway lines.
- If management measures are taken on the most dangerous section, where the main difficulties in train traffic occur.

In this case, at the beginning of traffic disruptions, on the approaches to the most dangerous place, there are long delays of freight trains, which subsequently spread over a long range. At the point of main difficulties, delays in freight trains continue due to restrictions in capacity or deterioration in the operational work of final destinations.

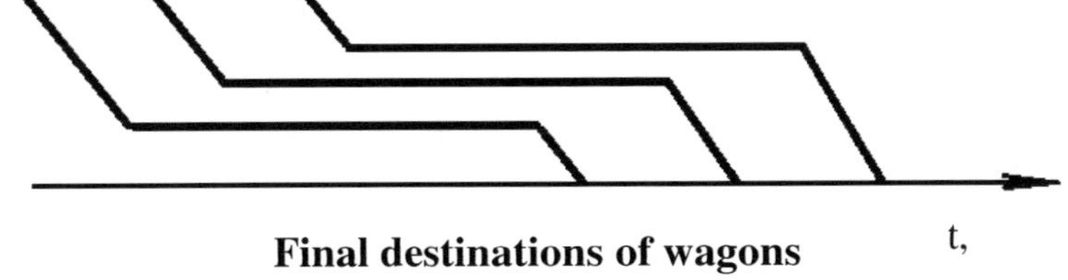

Figure 1. Scheme of delays for freight trains at the point of major initial disruptions in traffic at destination stations on the continent of Eurasia

Fig. 1 shows that in the event of the appearance of restrictions, for example, at points of unloading in the eastern part of the continent of Eurasia, further trains pass through with their long delays along the route.

Improving the options for the passage of trains as a whole throughout the entire railway network of a large country will possibly be carried out by the integrated management of the passage of trains throughout the railway network of a large country. In this case, in front of the tracks located in the central geographic point of the country, at the cargo-handling point, in the Kuzbass region in Russia, it will be necessary to delay the trains of freight trains going to the points of major traffic disruptions before the tracks being repaired. In this case, freight trains detained before being repaired at a freight station on the continent of Eurasia will then reduce the time they spend in front of their destination. In this case, a more stable passage of freight trains in general along the entire main railway line will be provided.

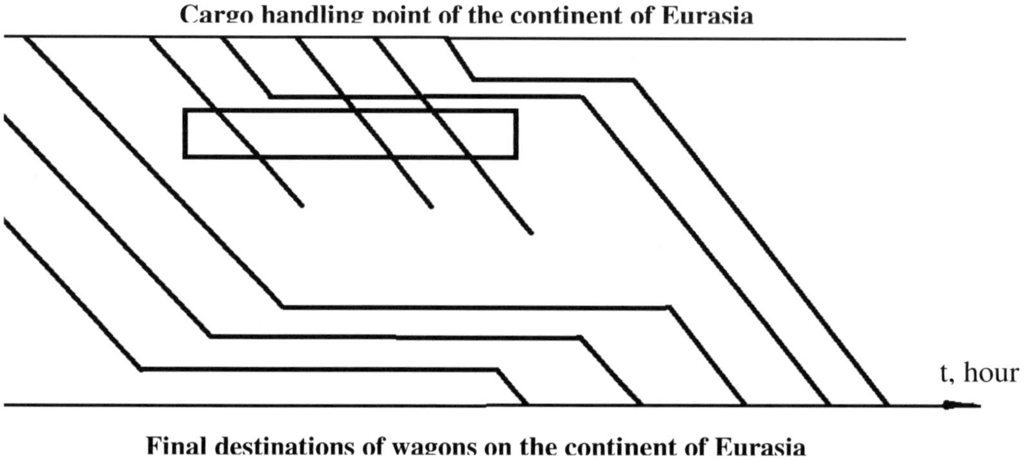

Figure 2. Scheme of the passage of freight trains when driving them on the main railway line.

The above conventional example shows that the best result in managing the operational work of the railway network as a whole is not achieved when management measures are carried out only at the end points of the beginning of the main disruptions in traffic. This option takes place in the presence of private railways in the country and their isolated traffic control only on their own sections. The greatest management effect is achieved when rational measures for the management of the country's railway network are carried out in far-removed points located in the central geographic center of the Eurasia continent with the largest volume of wagon loading.

In general, for all railway transport, measures for the control of train traffic are provided by the following tracks.

Major train disruptions occur on the outskirts of a large country. At the same time, rational measures to control the movement of trains find themselves in far-removed parts of the main lines located in the central geographic center of the country. This example also shows that private railroad management, as is the case in the United States, does not guarantee good economic results if trains are scheduled to run only on their own trunk lines. Only a comprehensive regulation of the passage of trains and the organization of management from a single center according to rational technology, taking into account the delays of trains along the route, and first of all during the overhaul of the track, makes it possible to ensure a more rational achievement of success in management than private management of the order of passage of trains on individual roads big country.

In general, for the entire railway transport of the continent of Eurasia, the performance parameters will be evaluated as follows.

The most dangerous points in the management of train access in railway transport are the final points of unloading and border points, usually located on the outskirts of the continent of Eurasia. At the same time, the best management results will be achieved when rational measures for the management of train traffic are carried out primarily in the central geographic center of the continent of Eurasia. This region is located in Kuzbass, where the largest volume of cargo is loaded on the continent as a whole.

4. Water transport, management of the operation of a complex of hydroelectric power plants

There are certain analogies with the railway transport of the continent of Eurasia.

In an extended river basin, several hydroelectric power plants are usually installed, the so-called cascade of hydroelectric power plants, for example, on the Angara River. For a cascade of hydroelectric power plants, the use of methods of operational work of railway transport is possible by choosing options that ensure the achievement of higher economic results, compared with the options for isolated management of the operational work of only individual hydroelectric power plants. Operational work management options will be established as follows.

The lack of integrated management in the operational work of a number of hydroelectric power plants on long rivers leads to frequent floods. For example, in 2019, catastrophic floods occurred twice on the Angara River in Tulun, with four large hydroelectric power plants in a cascade on the Angara River. These floods occurred with an interval of only one month in June and July. On the Amur River, floods with catastrophic consequences occur regularly due to the increased water level due to the lack of rational water level management.

At the same time, due to the irrational system of using the water of the cascade of hydroelectric power stations in 2019, the Volga River became shallow. Failures in the use of water resources on long rivers occur due to the lack of scientific foundations for the influence of hydroelectric power plants on each other. The peculiarity of the operation of hydroelectric power plants located both in the lower reaches or at the mouth, as well as in the upper reaches at the sources of rivers was not taken into account. The peculiarity of the location of hydroelectric power plants at different heights above sea level was not taken into account.

The probability of occurrence of the most dangerous places of an extended river basin for the possibility of emergence of critical situations with floods was not established. The points for the regulation system were not established, the use of which in practice at the most important points allows to ensure the achievement of the best results in the management of water energy resources in a number of countries of the world.

To obtain the best results in ensuring the achievement of a faster discharge of water in the basins, in order to prevent the occurrence of destructive floods on individual rivers of each extended river basin, it is necessary to take into account that in the presence of a cascade of hydroelectric power plants, different power plants are located at different heights above sea level. Therefore, each of them has a different potential in generating electricity. This feature will have a significant impact on the procedure for establishing a rational method of management in their use, both to ensure greater generation of electricity and to prevent the occurrence of catastrophic floods on large rivers of the river basin.

In managing the operation of a cascade of hydroelectric power plants, it is necessary to solve two main problems in operational conditions:

- to provide a temporary reduction in water flow for the most dangerous location in the entire cascade of hydropower plants;
- to provide high potential in electricity generation for existing situations of fluctuations in the water flow and for its reduction in winter, in general for the entire river basin of a large extent.

Each of these tasks, taking into account the presence of a cascade of hydroelectric power plants located both at the mouth and in the upper reaches of the rivers of the river basin, will be solved in the following way.

It is necessary to assess the situation separately: both for the lower hydroelectric power plants located both at the mouth of the river and in the upper reaches of rivers located at high altitudes above sea level.

During forced floods, with a large amount of snow melting, an excess of water appears, which can lead to floods. Also, during the period of melting of glaciers or snow in the mountains or during the period of prolonged heavy rains, devastating floods may occur.

or the upper hydroelectric power plants there is a small flow of water from a limited area. For these hydroelectric power plants, it is possible to both retain and discharge water. In any case: either water is discharged or retained, it in any case remains in the sphere of the hydroelectric power station of the entire given river basin. During floods (or hazardous situations) in the lower part, the retention of water in the upper hydropower plant allows for a low risk of flooding downstream of the river. However, the possibilities for flood prevention in this case will be limited.

In the future, as we move downstream for the analysis of the operation of the hydroelectric power station, an increasing volume of water flow is recorded. This makes it possible to reduce the volume of water, both by increasing power generation and by dumping. The flood prevention capabilities of these hydropower plants are increased compared to those located in the upper reaches of this long river basin.

It should be taken into account that in the event of an increased discharge of excess water from the upper hydroelectric power plants, as we move downstream of the river, the possibility of floods is created, since the water level in the hydroelectric power stations located downstream for this part of the river will increase. In general, for all hydroelectric power stations, except for the lower one, this water will remain for a certain time within the considered river basin of a large extent and with an increased total water supply.

A completely different situation will be with the operation of the lowest hydroelectric power station for this river basin. In this case, it is possible to quickly discharge the increased volume of accumulated water from the entire river basin, and not only from its part, as it was for higher-located hydroelectric power plants. In addition, the water leaves this river basin forever and is the greatest prevention of huge floods with catastrophic consequences.

In addition, from the lower hydroelectric power plants, water immediately enters the ocean or sea at a constant level. Therefore, the water flowing out of the lower hydroelectric power station does not create additional potential for increasing the level of electricity generation as a whole in the entire cascade of hydroelectric power stations of this large-area river basin with abundant precipitation.

Therefore, in order to get rid of excess water in a given river basin, it will be necessary first of all and to the greatest extent to get rid of excess water by means of its increased release from the lower hydroelectric power station.

The accumulation of water potential for the generation of increased volumes of electricity can be carried out as follows.

To ensure the achievement of higher success in the use of powerful resources, it is necessary to control the cascade of hydroelectric power plants in such a way that, with the available flow of water, a large generation of electricity is provided, and primarily throughout the entire year.

When assessing the options for the operation of the hydroelectric power station, it is necessary to take into account that the external inflow of water in the winter period of the year is very small. At the same time, there is an increased flow of water in spring.

The generation of large volumes of electricity will be achieved when water enters the turbines from a high altitude of each reservoir. To achieve this goal, it will be necessary that the water level at all hydropower plants in the river basin has an increased value.

nly in spring, and at certain times of the year during heavy torrential rains, the water level is dangerous and can lead to catastrophic floods. However, in the rest of the year, if all units of each hydroelectric power station are in operation, it will lead to a rapid loss of the level of accumulated water in the reservoirs. Therefore, during the period of reduced water intake, there should be a smaller number of units that generate electricity. This will make it possible to maintain a high level of operation of the hydroelectric power station and ensure an increased generation of electricity during the day and throughout the year.

In order to accumulate more potential, one should bear in mind, first of all, the upper hydroelectric power stations. It is located at the highest altitude above sea level. However, in practical terms, the water accumulated at the uppermost hydroelectric power station has the greatest energy potential. Therefore, this hydroelectric power station should have improved results during the period of filling the reservoir.

In addition, the uppermost hydroelectric power station will be located in the mountains. The surface of the reservoir will be small. Therefore, if a certain amount of water is filled in a short period, the level of the river reservoir rises to a great height. Therefore, the upper hydroelectric power station will have to include more than the average level of the number of units in order to provide a large amount of electricity generation at low water consumption.

With a small external water supply, the units of each hydroelectric power station will have to have a large reservoir height. This large height should be formed by reducing the number of operating units of the hydroelectric power station. In this case, with a high level of reservoirs in general at all hydroelectric power plants during the winter period, the number of units of operating hydroelectric power plants will gradually decrease.

In spring, the amount of incoming water begins to increase at a fast time. Compared to the winter period, special conditions for the operation of the hydroelectric power station must be created.

- Preparation period. This is for a month of massive snow melting or increased water inflow into the river basin. In this case, it will be necessary to ensure the preparation of reservoirs for the unhindered passage of water flows, and at this time there will be a high water level in the reservoirs.
- Period of massive snow melting, when the accumulation of water reserves for the coming season occurs first.
- Period of increased electricity generation throughout the summer.
- Winter period of gradual decrease or reduction of water reserves in reservoirs. Preparing for massive snow melt.

The accumulation of water and the degree of filling allows a higher level of power generation to be achieved in a faster time. The amount of electricity generated does not only depend on the volume of water. The degree of power generation depends to a large extent on the height of the hydroelectric power station: the lower or upper hydroelectric power station.

The higher the water is, the greater the potential will be in generating increased flows of electricity. Therefore, it will be necessary to use the obtained potential to a greater extent to generate increased volumes of electricity generation while reducing the volume of accumulated water.

In the absence of rational regulation (private ownership of hydroelectric power plants), the most important and most dangerous points in the operation of the entire cascade will be established as follows.

In the article under consideration, we will assume that the increased water flows will gradually shift: from the sources - up to the mouth. At the same time, the likelihood of flooding will be assessed in the absence of operational control of the operation of the hydroelectric power station and the absence of centralized regulation of the flow of water in various parts of the river of a long river basin.

In case of heavy rains, only the upper reaches of the river basin will have the largest area. Such a flow of water from a small area can be relatively easily discharged below or become easily directed to create increased water reserves in the upper part. In this case, the largest volume of water discharge (over the capacity of the units) will not lead to disastrous consequences in the lower part of the longest river.

In the case of a shift of the increased water flow downstream, the discharge area increases and the volume of water that can be released downstream increases.

Subsequently, as the increased inflows of water move downstream, the possible volume of excess water increases. This will increase the volume of water and lead to the likelihood of flooding or catastrophic flooding.

The further downstream of the extended theca, the more water can be released downstream with disastrous consequences.

However, even in the absence of centralized regulation, liquidation of catastrophic consequences will change according to completely different laws. Here the movement will be carried out the other way around: from the mouth - to the sources of an extended river.

In the case of increased flows of water at the mouth, released in large volumes, they will quickly fall into the ocean. In this case, there is no accumulation of water and usually there is practically no deterioration in the living conditions of people near the ocean level. At the same time, it is necessary to prevent flooding of houses, since as you move up, the height of the river above sea level increases.

Table 1. Comparison of the performance indicators of railway transport with the operation of a cascade of hydroelectric power plants in an extended river basin under conditions of seasonal fluctuations in water flows

Indicators	For rail transport	For the operation of the hydroelectric power station
Lack of a single control center for the operation of a hydroelectric power station or isolated control of train traffic on each road	There are long delays of freight trains along the route	Leads to floods or reduced electricity generation and deterioration of navigation in river basins
The presence of a single control center with rational network technology	Allows to significantly reduce delays of freight trains en route, ensures stability in the operation of railway junctions	Makes it possible to prevent the occurrence of large floods, causes increased electricity generation throughout the year
The most dangerous points where long-term failures in their work most often occur	Final destinations at unloading stations on the continent of Eurasia	The central part of the river basin, where increased water flows are concentrated
The most important points in managing the operation of technical objects	The central geographic part of the continent of Eurasia with an increased volume of loading, where the cargo point is located	Outlying hydroelectric power plants of the river basin: the initial one - at the mouth and the upper hydroelectric power station at the source of a large river
Management of train traffic and power generation at hydropower plants in river basins	The best results are achieved when rational measures are applied in the central geographic part of the continent of Eurasia, located in Kuzbass, Russia	The best results are obtained in the operation of hydropower plants located at both ends of the river basin: at the upper and lower hydropower plants. Upper hydroelectric power plants serve to store water or energy potential, as well as to increase electricity generation. The use of lower hydroelectric power plants allows to remove excess water in a faster time in order to prevent large catastrophic floods in river basins

5. Conclusion

- Methods of centralized control over the passage of trains on railway transport, as well as the management of the operational work of a cascade of hydroelectric power stations on long rivers, can improve the economic indicators of the country's energy economy.
- In railway transport, as well as for river basins, the operation of hydroelectric power plants must take into account their outskirts and central part.
- For railway transport, the most dangerous points in operational work are the outskirts of the continent of Eurasia or border points between states. For the energy sector, the most dangerous points causing catastrophic floods are the central part of long river basins.

- In the management of train traffic on railway transport, the greatest effect is achieved by regulating the sequence of passage of freight trains in the central geographic part of the continent of Eurasia. For a river basin, the greatest effect is achieved for hydroelectric power plants located at the mouth of the river, as well as at its source.
- The lower hydroelectric power plants at the estuary make it possible to prevent destructive floods with an excess of water in the reservoirs. Upper hydropower plants located high above sea level create potential for high-volume power generation.

References

[1] Veselov F V and Solyanik A. I. 2020 Midterm Development Conditions for the Electric Power Industry of Russia under Hard Price Constraints. *Studies on Russian Economic Development* **31** 64-70.

[2] Sanzhapov B Kh and Stulova N V 2014 Decision support model in the problem of analyzing the ecological risk of urban environment pollution by objects of the fuel and energy complex *Internet Bulletin of VolgGASU. Ser.: Building Informatics* **11 (32).**

[3] Makarov A, Mitrova T and Kulagin V 2019 Scenario forecast of energy development in Russia, *Regional energy and energy saving* **4**

[4] Dmitrenko A V, Lesnykh E V, Bekher S A, Dmitrenko I Y and Buryanina N S 2020 Choosing rational locations for placing wind farms in Eurasia with regard *VIII International Scientific Conference Transport of Siberia.*

IOP Conf. Series: Materials Science and Engineering 1151 (2021) 012014 doi:10.1088/1757-899X/1151/1/012014

Diagnostics of tension in long rods with the help of traveling bending waves

V M Bardakov, N N Klimov, A A Ermakov and D Yu Pomerantsev

Irkutsk State Transport University, 15, Chernyshevsky St., Irkutsk, Russia

E-mail: klinn42@mail.ru

Abstract. The article considers the possibility of measuring the tension of long thin structures by measuring the speed of traveling bending waves. Earlier, with the participation of the first two authors of the work, a theoretical consideration of the possibility of determining the stress of a continuous-welded railway track by measuring the velocity of bending waves was carried out. For verification, the velocity of propagation of bending waves was measured, presumably during the compression of a continuous-welded track. This is possible when the temperature of the rails is higher than the laying temperature of the continuous track. In the same consideration, it is taken into account that bending waves exist in an extended product of finite thickness without tension, and the results are presented for the case when the continuous track is in a tensioned state. This consideration takes into account the fact that in a long workpiece of finite thickness there are also bending waves without tension. These waves, unlike the dispersion-free waves of tension, possess dispersion. A method of measuring the speed of these waves is proposed.

1. Introduction

There are known methods to determine mechanical tensions based on measuring the velocities of longitudinal and transverse waves and turning the plane of their polarization [1-3]. A method for determining mechanical tensions based on the application of the acoustoelasticity effect [4] is also used. In particular, on the basis of this method, the engineering firm "Inkotes" developed a device named IN-5101A, the principle of which rests on the generation of the ultrasonic probing pulses and the recording of the speed of reflected elastic waves excited in the material of the controlled object.

As part of this work, the possibility of measuring the tension of long metal structures (smooth track rails, high-pressure vessels, pipelines) is considered by measuring the increase in the speed of the propagation of the bending waves of the audio range in the workpieces when they are stretched, excited by a calibrated impact.

Diagnostics of the tension force of thin strings using traveling bending waves is simple and clear. It is necessary to quickly create an initial excitation for a long stretched string in the form of the string's local length deviation from the equilibrium state. It can be a quick impact or a quick plucking. The initial excitation, splitting in two, moves without changing the shape in both directions at a speed of $C_T = \left(\dfrac{T}{\rho_{lin.}}\right)^{1/2}$, as the traveling waves in a thin stretched string are dispersion-free (T is the force of

Content from this work may be used under the terms of the Creative Commons Attribution 3.0 licence. Any further distribution of this work must maintain attribution to the author(s) and the title of the work, journal citation and DOI.
Published under licence by IOP Publishing Ltd

string tension per unit area, $\rho_{lin.}$ is the linear string density). Measuring the speed of C_0, we find a tension T.

Diagnostics of tension in a stretched rod of finite thickness is complicated by the fact that bending waves in the thin rod also exist without tension. And these waves are dispersion waves, initial excitations are diffused. The additional tension of the rod changes the process of spreading bending waves along the rod. One needs to take into account this change in the pattern of the excitation spread and determine how to measure the tension T in the rod of finite thickness. In works [5, 6], we measured the velocity of flexural waves to diagnose the stress state of a continuous welded track, and the proposed article is a further development of the method.

2. Theoretical consideration

The dispersion ratio for the bending waves of the rod of finite thickness, according to Lamb [7], looks like this:

$$\omega = C_T k + C_0 \frac{k\chi}{\sqrt{1+k^2\chi^2}}, \qquad (1)$$

where $C_0 = \sqrt{\frac{E}{\rho}}$ is the velocity of longitudinal waves in the media with Young's modulus E, ρ is the density of the media. The group velocity for thin waves is as follows:

$$\upsilon_{gr} = \frac{\partial \omega}{\partial k} = C_T + C_0 \frac{2k\chi + k^3\chi^3}{\left(1+k^2\chi^2\right)^{3/2}}. \qquad (2)$$

Here k is the wave number, and χ is the rod cross-section radius of inertia. Besides, depending on the direction of the bending excitations relative to the plane, the value of χ changes.

If we take the cross-section of the rod in the form of a rectangle, we can talk about two bending planes:

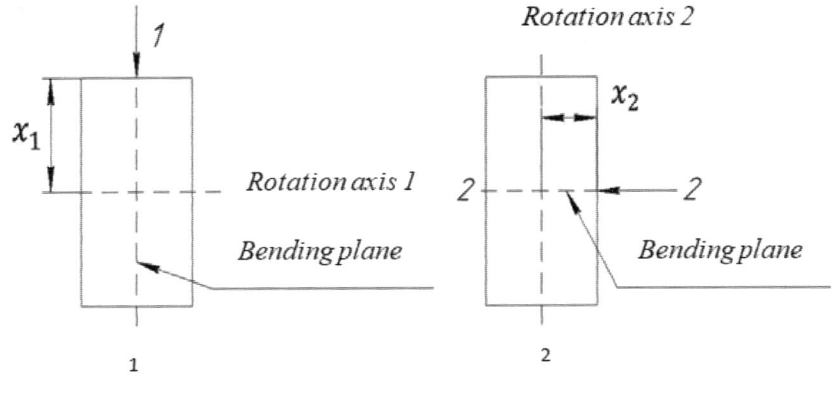

Figure 1. Rod cross-section radii of inertia.

Figure 1 depicts two directions of external impact (impacts) on the rod (1 – 1) and (2 – 2), which correspond to their own axes of rotation of bending excitations and their own inertia radii χ_1 and χ_2.

Further, let us keep in mind that, depending on the direction of external impact, it is necessary to use either χ_1 or χ_2 for the radius of inertia.

The phase velocity of the waves from (1) will be as follows:

$$\upsilon_{\text{ph}} = C_{\text{T}} + C_0 \frac{k\chi}{\sqrt{1+k^2\chi^2}}. \tag{3}$$

From the comparison (3) and (2) it is clear that the group velocity at all k is more than the phase velocity. Besides, at $k\chi >> 1$ $\upsilon_{\text{ph}} \to \upsilon_{gr}$. At $k\chi << 1$, but provided that $2C_0 k\chi >> C_{\text{T}}$, the group velocity is maximally different from the phase velocity, twice exceeding it. And finally, when $k\chi$ are so small that $2C_0 k\chi < C_T$, the phase velocity tends to the group velocity, which tends to the value C_{T}. In these comparative characteristics, we implied that $C_T << C_0$.

The development of the bending excitation of the rod can be broken down into two stages:

1) the formation of the initial excitation under the influence of external impact;
2) the scattering of the initial excitation in the form of a wave process in both directions from the initial excitation.

Let us consider both stages. When considering the first stage, let us take into account that the external impact is an impact by the body of some mass M and size along the *x*-axis (along the rod) of the magnitude $2\Delta x_{sp}$.

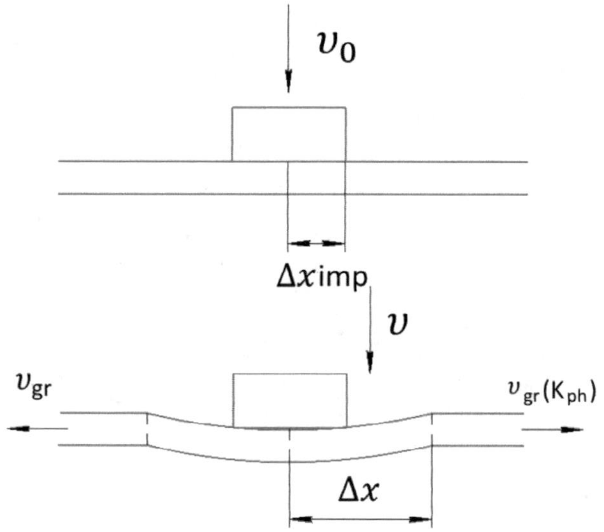

Figure 2. Formation of the initial excitation

The area of initial excitation Δx is increasing with the velocity of $\upsilon_{gr}(k_{ph})$, where k_{ph} is the maximum wave vector of excitation at the time when it was of the size of Δx. Let us accept that $k_{ph} = \dfrac{2\pi}{\Delta x}$. With an increase of Δx, the wave number decreases, the group velocity changes according to the formula (2). At the initial moment of the impact, $k_{phimp} = \dfrac{2\pi}{\Delta x_{imp}}$, and $k_{phimp}\chi >> 1$. Then $\upsilon_{gr}(k_{phimp}) \approx C_0$. Over time, $k_{ph}\chi$ decreases during the impact and, most certainly, even before the end of the impact, the parameter of $k_{ph}\chi$ will be much smaller than one. Let us denote the magnitude of $k\chi$ at the end of the impact $k_0\chi << 1$, and besides, $k_0 = \frac{2\pi}{\Delta x_0}$, where Δx_0 is the half-length of the initial bending excitation. The moment of the end of the impact can be identified with the moment when the body of mass M bounces away from the rod. Careful consideration of the formation of the initial bending excitation is a separate complex problem and is of only academic interest. For the purposes of diagnosis of tension T, the details of the formation of the initial excitation are not important. A more powerful impact (with a greater mass M) gives an increase of Δx_0.

Let us now consider the second stage. The magnitude of $k_0 = \frac{2\pi}{\Delta x_0}$ will be the maximum wave number for the initial excitation. Since the problem is linear, the composition by wave numbers at the stage of wave propagation evolution will not change in excitation. The edge of the wave excitation zone at the second stage will move with a velocity of:

$$\upsilon_{gr0} = C_T + C_0 2k_0\chi \tag{4}$$

Let us assume that $C_0 2k_0\chi >> C_T$ and

$$\upsilon_{gr0} \approx C_0 2k_0\chi . \tag{5}$$

The phase velocity of the wave excitation humps just behind the edge of the wave zone will be equal to $\upsilon_{ph0} \approx C_0 k_0\chi$. This effect can be tested experimentally (see the next section).

Sensors need to be positioned at points along the *x* further than Δx_0, although Δx_0 is not known in advance and needs to be specifically identified. The method of determining Δx_0 is in the next section.

Knowing the distance between the sensors Δ_D, by delaying the arrival of the wave zone boundary Δt_{deltim} to the second sensor, we determine the group velocity:

$$\upsilon_{gr0} = \frac{\Delta_D}{\Delta t_{deltim}} . \tag{6}$$

If there is no tension T, then $C_T = 0$, and there is no clear second boundary of the wave packet, so the lower boundary of the wave packet moves from the side of $\upsilon_{gr\min}$ which is equal to C_T. If $C_T \neq 0$, then the speed C_T is determined by the delay Δt_{lbwz} of the arrival of the lower boundary of the wave zone to the second sensor:

$$C_T = \frac{\Delta_D}{\Delta t_{lbwz}} . \tag{7}$$

For a better measurement, it is necessary that C_T is not much smaller than υ_{gr0}. And to reduce υ_{gr0}, one needs to increase the power of the impact.

IOP Conf. Series: Materials Science and Engineering 1151 (2021) 012014 doi:10.1088/1757-899X/1151/1/012014

3. Experimental Method

It is necessary to identify Δx_0 for the given mass of the striker M. To do this, you need to have three sensors, placing them as shown in Figure 3.

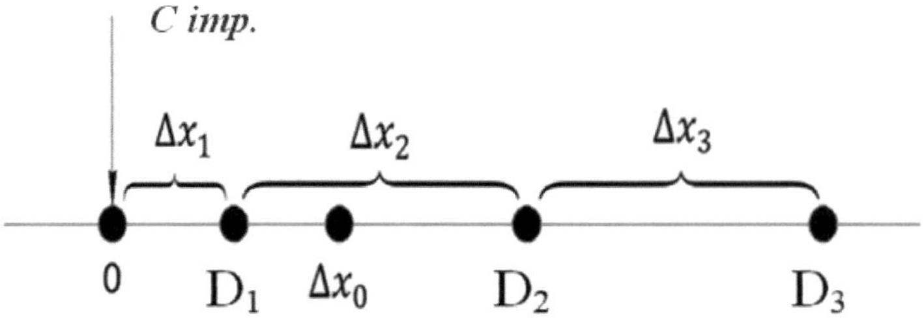

Figure 3. The location of measurement sensors.

It is advisable that the sensor D_2 is immediately behind Δx_0, and the sensor D_1 is in front of Δx_0. Then $\overline{\upsilon_1} = \frac{\Delta x_1}{\Delta t_1}$, the average speed of the front edge of the wave zone to the first sensor (Δt_1 is the time from the start of impact) $\overline{\upsilon_2} = \frac{\Delta x_2}{\Delta t_2}$ is the average velocity of the wave zone edge between the sensors D_2 and D_1. $\overline{\upsilon_3} = \frac{\Delta x_3}{\Delta t_3} = \upsilon_{gr0}$. If $\Delta x_1 < \Delta x_0$, and $\Delta x_1 + \Delta x_2 > \Delta x_0$, then $\overline{\upsilon_1} > \overline{\upsilon_2} > \upsilon_3 = \upsilon_{gr0}$.

By moving the sensor D_1 along x_1, we will find $\Delta x_1 = \Delta x_0$ such that $\overline{\upsilon_1} > \upsilon_2 = \upsilon_3 = \upsilon_{gr0}$ is fulfilled.

Let us consider the experiment methodology according to the proof that the phase velocity of humps in the wave zone, just behind the front edge, is half the size of υ_{gr0}.

Let us draw the supposed pattern of records from two sensors located at a distance of Δx_D.

If Δx_D is large enough, then at the second sensor, due to the fact that the phase velocity is $\upsilon_{ph} < \upsilon_{gr0}$, a new hump (zero) is generated and together with other humps moves in the edge system deep into the wave zone.

Conditions of the generation of the first hump are:

$$\Delta t_{ph} - \Delta t_{gr} > T_{ph}, \tag{8}$$

$\Delta t_{ph} = \frac{\Delta x_D}{\upsilon_{ph}}$, $\Delta t_{gr} = \frac{\Delta x_D}{\upsilon_{gr0}}$, T_{ph} is the time interval between the two humps in the image from a single sensor. Note that $\omega_0 = k_0 \upsilon_{ph} \rightarrow \frac{2\pi}{T_{ph}} = \frac{2\pi}{\Delta x_0} \upsilon_{ph}$ and $T_{ph} = \frac{\Delta x_0}{\upsilon_{\phi}}$

Then from (8) we obtain:

$$\Delta x_D > \Delta x_0 \frac{\beta}{\beta - 1}, \tag{9}$$

where $\beta = \frac{\upsilon_{gr0}}{\upsilon_{ph}}$. At $\beta < 2$ magnitudes of $\frac{\beta}{\beta - 1} > 2$.

If, on the contrary, we set Δx_D in such a way that (9) is not fulfilled, that is, for example,

$$\Delta x_D < \Delta x_0 \tag{10}$$

there will be no generation of additional humps of wave excitations, and then the humps on both sensors will be easily identifiable.

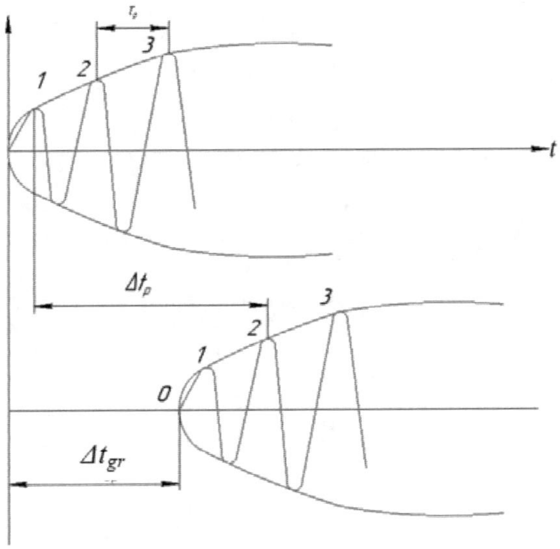

Figure 4. The supposed pattern of the observation of the signal on two sensors.

In this case, tracking the delay of Δt_{ph1} between the occurrence of the first hump at the second and the first sensor, we will obtain:

$$\upsilon_{ph} = \frac{\Delta x_D}{\Delta t_{ph1}} \tag{11}$$

$$\upsilon_{gr0} / \upsilon_{ph} = \frac{\Delta t_{ph1}}{\Delta t_{gr}} \tag{12}$$

4. Conclusion

The performed consideration of the formation of a bending wave arising from the impact of an extended workpiece in a tensioned state allows one to continue measuring the velocity of bending waves in order to determine the stressed state of a continuous-welded railway track. The above technique makes it possible to determine the tensile forces for long structures by increasing the velocity of bending waves.

References

[1] Gushcha O I, Guz A N, Lebedev V K, Makhort V G and Trotsenko V P 1975 Acoustic method of controlling stress in solid media *Bull. of inventions* **41**

[2] Thompson R B and Alers G A 1978 US, Patent No. 40808836 *Method of measuring stress in a material.* Int.Cl. 73\597 (G 01 B7/16).

[3] Gushcha O I and Makhort V G 1983 Acoustic method of controlling stress in solid media *Bull. of inventions* **36**

[4] Nikitina N E 2005 *Acoustic elasticity. Practical experience* (Nizhny Novgorod: TALAM Publ.)

[5] Bardakov V M, Klimov N N, Dudakov S V, Lopatin M V, Muratov V I, Kutsenko S M and Filatov E V 2010 On the possibility of acoustic diagnostics of the stress state of a continuous welded track.

Problems and Prospects of Research, Design, Construction and Operation of Railways: Proceedings of the IV All-Russian Scientific and Practical Conference with International Participation. V.1 (Irkutsk: IrGUPS) pp. 341 - 354.

[6] Klimov N N, Zubkova D A, Kutsenko S M and Dudakov S V 2011 Automation of data processing of acoustic diagnostics of the stress state of a continuous welded track *Pecs Modern technologies. System analysis. Modeling* **4(32)** 209-214

[7] Lamb G 1960 *The Dynamical Theory of Sound.* State Publishing (House of Physics and Mathematics, FIZMATGYZ Publ)

IOP Conf. Series: Materials Science and Engineering 1151 (2021) 012015 doi:10.1088/1757-899X/1151/1/012015

Evaluation of central tendency measures in relation to the parameters of contact network feeders controlled by relay protection

R S Trifonov, D A Yakovlev, S A Filippov and O A Solovyeva

Transbaikal Institute of Railway Transport, 11, Magistralnaya Str., Chita, 672040, Russia

E-mail: rttrif@rambler.ru

Abstract. The paper considers the importance of possible input values of standby stages of remote protection of contact network feeders from the point of view of their influence on the reliability of protection. The measures of the central tendency of parameters controlled by relay protection of contact network feeders were evaluated based on the theory of statistics, using the example of the Trans-Baikal Railway for 2018-2019. In total, 14 parameters were measured according to GOST 32144-2013. Electric power. Electromagnetic compatibility of technical facilities. Standards for the quality of electric energy in general-purpose power supply systems, in 2020 on the feeder of the contact network of the Sokhondo traction substation of the Trans-Baikal Railway against 88965290 obtained values, which made it possible to select significant values from all possible parameters on the basis of mathematical analysis. Statistical data on the following parameters were considered and analyzed: effective voltage value, effective current, peak voltage amplitude factor, peak current amplitude factor, harmonic voltage distortion factor, harmonic current distortion factor, phase shift factor, load resistance, angle between the first voltage harmonic on substation buses and feeder current. The reasonable selection of significant parameters proposed by the authors allows improving the operation of standby stages of remote protections of contact system feeders in heavy traffic conditions and increasing their selectivity.

1. Introduction

Russian Railways is a dynamically developing company with a number of strategic objectives specified in [1].

The analysis of the current state of the infrastructure performed in block 5.3 [1] indicates the following key problems of the company at the time of strategy development:

- non-compliance of the level of carrying capacity in certain areas with the transport market demand, presence of about 10.2 thousand km of "bottlenecks";
- high wear and tear of certain infrastructure elements;
- lack of opportunities for breakthrough growth of reliability, speed and accuracy.

In the same section [1], the key tasks of the company's infrastructure development include the following:

- development of infrastructure in order to switch to heavy-weight traffic with weight standards of trains of 9 thousand tons or more;

Content from this work may be used under the terms of the Creative Commons Attribution 3.0 licence. Any further distribution of this work must maintain attribution to the author(s) and the title of the work, journal citation and DOI.
Published under licence by IOP Publishing Ltd

- optimization of expenses due to increase of energy efficiency and labor productivity, rational organization of repair and maintenance works;

- improvement of transport safety and mitigation of risks associated with the influence of the "human factor" on the production process.

Since the launch of the development strategy many things have already been achieved by 2020. Thus, for more than five years double trains of increased weight reaching 12,600 tons have been running along the Trans-Baikal Railway using performance charts, which generally corresponds to the strategic objectives and goals outlined in [1].

However, in some cases the increase in the weight of trains has a very negative effect on the operation of directional remote protection of contact network feeders (DRP CNF), in particular the third backup stage. The increase in the weight of train sets while reducing the inter-train interval in order to optimize the economic performance has led to the fact that the maximum operating current tends to the maximum permissible value in the operating mode, which in turn leads to a decrease in the voltage of the contact network feeder, and in some cases to the minimum permissible value, especially in sections with a complex track profile, which also include the Trans-Baikal Railway, and the undervoltage may lead to the shutdown of trains. Current stepup – to contact wire burns and spurious tripping – again shutdown, wear and tear of switching equipment.

Mathematically, this can be generally represented as follows. The variety of factors affecting the operation of the standby stage of remote protection of the contact network feeders can be conditionally divided into two classes – external (*endogenous*) and internal (*exogenous*):

$$\mathbf{O}_{\mathbf{I}}(t) = \mathbf{S} \cup \mathbf{T}, \tag{1}$$

where $\mathbf{O}_{\mathbf{I}}(t)$ – m-dimensional vector-function characterizing a set of external primary influencing factors;

$\mathbf{S}$ – m-dimensional vector determining the state of external power supply system (EPSS);

$\mathbf{T}$ – m-dimensional vector describing the state of the traction power supply system (EPSS) associated with the parameters of the traction network of the protected feeder zone.

EPSS parameters are determined by the state of the vector $\mathbf{S}$ depending on the following values:

$$\mathbf{S} = [\mathbf{P}, \mathbf{Q}, \cos\varphi, \mathbf{U}, \mathbf{X}_{\mathbf{s}}]^{T}, \tag{2}$$

where $\mathbf{P}$ – m-dimensional vector-function characterizing the active power transmitted through the EPSS;

$\mathbf{Q}$ – m-dimensional vector-function determining the reactive power transmitted through the EPSS;

$\cos\varphi$ – active power factor of the EPSS;

$\mathbf{U}$ – m-dimensional vector-function representing the level of the EPSS voltage;

$\mathbf{X}_{\mathbf{s}}$ – m-dimensional vector-function describing the EPSS resistance to the traction substation on which the protection is installed.

In turn, the EPSS parameters are determined by the state of the vector $\mathbf{T}$ depending on the following variables:

$$\mathbf{T} = [\mathbf{K}, \mathbf{F}, \mathbf{W}, \mathbf{Q}, \mathbf{M}, \mathbf{P}]^{T}, \tag{3}$$

where $\mathbf{K}$ – m-dimensional vector characterizing the type of the traction network of the protected feeder zone;

$\mathbf{F}$ – m-dimensional vector describing the EPSS state associated with motion parameters on the protected feeder zone;

$\mathbf{W}$ – m-dimensional vector defining parameters of traction substations;

$\mathbf{Q}$ – m-dimensional vector characterizing the parameters of electric motive power in the protected section;

$\mathbf{M}$ – m-dimensional vector representing weight parameters of trains running on the protected section;

$\mathbf{P}$ – m-dimensional vector characterizing the path profile of the protected section.

Many external (*endogenous*) secondary factors characterize the parameters of the mode, in which the protected system is located, and are divided into two subsets – $\mathbf{O}_{\mathbf{II}н}(t)$ and $\mathbf{O}_{\mathbf{II}к}(t)$ characterizing the current mode of the system.

The set of the normal mode parameters are characterized by the state of vector $\mathbf{O}_{\mathbf{II}н}(t)$:

$$\mathbf{O}_{\mathbf{II}н}(t)=[\mathbf{I}_{н}(t),\mathbf{U}_{н}(t),\lambda,\mathbf{V}_{\mathbf{I}},\mathbf{V}_{\mathbf{U}},\boldsymbol{\varphi}_{н\mathbf{I}_k},\boldsymbol{\varphi}_{н\mathbf{U}_k}]^T, \tag{4}$$

where $\mathbf{O}_{\mathbf{II}н}(t)$ – m-dimensional vector-function characterizing a set of external secondary influencing factors in a normal mode;

$\mathbf{I}_{н}(t)$ – m-dimensional vector-function determining the current of the protected feeder in a normal mode;

$\mathbf{U}_{н}(t)$ – m-dimensional vector-function describing the voltage in the traction network of the protected feeder in a normal mode;

λ – system power factor;

$\mathbf{V}_{\mathbf{I}}$ – m-dimensional vector-function characterizing harmonic current of the traction network;

$\mathbf{V}_{\mathbf{U}}$ – m-dimensional vector-function characterizing harmonic voltage of the traction network;

$\boldsymbol{\varphi}_{н\mathbf{I}_k}$ – m-dimensional vector representing the current angle of the k harmonic in a normal mode;

$\boldsymbol{\varphi}_{н\mathbf{U}_k}$ – m-dimensional vector characterizing the voltage angle of the k harmonic in a normal mode.

The set of the emergency mode parameters are characterized by the state of vector $\mathbf{O}_{\mathbf{II}к}(t)$:

$$\mathbf{O}_{\mathbf{II}к}(t)=[\mathbf{I}_{к}(t),\mathbf{U}_{к}(t),\boldsymbol{\varphi}_{к\mathbf{I}_k},\boldsymbol{\varphi}_{к\mathbf{U}_k},\mathbf{S},\mathbf{T},\mathbf{W}]^T, \tag{5}$$

where $\mathbf{O}_{\mathbf{II}к}(t)$ – m-dimensional vector-function characterizing a set of external secondary influencing factors in the emergency mode;

$\mathbf{I}_{к}(t)$ – m-dimensional vector-function describing the current of the protected feeder in the emergency mode;

$\mathbf{U}_{к}(t)$ – m-dimensional vector-function determining the voltage in the traction network of the protected feeder in the emergency mode;

$\boldsymbol{\varphi}_{к\mathbf{I}_k}$ – m-dimensional vector representing the current angle of the k harmonic in the emergency mode;

$\boldsymbol{\varphi}_{к\mathbf{U}_k}$ – m-dimensional vector characterizing the voltage angle of the k harmonic in the emergency mode.

External factors are interconnected: $\mathbf{O}_{\mathbf{II}н}(t)\subseteq\mathbf{O}_{\mathbf{I}}(t)$, $\mathbf{O}_{\mathbf{II}к}(t)\subseteq\mathbf{O}_{\mathbf{I}}(t)$. A different combination of primary factors leads to various changes within a limited region of secondary values for each mode: $\mathbf{O}_{\mathbf{II}н\min}(t)\leq\mathbf{O}_{\mathbf{II}н}(t)\leq\mathbf{O}_{\mathbf{II}н\max}(t)$, $\mathbf{O}_{\mathbf{II}к\min}(t)\leq\mathbf{O}_{\mathbf{II}к}(t)\leq\mathbf{O}_{\mathbf{II}к\max}(t)$.

Change of parameters $\mathbf{F}$ and $\mathbf{M}$ is critical for the operation of the standby stage of remote protection of the contact network feeders. So, if $\mathbf{F}\to\min$ (packet traffic schedule), and $\mathbf{M}\to\max$ (handling of heavy trains), then the state of the vector $\mathbf{O}_{\mathbf{II}н}(t)$ changes so that it takes the value $\mathbf{O}_{\mathbf{II}н}(t)=\sup\mathbf{O}_{\mathbf{II}н}(t)$ and tends to the region $\mathbf{O}_{\mathbf{II}к}(t)$.

If in this case the vector $\mathbf{O}_{\mathbf{II}к}(t)$ takes the value $\mathbf{O}_{\mathbf{II}к}(t)=\inf\mathbf{O}_{\mathbf{II}к}(t)$, then $\mathbf{O}_{\mathbf{II}н}(t)\cap\mathbf{O}_{\mathbf{II}к}(t)$, which dramatically increases the possibility of false response and hence non-selective operation of the standby stage of remote protection.

The negative impact on the operation of the third stage of DRP CNF is the decrease of protection reliability manifested as the increase in the number of non-selective actuations.

This statement can be illustrated by the analysis of the number of outages at the Trans-Baikal Railway for 2018 and 2019, presented in Figures 1 and 2.

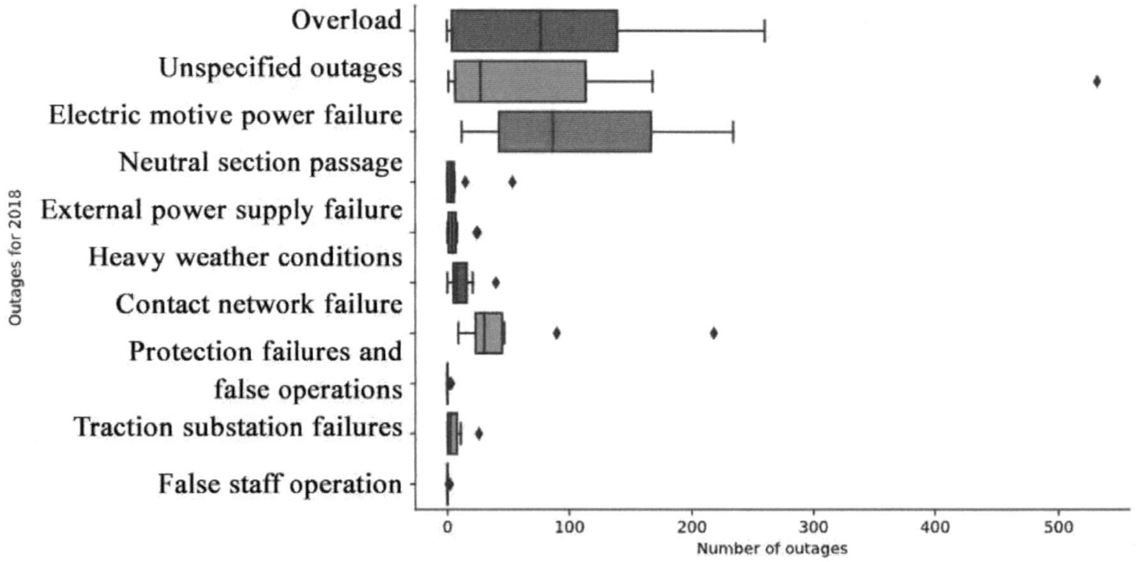

Figure 1. Statistics of relay protection outages at the Trans-Baikal Railway for 2018

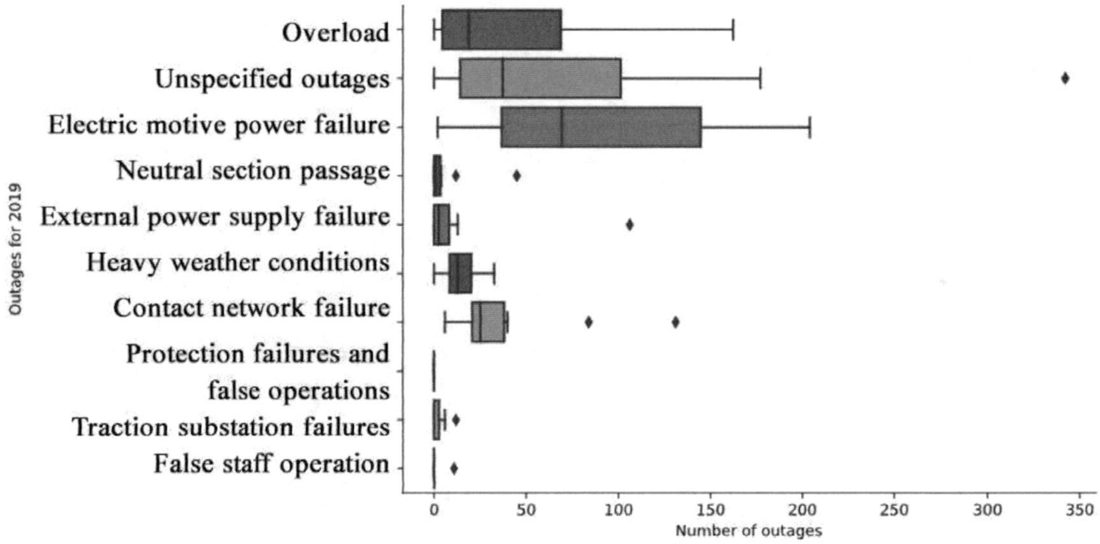

Figure 2. Statistics of relay protection outages at the Trans-Baikal Railway for 2019

Figures 1 and 2 show the total number of actuations of contact network feeder protections by types of outages for all power supply distances of the Trans-Baikal Railway. The analysis of diagrams shows that the largest number of protection actuations is observed for the following types of outages:

- unspecified outages;
- electric motive power failure;
- overload.

Table 1 shows the calculation of central tendency measures for frequent types of outages for 2018 and 2019 for the Trans-Baikal Railway as a whole.

Table 1. Central tendency measures for the most frequent types of outages for 2018/2019

Central tendency measures	Overload 2018/2019	Unspecified outages 2018/2019	Electric motive power failure 2018/2019
Mean	94.00 / 47.50	95.80 / 79.80	104.70 / 86.8
Standard deviation	101.24 / 60.11	163.97 / 107.67	78.48 / 72.26
min	0.00 / 00.00	1.00 / 0.00	12.00 / 2.00
25%	4.00 / 4.50	6.50 / 14.25	42.75 / 37.00
50%	77.50 / 19.00	27.50 / 37.50	87.00 / 69.00
75%	139.25 / 68.50	113.50 / 101.00	166.75 / 144.50
max	260.00 / 162.00	532.00 / 342.00	234.00 / 204.00

The analysis of the table shows that according to the specified indicators, 2019 in general is characterized by the decrease in the total number of outages compared to 2018, but their number remains significant, since the fact of each outage bears a risk for the reliability and continuity of the transportation process. It is also necessary to clarify that for a number of traction substations the situation is more critical, which, in particular, is indicated by a significant margin of standard deviation and as a result of a strong variation of outage values relative to the mean value, as well as the presence of upward trends indicated as separate points in Figures 1 and 2.

This situation is largely caused by the fact that after the transition to the microprocessor base for DRP CNF terminals the algorithms forming the basis of protection remained unchanged.

The fundamental disadvantage of these algorithms is that they do not make it possible to unambiguously distinguish between normal and emergency operating modes of the system during the movement of heavy trains. When the algorithms were developed, they showed a significant increase in the EPSS reliability with weighting standards and intervals of ordinary freight traffic. However, in modern conditions, they do not provide a proper level of reliability, which is confirmed by outage statistics.

The modern microprocessor element base allows utilizing a variety of possibilities in applying new approaches to the development of remote protection algorithms. The first step in developing new algorithms and generally improving the technical perfection of relay protection system of contact network feeders is to understand the processes within the traction power supply system of heavy trains.

To this end, experimental measurements of electrical energy quality indicators of the contact network feeder No. 4 (CNF-4) were made at the Sokhondo traction substation of the Trans-Baikal Railway. The purpose of this paper is to analyze the results of measurements in order to understand the processes and nature of changes in values for heavy trains.

2. Description of conditions and measurement results

Experimental measurements were carried out at the CNF-4 of the Sokhondo traction substation of the Trans-Baikal Railway using an electrical energy quality analyzer PQM-700 by Sonel.

According to its type, the Sokhondo traction substation is a transit substation. The choice of the feeder for the installation of the quality analyzer is caused by the fact that in case of heavy trains, regular false actuations of the third stage of the DRP CNF are observed. According to the substation report, in 2019 the largest number of outages on the fourth feeder was 89, which is 33% more than in 2018.

This feeder has a microprocessor protection terminal CZA-27.5 kV equipped with all required types of protection according to the current norms and rules [2].

The measurements were carried out from 2020-09-08 04:22:49.242 to 2020-09-19 18:03:11.715, which amounted to a total duration of 1 week, 4 days, 13 hours, 40 minutes and 22 seconds.

The parameters were measured according to [3] in the amount of 14 items. The sampling rate was 1 second.

For further study, the interesting parameters from the point of view of the DRP CNF operation were selected both for the analysis of current algorithms and for the development of new approaches:

- U – effective voltage, kV;
- I – effective current, A;
- CF_U and CF_I – peak amplitude factors of voltage and current;
- THD_U and THD_I – harmonic distortion factors of voltage and current;
- U_{hx} and I_{hx} – voltage and current harmonics, kV and A;
- $\cos(\varphi)$ – phase shift factor.

The final dataset form for the study made 89 variables and 999610 observations totaling to 88965290 values. There are no missing and duplicate values.

3. Analysis of central tendency measures

Before further data analysis, it is necessary to indicate the secondary controlled parameters of the DRP CNF, which are not measured directly by the quality analyzer, but were calculated additionally.

The protection calculates secondary controlled setpoints as follows:

$$\mathbf{Z}(nT) = [\mathbf{U}(nT), \mathbf{I}(nT), \boldsymbol{\varphi}_\mathbf{Z}]^T = \frac{\mathbf{U}(nT)}{\mathbf{I}(nT)} \tag{6}$$

$$\boldsymbol{\varphi}_\mathbf{Z} = \arg[\mathbf{Z}(nT)] = \arg\left[\frac{\mathbf{U}(nT)}{\mathbf{I}(nT)}\right] \tag{7}$$

$$THD_I = \frac{\sqrt{\sum_{h=2}^{9} \mathbf{I}_h^2(nT)}}{\mathbf{I}(nT)} \cdot 100\%, \tag{8}$$

where $\mathbf{Z}(nT)$ – m-dimensional vector function characterizing the resistance of the protected feeder;

$\boldsymbol{\varphi}_\mathbf{Z}$ – m-dimensional vector defining the resistance angle;

$\mathbf{U}(nT)$ and $\mathbf{I}(nT)$ – discrete values of the vector function of voltage and current after filtration;

THD_I – harmonic factor of the feeder current;

n – number of samples;

T – sampling period.

Table 2 shows the calculation results of central tendency measures for measured parameters.

Table 2. Calculation of central tendency measures

Central tendency measures	U,кВ	I,A	CF_U	CF_I	THD_U	THD_I	$\cos(\varphi)$	Z, Ohm	φ,°
Mean	25.15	355.06	1.59	1.59	13.51	24.11	0.47	135.28	25.15
Standard deviation	0.57	186.71	0.06	0.26	3.36	10.18	0.27	259.60	0.57
min	21.93	5.59	1.42	1.25	2.80	8.08	-1.00	22.41	21.93
25%	24.79	209.57	1.55	1.43	11.15	19.91	0.42	50.90	24.79
50%	25.18	339.52	1.59	1.52	13.50	22.37	0.55	74.06	25.18
75%	25.57	488.85	1.63	1.66	15.95	25.73	0.62	121.31	25.57
max	26.74	1033.10	1.97	5.95	26.22	327.67	1.00	4557.87	26.74
Shift	-0.42	0.33	0.63	2.89	-0.03	10.94	-2.23	7.03	-0.42
Excess	0.39	-0.52	0.90	12.23	-0.39	208.23	5.91	58.52	0.39

The analysis of descriptive statistics of the effective voltage value of the CNF-4 of the Sokhondo traction substation shows that the mean value equal to 25.15 kV is almost equal to the nominal effective value specified by [3] and equals 25 kV.

The standard deviation of 0.57 indicates that the values have a slight variation relative to the mean value, and this also confirms the value of the median or 50% percentile, which equals 25.18 kV. The excess has a value of 0.39, which indicates an "acute" distribution of voltage values, while observing a slight negative shift of -0.42 indicating that in general the values are shifted upwards. From the point of view of the DRP CNF operation, the minimum voltage is 21.93 kV, which is close to the minimum permissible value specified by [3]. Figure 3 shows the distribution presented as a violin diagram (left) and the distribution function (right) of voltage values illustrating the above conclusions.

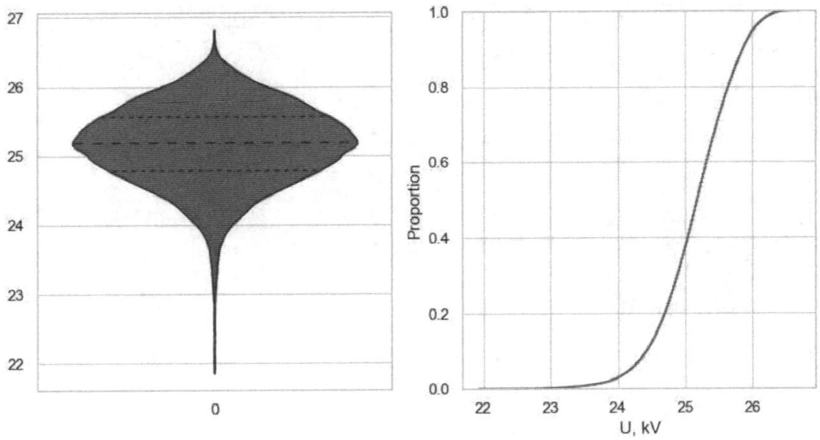

Figure 3. Violin diagram and voltage distribution function diagram

The analysis of the descriptive statistics of the current distribution diagrams presented in Figure 4 shows that the mean current equal to 355.06 A is generally characteristic of the normal mode in the conditions of freight traffic. However, the standard deviation of 186.71 A indicates that the current values have a significant variation relative to the mean. The median value of 339.52 A is slightly different from the average, which indicates a shift to the left relative to the average, and this is also confirmed by the asymmetry value equal to 0.33.

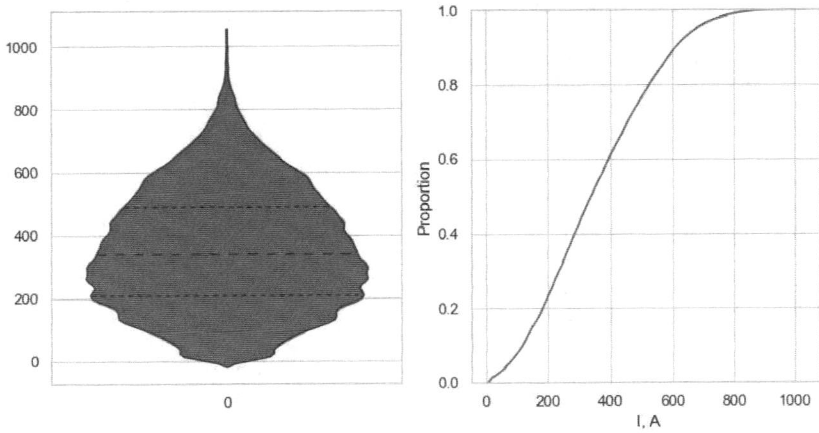

Figure 4. Current distribution

Figure 5 shows the distribution of the peak voltage factor. The analysis of the central tendency measures shows that the mean value equals 1.59, which is quite close to the normal value of the sinusoidal signal equal to 1.414, however, there are significant voltage distortions from the variable load and a number of other factors characteristic of the traction power supply system, which is clearly seen by the shape of the distribution function diagram. The standard deviation equals 0.06, which generally indicates a slight variation in values relative to the average, and this is also confirmed by the fact that the median value coincides with the mean value. The data have a normal distribution with a slight shift towards smaller values since the shift value is positive and equals 0.63. The excess value of the peak voltage amplitude factor is 0.90, which indicates the presence of an "acute" distribution vertex, which is also confirmed by the violin diagram in Figure 5.

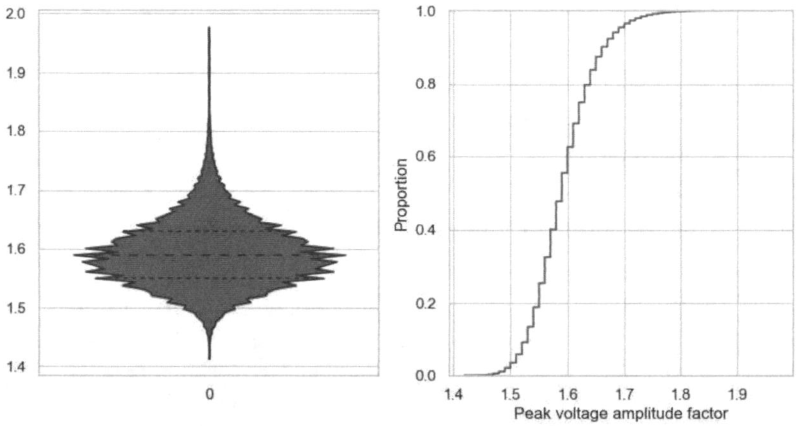

Figure 5. Distribution of peak voltage amplitude factor

The distribution of the peak current amplitude factor is shown in Figure 6. The joint analysis of Table 2 shows that, like the voltage, the current on average has a shape close to the ideal sinusoid, since the mean value is similar to the value of the peak voltage amplitude factor and equals 1.59. However, there is a significantly higher standard deviation of 0.26, which indicates a shift in values relative to the average, and this is also confirmed by the fact that the median is different from the average and equals 1.52. A significant positive asymmetry of 2.89 indicates the distribution shift towards smaller values, while the shift value itself is much greater than 1, which indicates a distribution of data different from normal. The maximum value of the peak current amplitude factor is 5.95, and the excess equals 12.23, which indicates significant peak current values of about 5-8% of the total number of observations according to the distribution function diagram.

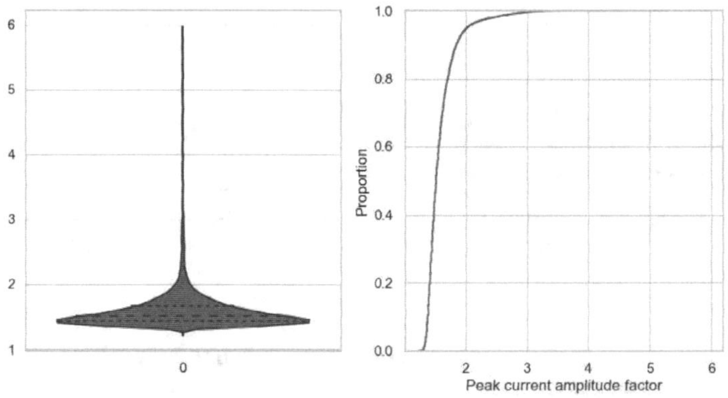

Figure 6. Distribution of peak current amplitude factor

IOP Conf. Series: Materials Science and Engineering 1151 (2021) 012015 doi:10.1088/1757-899X/1151/1/012015

Figure 7 shows the distribution of the harmonic voltage distortion factor for the main harmonic of 50 Hz. The average value equals 13.51, which indicates the significant presence of higher harmonic components in the voltage signal, which is generally a significant factor in terms of the DRP CNF operation, since signal distortions significantly affect both the measurement results and the protection. The standard deviation equals 3.36, which indicates a slight shift in observations relative to the average, which is further confirmed by the fact that the median value is almost equal to the average and makes 13.50. The maximum value of the harmonic voltage distortion factor equals 26.22, which indicates a significant distortion of the waveform, while the number of such observations according to the distribution function diagram ranges from 3 to 5% of the total number of observations. The shift value is -0.03, and the excess value is -0.39, which indicates data distribution close to normal.

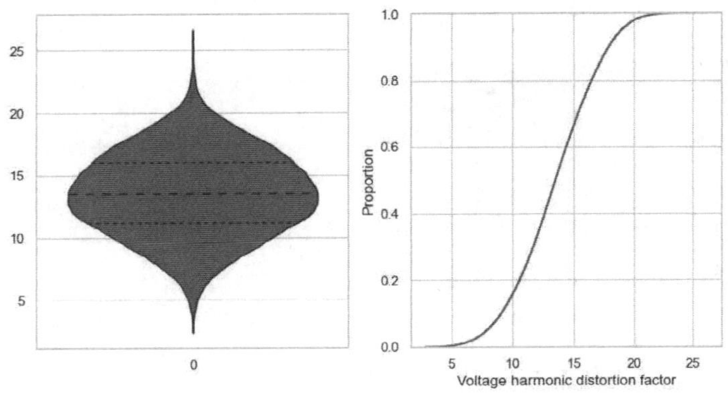

Figure 7. Distribution of voltage harmonic distortion factor

The distribution of the harmonic current distortion factor is shown in Figure 8. The analysis shows that the average value of the factor is 24.11, which indicates even more “dirt” current signal with higher harmonic components compared to the voltage. This circumstance also has an extremely negative impact on the DRP CNF operation. A median of 22.37 indicates a shift of values to the left relative to the average, with an asymmetry value of 10.94, which indicates a strong shift towards smaller values and that the structure of data distribution is not normal. A significant excess of 208.23 indicates that the distribution structure has a significant sharp vertex and long tails. The analysis of the violin diagram shows that the right tail of the distribution is significant caused by the relatively rare heavy trains in relation to the number of other types of trains. The maximum value of the harmonic current distortion factor is 327.67 and the number of such observations varies from 3 to 5% of the total number of observations, which is generally typical for heavy trains. Such high maximum value indicates that in case of heavy trains the current shape is significantly distorted.

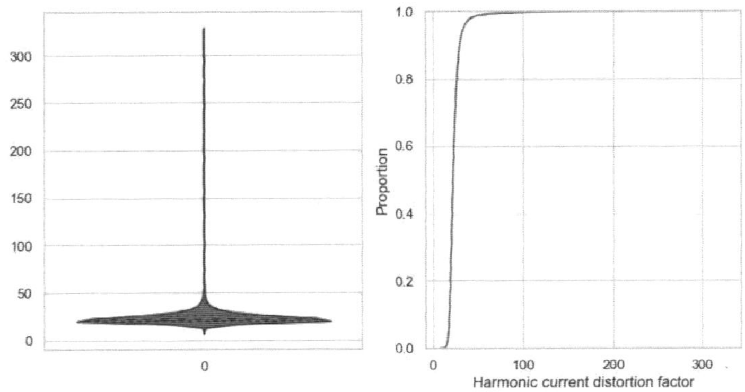

Figure 8. Distribution of harmonic current distortion factor

IOP Conf. Series: Materials Science and Engineering 1151 (2021) 012015 doi:10.1088/1757-899X/1151/1/012015

Figure 9 shows the distribution of the phase shift factor. The average value of the factor is 0.47, which in general indicates the unsatisfactory quality of electric energy. The median value equals 0.55, which indicates a shift in observations relative to the average value towards large values. The minimum and maximum values vary from -1 to 1, which indicates a change in the nature of the load from capacitive to inductive. Besides, according to the diagram of the distribution function, most of the load is inductive and a long tail indicates a shift towards the capacitive load. From the point of view of the DRP CNF operation, this circumstance is a complicating factor because it may lead to distortion of remote measurements and, as a result, to failures or false protection operations. A negative displacement of -2.23 indicates a shift towards larger values, and a high excess value of 5.91 indicates that there are more values at the distribution edges than around the average.

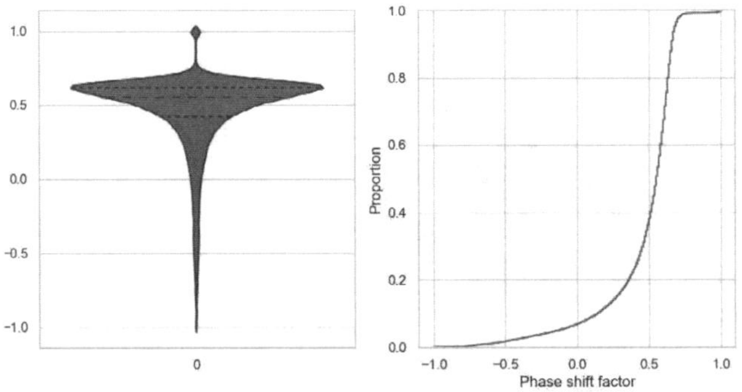

Figure 9. Distribution of phase shift factor

Figure 10 shows the distribution of load resistance. The average value is 135.28 Ohm, which is generally normal for freight traffic, since according to [4, 6] the values Z > 30 Ohm belong to the normal mode according to the resistance criterion. The standard deviation of 259.60 Ohm indicates that the values are strongly displaced from the average, which is also confirmed by the shape of the violin diagram and the significant difference between the median and the average equal 74.06 Ohm. The minimum resistance was 22.41 Ohm, which, according to the criteria of the boundary conditions presented in [4], indicates the load value of heavy trains below the limit of the accepted normal mode. The maximum value of load resistance equal to 4557.87 Ohm is derived from the minimum current of 5.59 A, which indicates the possible presence of equalizing currents in the studied inter-substation zone, which is also a complicating factor for the DRP CNF operation. A significant positive asymmetry value of 7.03 Ohm indicates a shift in the distribution towards smaller values. The value of the excess resistance of 58.52 Ohm indicates the presence of long tails, i.e. there are higher values at the distribution edges than around the average.

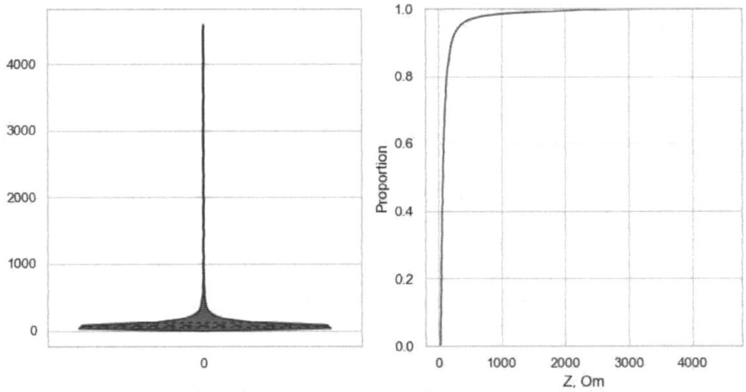

Figure 10. Load resistance distribution

Figure 11 shows the angle distribution between the first harmonics of the voltage on substation buses and feeder current. The average angle is 25.15°, which from the point of view of the DRP CNF operation is a positive factor, since according to [4] the boundary condition for the angle is $\varphi < 40^{\circ}$. A relatively small standard deviation of 0.57, as well as the median value almost coinciding with the average of 25.18 and the negative asymmetry value of -0.42 indicate a slight spread of values relative to the average and their shift towards large values. A positive angle excess of 0.39 indicates a sharp shape of the distribution, as confirmed by the shape of the violin diagram.

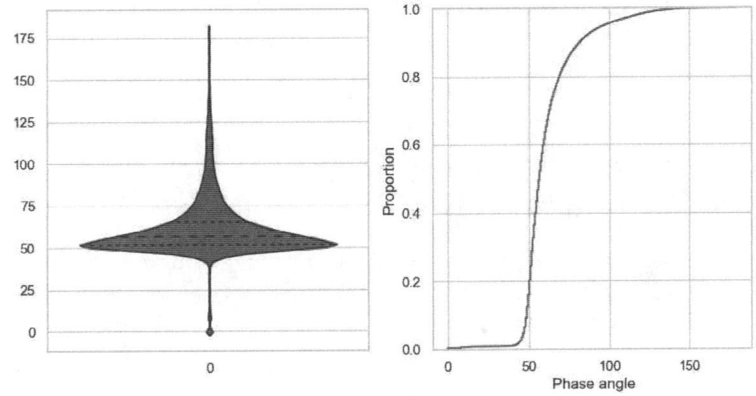

Figure 11. Distribution of phase angle between first harmonics of voltage on substation buses and feeder current

4. Conclusion

In this work, the main central tendency measures of parameters controlled by relay protection of contact network feeders during the operation of heavy trains were studied using the example of the fourth feeder of the contact network of the Sokhondo traction substation of the Trans-Baikal Railway.

The results of the study confirm the authors' hypothesis that the operation of heavy trains on a regular basis leads to a decrease in voltage in the contact network feeder to the boundary values of the minimum permissible value specified in [3], which is presented in Figure 3 in the form of a tail with a range of values from 23 to 22 kV and below.

Besides, the values of the feeder current also tend to the maximum permissible values and go beyond the limits of the normal mode equal to 850 A specified in [4]. This fact is presented in Figure 4 as a current distribution tail.

The operation of heavy trains does not best affect the quality of electrical energy and leads to significant distortions in current and voltage waveform, which is confirmed by high values of peak amplitude factors of voltage and current, reaching 1.97 and 5.95 at their maximum, respectively. The operation of heavy trains causes a strong "contamination" of the traction network with the highest harmonic components, which is confirmed by high harmonic distortion factors of voltage and current reaching 26.22 and 327.67 at their maximum, respectively. It is worth noting an extremely low average value, which generally indicates an unsatisfactory quality of electric energy. The above facts have a very negative effect on the quality and accuracy of the DRP CNF operation and, as a result, on the reliability of the transportation process as a whole.

The analysis and evaluation of load resistance distribution and angle between the first harmonics of voltage on the substation buses and feeder current confirms the authors' hypothesis that the operation of heavy trains may lead to overlapping of normal and emergency modes, thus causing ambiguity in the assessment of the DRP CNF mode and, as a result, an increase in the number of protection actuations due to overload and unspecified causes.

The general results of the analysis as a whole confirm the key thesis of the authors, that on the one hand the existing element base allows implementing fundamentally new protection algorithms, but the classical approaches used today to a great extent: first, do not unlock the full potential of the modern element base, and second, do not allow unambiguously distinguishing between normal and emergency operation modes and, as a result, failure to fulfill key requirements for relay protection such as selectivity, reliability and stability.

5. Acknowledgements

The team of authors expresses deep gratitude to the deputy head of the Trans-Baikal Directorate on Energy Supply of Traction Substations and Technological Networks, a structural division of Transenergo 2, a branch of Russian Railways, Denis Yu. Yankin for his comprehensive assistance and help in organizing and conducting experimental measurements at the existing substations of the Trans-Baikal Railway.

References

[1] Minutes No. 19 of the JSC Russian Railways Board of Directors 2013 *Development strategy of Russian Railway holding until 2030*, available at: https://ar2016.rzd.ru/ru/strategy/development-strategy-2030

[2] Order of the Russian Railways No. 1351r 2015 *Russian Railways Standard 07.021.1-2015. Protection of railway power supply systems against short circuits and overloads, Part 1. General principles and rules for the construction of protections, interlocks and automation in traction power supply systems*

[3] GOST 32144-2013. *Electric power. Electromagnetic compatibility of technical facilities. Standards of electric power quality in general-purpose power supply systems*. (Moscow: Standartinform)

[4] Russian Railways Order No. 615r 2016 *Russian Railways Standard 07.021.5-2018. Protection of railway power supply systems against short circuits and overloading, Part 4. Procedure for selection of protection facilities in AC traction power supply systems*

[5] Trifonov R S 2015 I*ncreased selectivity of the standby stage of remote protection of the contact network feeders based on adaptive identification:* Extended abstract of the dissertation of the Candidate of Technical Sciences, Chita.

[6] Figurnov E P 2009 R*elay protection: textbook.* (Moscow: Educational and Methodological Center for Education at Railway Transport)

Rig for testing active safety system for two-wheelers

A S Melnikov, M L Piatrenka and A U Yushkevich

Belarusian-Russian University, 43, Prospekt Mira, Mogilev, 212000, Republic of Belarus

E-mail: mishapetren@gmail.com

Abstract. The paper deals with the measuring equipment used to conduct two-wheeler braking system testing. When in motion, two-wheelers are initially less stable than cars, which necessitates creating more advanced active safety systems (ASS) that take into account the characteristics of their motion. Active safety systems available today operate on the basis of control algorithms that use kinematic parameters as sources of information, which requires performing approximate estimation of the relative slip in the contact patch between the wheel and the road. This leads to more complex control algorithms and increased cost of the entire system. To improve the safety of two-wheelers, an active safety system operating on the basis of control algorithms which use primary information about road conditions as sources of information must be developed. The primary information includes a number of forces, brake power, the magnitude of lateral forces, normal reactions in the contact patch between the wheel and the road. When creating and testing new anti-lock braking systems and their improved control algorithms, rig and road tests must be performed where testing conditions closely represent the real modes of operation of the active safety systems being developed. To conduct the rig tests under conditions close to real-world scenarios, a specialized rig for two-wheelers must be developed and created.

1. Introduction

The number of two-wheelers in the world is constantly increasing. According to the requirements set out in UN ECE Regulation No. 13 [1] and the laws adopted in 2012, all motorcycles having a cylinder capacity of 125 cm^3 and above must be fitted with anti-lock braking systems. Improvement of control algorithms based on kinematic parameters [2] implies a more sophisticated information processing unit, which results in an increased cost of the active safety system to be installed on a two-wheeler. This, in turn, leads to a significant increase in price of motorcycles with a cylinder capacity of 125 cm^3 and a decrease in competitive advantages. As an alternative, information sources based on primary factors, such as brake power, the magnitude of lateral forces and normal reactions in the contact patch between the wheel and the road can be developed [2, 4, 5, 6, 7] and the systems operating on the basis of sources of information about forces acting on the wheel should be improved.

In order to confirm the results of the development of active safety systems for two-wheelers, full-scale tests have to be performed.

2. Rig for active safety system testing

Full-scale tests are an integral part of the development, creation, refinement of new systems, components and mechanisms of vehicles. When performing full-scale tests, rig and road tests have to be conducted, and the results of computer simulation, rig and performance tests should be compared.

Content from this work may be used under the terms of the Creative Commons Attribution 3.0 licence. Any further distribution of this work must maintain attribution to the author(s) and the title of the work, journal citation and DOI.
Published under licence by IOP Publishing Ltd

The small dimensions of two-wheelers make it difficult to install measuring equipment on motorcycles and, as a result, to measure the parameters of the components being tested and the motorcycle controls.

When studying the active safety system, brake system and operating brake mechanisms, a number of parameters have to be measured. During rig tests, the brake torque, the actuating force acting on the brake pads of the brake mechanism being tested or the pressure in the hydraulic drive of the brake mechanism, the temperature of brake linings, the speed of rotation of the brake disc are determined. To identify, measure and detect the parameters of the brake system and the active safety system, the motorcycle and the rig are equipped with a number of sensors.

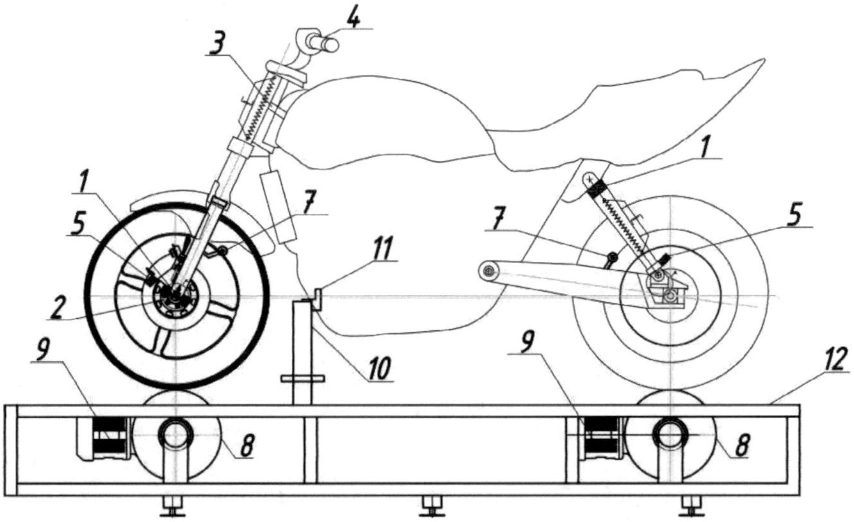

Figure 1. Measuring rig used to test brake systems, mechanisms and active safety systems: 1 – compression force sensor; 2 – bending moment sensor; 3 – tilt angle sensor; 4 – actuating force sensor; 5 – temperature sensor; 6 – motorcycle holder ; 7 – wheel rotation speed sensor; 8 – roller; 9 – electric motor; 10 – guide; 11– frame.

In order to measure and detect parameters of the motorcycle (lateral forces in the contact patch between the wheel and the road, normal reactions, wheel rotation speed), as well as brake systems and active safety systems of a two-wheeler, the rig is fitted with the following sensors:

- a speed sensor which detects the wheel rotation speed;
- a temperature sensor which detects the temperature of the brake disc during braking;
- a force sensor mounted on the actuator handle and the pedal of the rear brake mechanism; it detects the impact force applied to the actuator of the brake mechanism and the brake torque produced by the braking mechanism;
- two compression force sensors installed in the tubes of the front fork on the wheel rotation axis; they detect changes in the normal forces acting on the fork shock absorber;
- a compression force sensor installed in the shock absorber bracket of the rear swingarm of the motorcycle; it detects the change in the normal forces acting on the rear wheel;
- a steering angle sensor detects the steering angle of the front fork of the motorcycle during curvilinear motion;

Due to the holder fixing the motocycle frame to the rig, the motorcycle can move, which makes it possible to detect the redistribution of normal loads during the motorcycle braking and acceleration.

Two electric motors connected to the inverter are installed on the rig; their speed is controlled by means of the software developed specifically for this purpose.

The motors can run in different modes: they can operate synchronously and be controlled simultaneously; they can operate and be controlled each in its own preset mode or in a combined mode when one motor runs and the other is operated by means of a motorcycle wheel rotating in a power-actuated mode.

The brake power arising on the rig roller during braking as well as the power and the torque acting on the roller, which operates in traction mode, are determined by means of the motor. The motor power is N=4.5 kW, the speed is n=1420 min^{-1}.

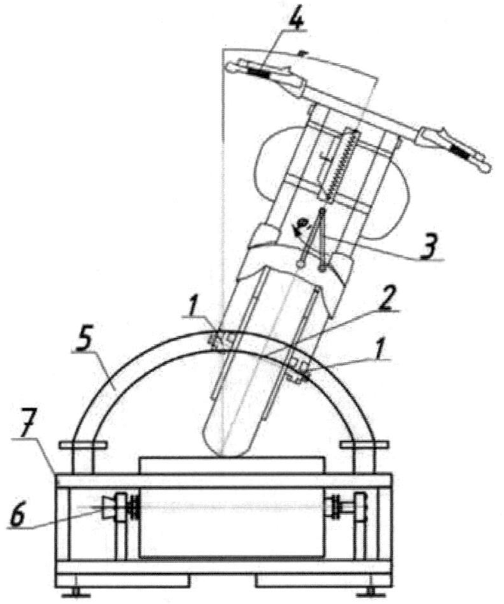

Figure 2. Measuring rig used to test brake systems, mechanisms and active safety systems: 1 – compression force sensor; 2 – bending moment sensor; 3– tilt angle sensor; 4 – actuating force sensor; 5 – guide; 6 – axial force sensor; 7 – frame

The developed rig is equipped with a measuring system for collecting and processing information that transmits the measurement results to a PC; the measuring unit can transmit data over distances of up to 100 m.

The module for collecting and processing information (MCPI) is the main part of the measurement and data processing unit. Its main functions are as follows:

1) to produce voltage levels needed to power the unit modules and sensors;

2) to process sensor readings;

3) to perform sensor reading filtering operations;

4) to implement the data exchange protocol between the MCPI and the PC via the UBS2.0 interface or via a wireless communication protocol based on radio modules.

A simplified block diagram of the MCPI is shown in figure 4.

The module has 16 analog inputs (A In) and 16 digital inputs (D In).

Analog input processing is performed by a precision high-speed and high-resolution analog-to-digital converter (ADC). To match the signal coming from the analog sensor to the ADC, a level matching module is used. The channel to be processed is selected by means of a high-speed analog multiplexer.

To process the digital sensor data, the built-in peripherals of the microcontroller are used. The data from all 16 digital inputs are processed by a high-speed multiplexer matrix.

The microcontroller program is based on a real-time operating system. Consequently, the microcontroller can perform data collection, processing, filtering and transmission operations simultaneously.

The data can be transferred via a high-speed USB2.0 interface (in FullSpeed or Hi-Speed mode) or via a wireless interface using a radio module. When using the radio module in an open area, the transmission range is 100 m.

The unit is powered from mains with voltage of 220 V/50 Hz, or, if it is not available, from a 12 V autonomous power supply.

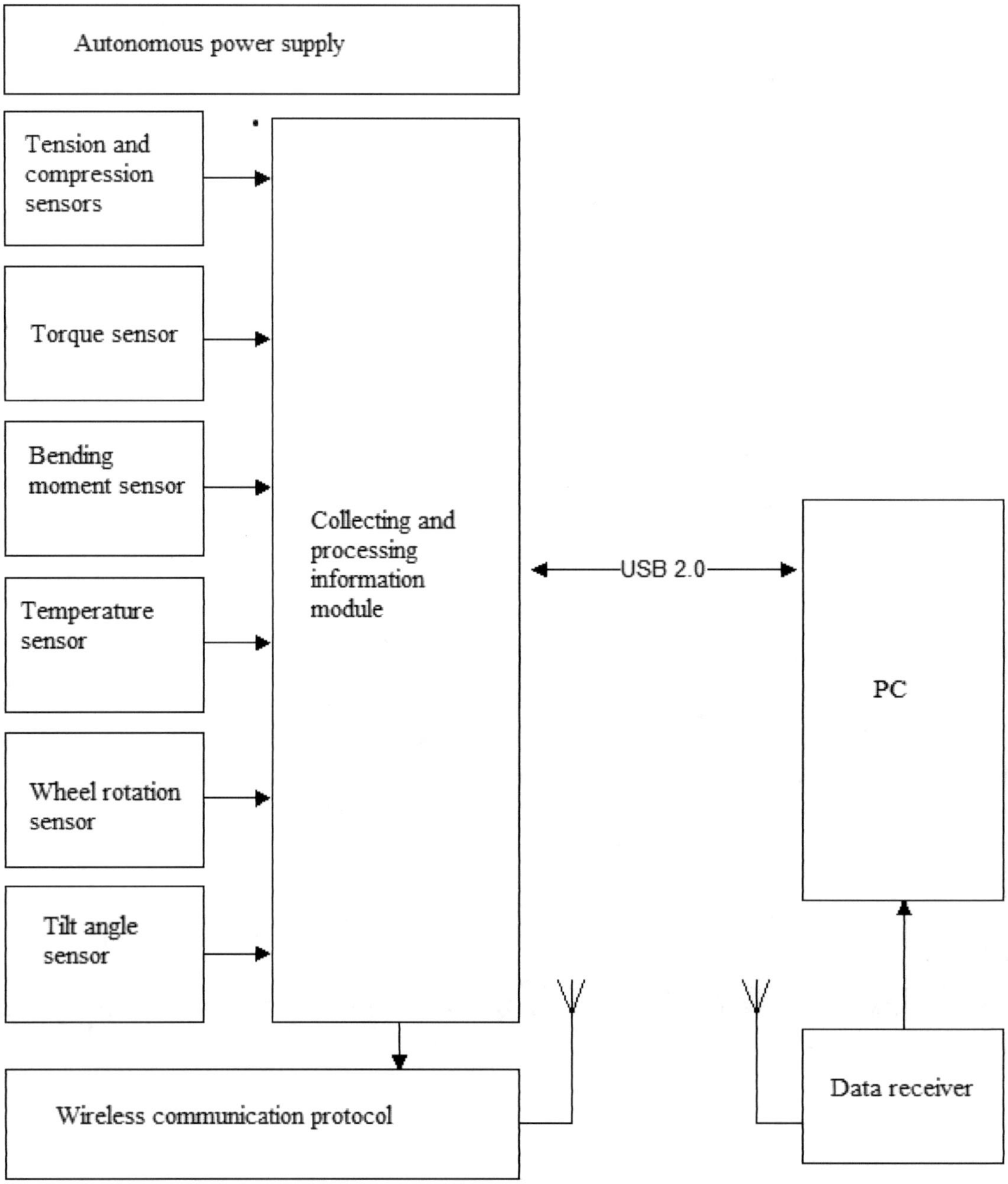

Figure 3. Block diagram of the measurement and data processing unit.

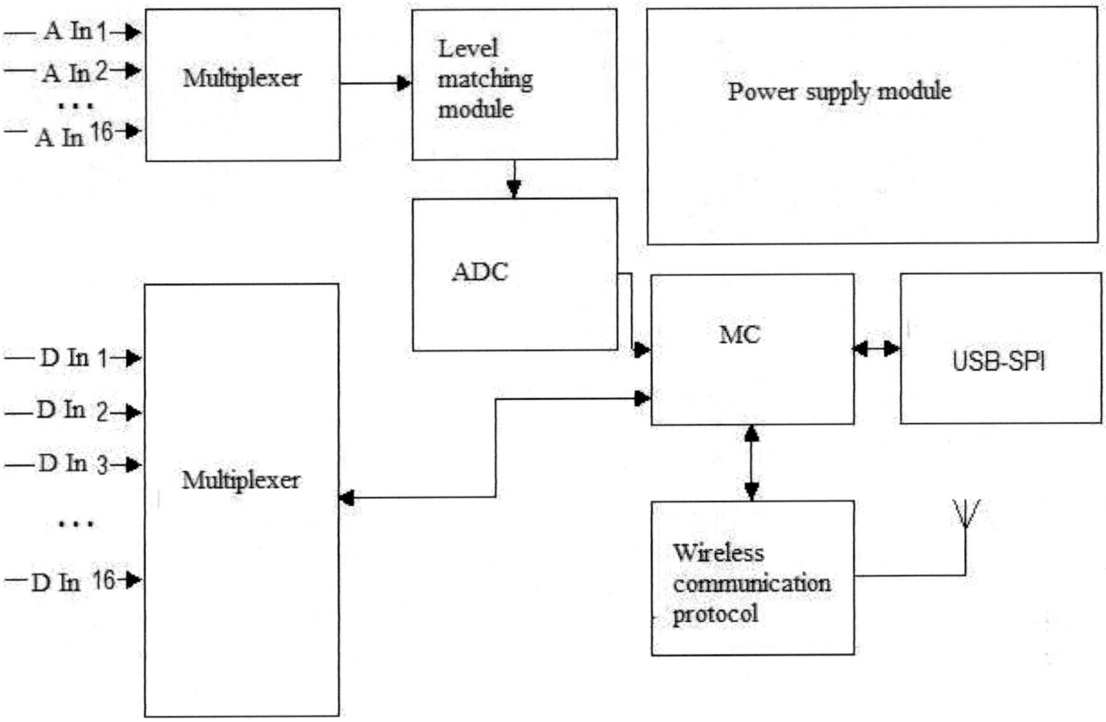

Figure 4. Simplified block diagram of the module for collecting and processing information.

The main purpose of the power supply module is to produce the voltage levels needed to supply power to the sensors and modules of the unit. Besides, the module monitors the state-of-charge levels of the autonomous power supply.

Tensiometers and piezometers are used as sources of information about the magnitude of lateral forces and normal reactions.

Table 1. Compression sensor characteristics

Parameter	Unit of measurement	Value
Measurement range	N	Up to 2000
Nominal output signal (n. s.)	mV/V	2
Nonlinearity (of n. s.)	%	0.15
Hysteresis (of n. s.)	%	0.1
Null balance (of n. s.)	%	± 2
Output signal temperature coefficient (of n. r./10 °C)	%	0.1
Null balance temperature drift (of n. s./10 °C)	%	0.05
Operating temperature range	°C	-10… +80
Permissible overload (of n. r.)	%	150
Recommended power supply	V	5
Protection class		IP 67

Table 2. Bending moment sensor characteristics

Parameter	Unit of measurement	Value
Measurement range	N·m	0-1500
Accuracy class	%	0.5
Protection class		IP 67
Refresh rate	Hz	150
Operating temperature range	°C	-10… +80
Recommended power supply	V	5

Modern anti-lock braking systems installed on motorcycles are sets of devices with complex control units and hydraulic brake system components. Rising material cost and advanced technologies used for their manufacture lead to a significant increase in the cost of the vehicles on which they are installed.

The high cost of control units used in active safety systems are caused by the use of complex computing devices operating on the basis of algorithms using kinematic parameters.

Application in control algorithms of sources of information about forces arising in the contact patch between the wheel and the road [4, 5, 6], namely the magnitude of lateral forces in the contact patch, normal reactions, actual torque can increase the speed of the control algorithm due to the use of primary sources of information that help improve the accuracy of decision-making of the generated control actions.

The rig used to test the motorcycle active safety systems is shown in figure 5.

Figure 5. Rig used to test active safety systems of the "MINSK" CX200 motorcycle: 1 – "MINSK" CX200 motorcycle; 2 – rig frame; 3 – roller; 4 – guide; 5 – MCPI data transmission cable; 6 - inverter control unit; 7 – MCPI data collection and processing unit; 8 – laptop; 9 – power supply; 10 – electric motor control inverter.

Figure 6. "Minsk" street motorcycle during brake dynamics testing with force measurement sensors placed on the rig: 1 – steering wheel angle sensor; 2 – front wheel-road contact patch normal load sensor; 3 – rear wheel-road contact patch normal load sensor; 4 – wheel-road contact patch lateral force magnitude sensor

The rig for active safety system testing is designed to measure the forces arising in the contact patch between the wheel and the road. The rig helps monitor a number of kinematic parameters, such as the wheel rotation speed, the roller speed, which provides accurate control of the moment when sliding occurs in the contact patch between the wheel and the road; the brake disc surface temperature, which helps control the efficiency of brake mechanisms and the temperature at the moment when the developed brake power is measured; the value of the force acting on the controls and the start of braking, which makes it possible to determine and measure the response time of the brake actuator of the system and evaluate the performance of brake actuators of various designs; forces that arise in the contact patch between the wheel and the road, such as brake power, the magnitude of lateral forces and normal reactions, due to which the wheel-road friction coefficient in the longitudinal and transverse directions can be calculated.

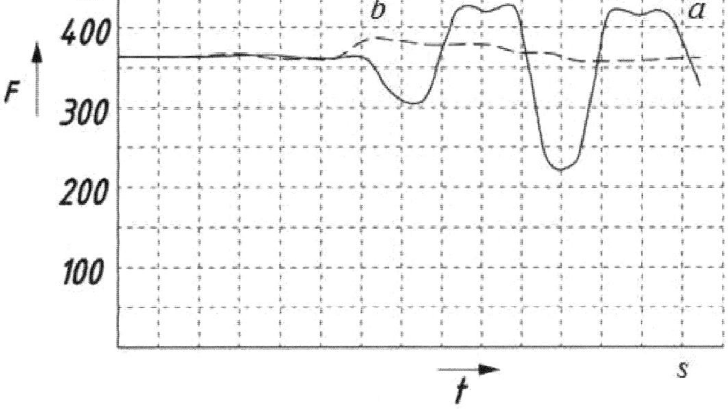

Figure 7. Lateral force graph obtained in "MINSK" CX200 motorcycle tests conducted for rectilinear and curvilinear motion: a – rectilinear motion; b – curvilinear motion with side slipping.

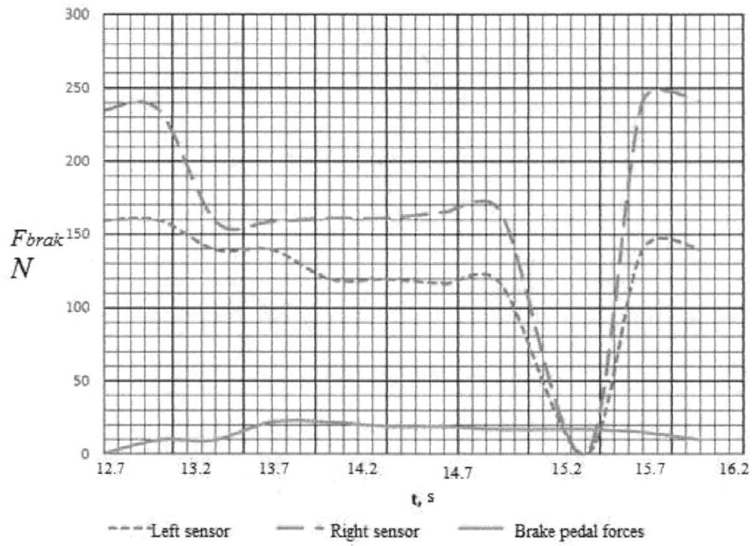

Figure 8. Braking force graph in the rear wheel contact patch obtained in "MINSK" CX200 motorcycle traction tests

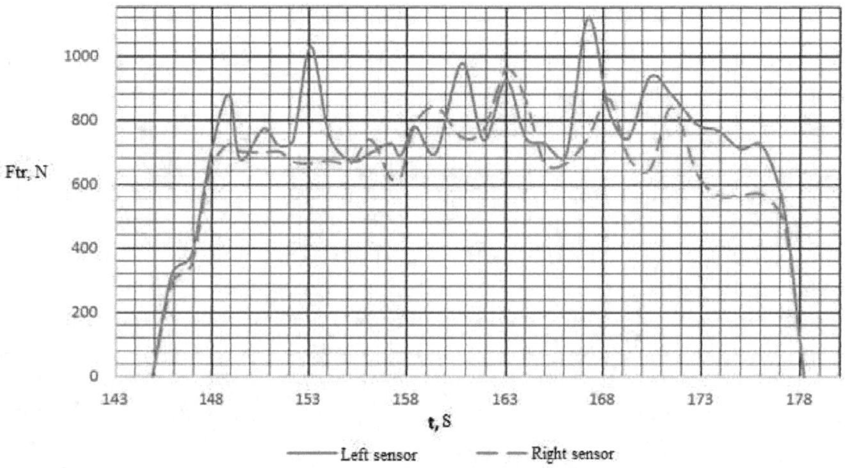

Figure 9. Tractive force graph in the rear wheel contact patch obtained in "MINSK" CX200 motorcycle acceleration tests

A rig that can be used to study and test active safety systems, braking systems and mechanisms of two-wheelers has been developed, which makes it possible to carry out tests and trials of braking systems and adaptive mechanisms for anti-lock braking systems.

3. Conclusion

1. The developed measuring system for collecting and processing information allows the data to be transmitted over distances of up to 100 meters, which in turn makes it possible to conduct road trials.
2. The developed rig can be used to conduct tests under conditions close to real road and traffic scenarios.
3. The rig can be used to test existing and new active safety systems.

References

[1] Regulation No 13 of the Economic Commission for Europe of the United Nations (UN/ECE) — (GOST R 41.13-99) Uniform provisions concerning the approval of vehicles of categories

IOP Conf. Series: Materials Science and Engineering 1151 (2021) 012016 doi:10.1088/1757-899X/1151/1/012016

M, N and O with regard to braking
[2] Kim V A 2003 *Methodology for creating adaptive active safety systems for motor vehicles based on force analysis* (Mogilev: Belarusian-Russian university)
[3] Zakharov A Y, Gorelov V A and Kotiev G O 2014 *Investigation of the dynamics of wheeled machines on a rig with chassis dynamometer* **7** 9–12 (Moscow: Mashinostroenie)
[4] Petrenko M L and Kim V A 2019 *Theoretical foundations of creation of tracking braking system of "Aist" bicycle Aktual'nye voprosy mashinostroeniya* **8** 21–25 (Minsk: OIM)
[5] Sazonov I S, Kim V A, Melnikov A S, Petrenko M L and Yushkevich AV Active safety system Patent BY 9363 IPC B 60T 8/00 Application no u20121031 applied on 2012.11.22.
[6] Yushkevich A V, Kim V A Theoretical justification of method of kinetic energy dissipation of "Minsk" motorcycle during braking. *Aktual'nye voprosy mashinostroeniya* **8** 26–29
[7] Ervinb R D, MacAdam C C and Watanabe Y 1978 *Measurement of Motorcycle Braking Performance by the Tow Method* 53-62

IOP Conf. Series: Materials Science and Engineering 1151 (2021) 012017 doi:10.1088/1757-899X/1151/1/012017

Methodology for assessing the competitiveness of options for the direction of the projected railway using SWOT analysis

Yu A Bykov and V A Fadeeva

Russian University of Transport, 9, Obraztsova st., Moscow, 127994, Russia

Abstract. The article presents a method for assessing the competitiveness of options for the directions of the projected railway at the pre-project stage. The first stage of the method involves the formation of an initial set of alternatives to directions using graph theory. The proposed approach to the formation of options for directions guarantees the consideration of all possible options for connecting the reference points defined by the design assignment. The second stage deals with the formation of an acceptable set of these alternatives using the SWOT analysis, method the use of which allows both quantitatively and qualitatively to assess the impact of internal and external environmental factors on the design object, that is, to take into account the weaknesses and strengths of each direction option, opportunities and threats of the external environment. The fourth stage of the considered method involves the determination of the integrated indicators of economic efficiency recommended for further design options for the direction of the railway.

1. Introduction

The principal direction of the projected railway, as well as the main parameters of its design (limiting slope, length of receiving and sending tracks, type of traction, etc.), in accordance with current regulatory documents should be established based on the results of technical and economic calculations for the future. This should take into account the savings in initial costs and ensure a further stepwise increase in its capacity as the required volume of traffic increases.

The interrelation of the issues of choosing the direction of the railway and its main parameters, taking into account the prospects for developing the capacity of the projected line, emphasizes the exceptional importance of a well-founded determination of the principal (main) direction and its possible options.

In general, the choice of the direction of the projected railway is influenced by various socio-economic, natural, technical and other factors, which in some cases are interrelated.

Socio-economic factors include, first of all, the purpose of the projected line, the location of the main cargo and passenger-generating centers, the availability of existing transport infrastructure and its interconnection with the new railway, etc.

In project practice, socio-economic factors determine the position of strong points, the mandatory passage of the route of the projected line. In some cases, reference points (abutment station of the projected line, the end point) can be defined by the design task.

Topographical, geotechnical, hydrological, climatic, seismic, and other design conditions are **natural** factors. These factors are determined by the location of fixed points through which it is desirable to pass possible variants of the railway route.

Content from this work may be used under the terms of the Creative Commons Attribution 3.0 licence. Any further distribution of this work must maintain attribution to the author(s) and the title of the work, journal citation and DOI.
Published under licence by IOP Publishing Ltd

Various combinations of connecting strong points and fixed points define a variety of options for the direction of the projected railway.

Evaluation of the competitiveness of the resulting set of options should be carried out taking into account **technical** factors, which primarily include the main design parameters: the limiting slope, the number of main paths, the type of traction, the length of the receiving and sending paths.

It becomes obvious that the variety of factors that influence the choice of the principal direction requires the adoption of balanced and reasonable design decisions, which largely determine the future investment and operating costs, as well as the economic efficiency of the projected railway.

The works of many scientists are devoted to solving this complex multi-variant and multi-criteria problem, among which we should note the works by A.V. Gorinov, I.V. Turbin, S.M. Goncharuk, A.V. Gavrilenkov, V. A. Podverbny, Yu.A. Bykov and others.

2. Materials and Methods

This article discusses the main provisions of the methodology for multi-criteria assessment of the competitiveness of the direction options of the designed railways, which includes four main stages:

1. Formation of the initial set of alternatives of the direction options:

$$x = \{x_1, x_2, \dots, x_m\}. \quad (1)$$

2. Formation of an acceptable set of alternatives of direction options:

$$y = \{y_1, y_2, \dots, y_n\},\ n < m. \quad (2)$$

3. Formation of an effective set of alternative direction options:

$$z = \{z_1, z_2, \dots, z_k\},\ k < n. \quad (3)$$

4. Determination by enlarged indicators of economic efficiency of design solutions for recommended options for further development of the direction from the set z.

The diagram shows the sequence of evaluating the competitiveness of options for the directions of the projected railway at the pre-project stage.

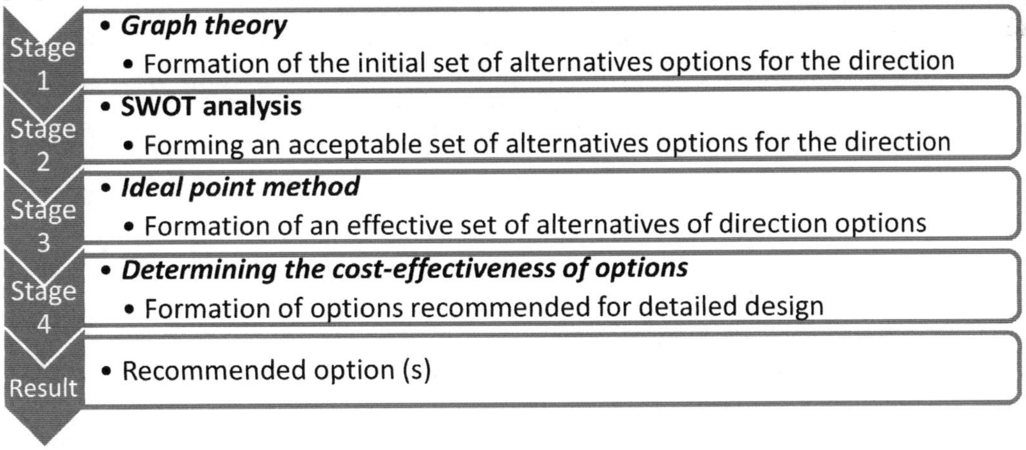

Figure 1. Scheme of sequential assessment of the competitiveness of options for the directions of the projected railway at the pre-project stage.

At the first stage of the method, graph theory methods are used to form the initial set of alternative direction options. They allow us to consider all possible direction options connecting strong points and fixed points that are nodes of the graph.

For the first time, the use of graph theory methods for forming a set of railway direction options was proposed by A.V. Gavrilenkov [1] and further developed in the works of Yu.A. Bykov.

It is known from graph theory that a subgraph that connects all nodes of graphs and does not contain closed contours is called a "complete tree". The number of such trees or the number of possible directions is calculated using the adjacency matrix determinant $d = |a_{ij}|$, a_{ij} – the number of edges with nodes i and j as endpoints, taken with a minus sign; a_{ii} – the number of edges adjacent to node i $(i = j = 1, \dots)$.

The described approach to the formation of the initial set of options for the direction of the projected railway allows us to consider all possible options, the assessment of the competitiveness of which is considered at the next stage of the methodology.

The second stage of the methodology is devoted to evaluating the competitiveness of the direction options from the set $x = \{x_1, x_2, \dots, x_m\}$ obtained at the first stage.

The method assumes the use of a modified SWOT analysis algorithm [2].

SWOT analysis allows both quantitative and qualitative assessment of the internal and external environment of the research object, while the internal environment is considered in terms of the **strengths (S)** and **weaknesses (W)** of the research object, and the external environment - in terms of the emergence of **opportunities (O)** and **threats (T)** for it [3].

A significant feature of the SWOT analysis method is the ability for the decision-maker to form well-known, but disparate and in some cases unsystematic ideas about the research object and its environment in the form of a logically consistent scheme of interaction of forces, weaknesses, opportunities and threats.

The authors of the article confirmed the possibility of using SWOT analysis to assess the competitiveness of options for the direction of projected railways, taking into account a large number of factors of the external environment (socio-economic, natural, environmental, etc.) and the internal environment (technical factors).

Methodological approaches to conducting a modified SWOT analysis of project decisions made at the initial (pre-project) stage by the example of choosing a direction option are considered in the work of the authors [4].

At this stage, a comprehensive analysis of internal and external factors that influence the choice of the direction of the projected railway allows us to "filter out" clearly non-competitive options and "narrow" the IMA to an acceptable set of options $y = \{y_1, y_2, \dots, y_n\}$, $n < m$.

The third stage of the method involves a multi-criteria assessment of the considered alternatives from the set of acceptable ones using the ideal point method.

As is known , the "ideal" method in this case is a conditional variant that has extreme (minimum or maximum) values for all the considered criteria.

The result of using the method is the definition of a generalized indicator in the accepted criteria space:

$$\varrho = \sqrt{c_1 r_1^2 + c_2 r_2^2 + \dots + c_j r_j^2} \quad (4)$$

where $r_1, r_2, \dots, r_j$ - normalized partial criteria;

$c_1, c_2, \dots, c_j$ - weight coefficients;

$$\sum_{j=1}^{j} c_j = 1, \quad c_j > 0. \quad (5)$$

The closer the option is to the "ideal" (less than ϱ), the better it is.

The use of the ideal point method for multi-criteria evaluation of variants of projected railways is considered in the work of the authors [5].

The result of the third stage of the method is the formation of an effective set of alternatives options for the direction $z = \{z_1, z_2, \dots, z_k\}$, $k < n$.

The fourth and final stage of the methodology under consideration offers a determination of the economic efficiency of options recommended for further, more detailed design based on aggregated indicators (required investments, income).

3. Conclusion
The proposed method multivariate and multicriteria evaluation of design decisions at the direction of the projected railway allows decision makers at each stage to fully assess the competitiveness of the options considered with the use of a large number of criteria (indicators), with both qualitative and quantitative evaluation, which significantly increases the objectivity of design decisions. The developed methodology can also be used to assess the competitiveness of options for passing international transport corridors through the territory of the Russian Federation.

4. References
[1] Gavrilenkov A V, Shivareva E A and Sholin V V 1991 *Transport construction* **6** 8-9
[2] Bystrov O F, Kainova D N and Tkachev G G 2010 *Scientific Bulletin of MSITI* **6** 65-70
[3] Vihansky O S 1998 *Strategic management*, pp. 20-38 (Moscow: Gardarika)
[4] Bykov Yu A, Fadeeva V A and Kozhevnikov V V, 2019 Designing the development of a regional railway network: collection of scientific papers. *FESTU* **7** 154-161
[5] Bykov Yu A, Vasiliev A B, Fadeeva V A, 2015 Transport of the Urals Yekaterinburg *USTU* **1 (44)** 53-56

Scheduling of construction works in the formation of technological processes of railway construction using a genetic algorithm

A V Polyanskiy[1]

[1] Russian University of Transport, 9, Obraztsova str., Moscow, 127994, Russia

E-mail: Polal_82@mail.ru

Abstract. The article is devoted to solving the task of scheduling construction works in the formation of technological processes of railway construction by the method of evolutionary modeling. Scheduling of construction works is the task of optimizing the amount of subcontracting involved to ensure maximum utilization of the construction contractor's own resources. The purpose of the task was to determine the sequence of work during the construction of railway track objects (roadbed, upper structure of the track, culverts). The solution of the task required taking into account project limitations: resource, technological, and information. The specifics of the activities of construction contractors and subcontractors required taking into account the time limit that set the time interval: early and late start of construction work. The set of restrictions and the dimension of the task contributed to the attraction of a genetic algorithm for its solution. A mathematical model was developed for this purpose. The model allowed us to assess the fitness function of individuals (solutions), taking into account the dynamic nature of the distribution of own and subcontracting resources for construction work. The results of solving the task were obtained in automated mode. Analysis of the results of the solution allowed us to determine the advantages of the genetic optimization method for obtaining the most effective solution to the task of scheduling construction works.

1. Introduction

The task of scheduling construction works within the framework of forming technological processes (TP) of railway construction belongs to the class of multi-criteria tasks for managing complex systems. The time to find solutions to such tasks increases exponentially depending on its dimension. Traditional methods of solving the task of finding a rational solution have a number of disadvantages. For example, combinatorial methods and mathematical programming methods require long calculations when studying large-dimensional tasks. In addition, these methods have difficulties in finding the optimum for the case of a dynamic constraint system that changes at each scheduling stage [1-3]. The application of the method presented in this paper takes into account the dynamic nature of the task; however, the search based on it is random and does not always allow you to find the optimal solution. At the same time, the use of evolutionary (genetic) optimization makes it possible to overcome the identified disadvantages of traditional methods, including significantly reducing the duration of the solution search to a quadratic dependence on the task dimension.

Content from this work may be used under the terms of the Creative Commons Attribution 3.0 licence. Any further distribution of this work must maintain attribution to the author(s) and the title of the work, journal citation and DOI.
Published under licence by IOP Publishing Ltd

2. Materials and methods

The general statement of the task of scheduling construction works can be formulated as the task of determining the sequence of works that meets the design requirements (restrictions) and the minimum time for their implementation. Project requirements that affect the work plan being developed will be resource, technological, and information constraints. Analysis of the activities of construction (contracting) organizations (CCO) requires taking into account the time limit, which sets the time interval: early and late start of construction work [4, 5].

The technological process consists of a set of complex technological processes (CTP). However, when forming a technological process (TP) from CTP, simple technological processes (STP) are distinguished, taking into account their resource supply (labor resources, technical means, materials). STP occupy a central position in the hierarchy of TP (Figure 1) and are the most optimal for the design of technology and organization of railway construction [6].

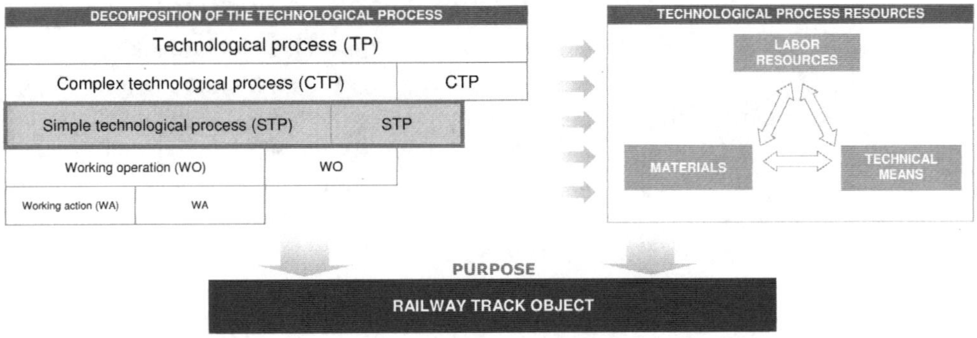

Figure 1. The structure of the technological process.

It should be noted that in some cases, when solving certain scheduling tasks, different sets of restrictions are taken into account, namely, all the above restrictions, with the exception of the time limit. However, special attention is paid to technological and resource constraints [3].

The objects of optimization are also different. The classical objective function (OF) in a scheduling task is to reduce the time required to complete work. In some works, OF is presented as the total time of deviations from the planned deadlines [4].

However, optimization of work performed by construction (subcontracting) organizations (CSO) has not yet been reflected in research activities, and this task is relevant for almost all CCO. Next, the activity of CCO will be considered and the task of calendar scheduling for the formation of TP will be defined, taking into account the performance of work by CSO.

We denote the scheduling period $t \in [0, T]$, where t - time point in the scheduling interval; T - duration of the scheduling period. The set of CTP is given: $CTP = \{CTP_i\}$, $i = \overline{1, N}$, where CTP_i - i -th CTP; N - number of CTP. The set of STP included in the CTP: $STP = \{STP_{i,j}\}$, $j = \overline{1, L_i}$, $\sum_{i=1}^{N} L_i = L$, where $STP_{i,j}$ j-th STP of i-th CTP; L - number of STP within the technological process. Construction work cannot be performed without the involvement of labor resources and technical means of CCO. To do this, the following symbols are used. The set of labor resources is available in the CCO: $KW = \{KW_r\}$, $r = \overline{1, R}$, where KW_r - number of people in the construction team r; R – number of construction teams in CCO. The set of labor resources of CSO will be considered unlimited in order to determine the allocated volume of third-party resources by attracting, if necessary, several specialized construction organizations. Thus, it allows you to remove the resource restriction of scheduling during the task solution.

The set of labor resources is required to perform STP of each CTP: $KWS = \{KWS(i, j, r)\}$, where $KWS(i, j, r)$ is the number of labor resources of the team to perform STP. Multiple durations of STP within TP: $KD = \{KD(i, j)\}$, multiple costs of performing STP per day: $CW = \{CW(i, j)\}$.

The set of technical means is available in CCO: $KT = \{KT_m\}$, $m = \overline{1, M}$, where KT_m – number of technical means in the set m; M - number of sets of machines in CCO. We will also assume that the set of technical means of CSO is infinitely large with conditions similar to the set of labor resources.

The set of technical means is necessary for performing STP of each CTP: $KTS = \{KTS(i, j, m)\}$, where $KTS(i, j, m)$ is the number of technical means included for performing STP. The set of durations of STP within TP: $KB = \{KB(i, j)\}$, the set of costs of performing STP per day: $CT = \{CT(i, j)\}$.

Plenty of time limitations began to perform STP: $TS = \{TSB(i, j), TSE(i, j)\}$, where $TSB(i, j)$ is the date of early run j-th STP i-th CTP; $TSE(i, j)$ – late start date to completion j-th STP i-th CTP. This restriction by the developer of organizational and technological documentation (OTD) includes technological and information restrictions if they appear for any STP within TP.

The controlled parameters in the task under consideration can be represented as a set of moments when STP start: $TB = \{TB(i, j)\}$. Now it is possible to solve the task of scheduling construction works taking into account the following optimization criteria:

1) minimization of the total cost of work performed by CSO:

$$F_1 = \sum_{i=1}^{N} \sum_{j=1}^{L_i} \left(CW(i, j) \cdot V(i, j) + CT(i, j) \cdot V(i, j) \right) \to \min \tag{1}$$

where $V(i, j)$ is the amount of work performed by CSO j-th STP i-th CTP;

2) minimization of the total downtime of the team and technical means of CCO:

$$F_2 = \frac{\sum_{t=0}^{T} \sum_{r=1}^{R} VF(t, r)}{T \cdot R} + \frac{\sum_{t=0}^{T} \sum_{m=1}^{M} MF(t, m)}{T \cdot M} \to \min \tag{2}$$

where $VF(t, r)$ – number of available resources of CCO construction team r at the moment t, $MF(t, r)$ – number of available technical means in the set of CCO m at the moment t.

Let us consider the following OF F that takes into account the above optimization criteria:

$$F = \varphi_1 \cdot \left(\frac{F_1^{In}}{F_1} \right) + \varphi_2 \cdot \left(\frac{F_2^{In}}{F_2} \right) \to \max \tag{3}$$

where φ_1, φ_2 – weight coefficients, $\varphi_1 + \varphi_2 = 1$; F_1^{In}, F_2^{In} – initial values of objective functions obtained by expert evaluation of the developer of OTD of the start dates of STP.

3. The use of Matlab and applications Genetic Algorithm and Direct Search Toolbox

To find the optimal construction plan (STP) it is advisable to use a genetic algorithm (GA). The steps of using GA involves the implementation of the following stages: 1) selecting the encoding method of the task solution (phenotype) in individual chromosome (genotype); 2) defining valuation method fitness function (FF) decisions; 3) description of the genetic operators; 4) determination of the laws of survival of the decision; 5) generation of initial population and GA.

Calculation of FF of individuals in a population can be performed in various ways described in [7-9], for example, by analytical methods, using simulation, artificial neural networks, and fuzzy logic systems. However, based on the works [10, 11], analytical methods for calculating FF are preferred for solving scheduling tasks taking into account resource constraints. Although this method is not without drawbacks. This approach does not allow us to take into account the dynamic nature of the behavior of the task being solved. However, the capabilities of the Matlab system (technical computing language) and the Genetic Algorithm and Direct Search Toolbox application allow you to simulate dynamic resource allocation for evaluating FF of a solution. At the same time, it is possible to limit the random search space and enhance optimization by taking into account the dynamically changing constraints of the scheduling task. The algorithm for applying the system and application during process optimization is shown in Figure 2.

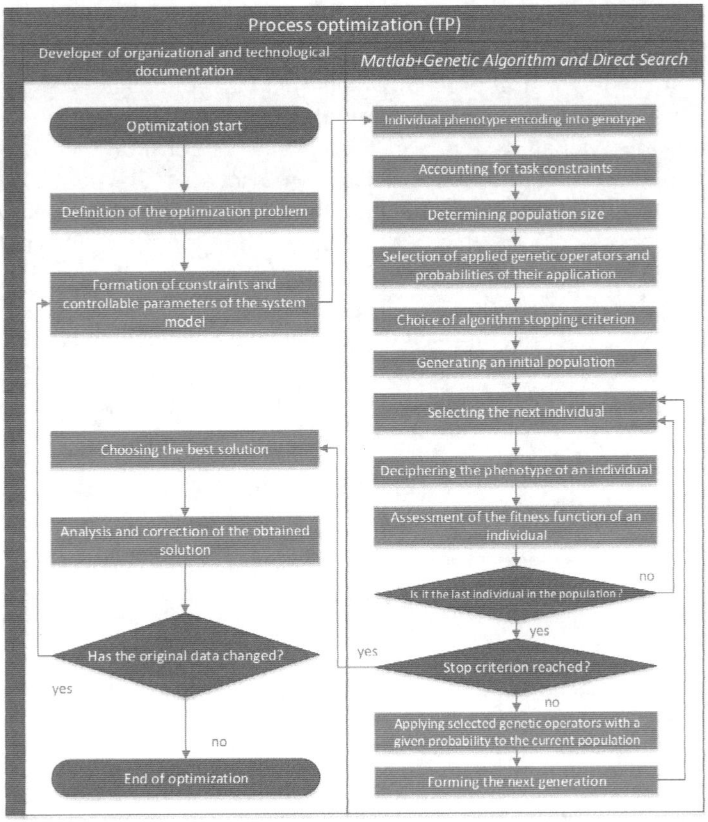

Figure 2. Algorithm for genetic optimization of the technological process using the Matlab system and the Genetic Algorithm and Direct Search Toolbox application.

The genetic optimization method used has the following features:

1) integration of expert, conceptual and evolutionary approaches;

2) focus on a wide class of scheduling tasks that allow you to configure the ready-made solution of the task for specific conditions by setting the parameters of GA;

3) support for a specific method for solving the scheduling task using GA based on the description of encoding the phenotype of an individual into a genotype.

The practical implementation of the method of genetic optimization using the Matlab system is carried out to solve the task of scheduling CTP in the framework of technological modeling in CCO during the construction of railway tracks objects (roadbed, upper structure of the track, culverts) to plot the Big ring of the Moscow railway (Moscow region, Russia). The purpose of scheduling is to minimize the downtime of teams, as well as to minimize the cost of work performed by CSO, as part of TP. The development of an evolutionary model for solving this task required the study of various options for encoding the solution into an individual chromosome: encoding a sequence of STP [10, 11], encoding priorities of STP [10], encoding rules for assigning priorities to STP [11], encoding calendar start dates [11]. As a result, the method of encoding start dates was chosen, due to the lack of redundancy in encoding and the ability to take into account time constraints.

Assessment of FF for solving the task of executing STP within TP using the formula (3) was performed in the Genetic Algorithm and Direct Search Toolbox application, taking into account the dynamic distribution of own and subcontracting resources for STP. Controlled parameters (input data) of the model represent the start dates of STP. The following parameters were set for GA during genetic optimization. First, the population size is 10 individuals. Second, individual size – 175 genes (5 genes for encoding each of the 35 simple technological processes under consideration); third, applied genetic operators – roulette-based reproduction, five-point crossover, five-point mutation with a probability of 10%, inversion with a probability of 5% [11]; fourth, algorithm stop criterion – 10 generations change; fifth, generation of a random initial population.

As a result of using the Matlab system and the Genetic Algorithm and Direct Search Toolbox application for genetic optimization, the values of FF of individuals in the population and the values of OF of the task depend on the change of generations. Figure 3 left shows the change in the minimum objective function value of task (1) under genetic optimization, and the right – change the maximum value of FF individuals of GA (3) in the course of its implementation. The graphs show that the best solution was achieved in the 7th generation.

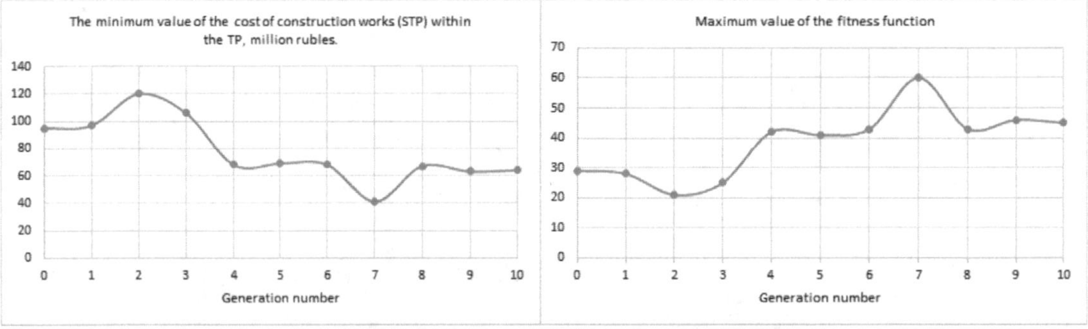

Figure 3. Dependence of the target function of the scheduling task and the fitness function of the individual of the genetic algorithm on the change of generations.

Histograms of OF (1) and (2) of the task solutions obtained using the genetic optimization method are shown in Figure 4 in comparison with OF of the original construction plan drawn up by the developer of OTD based on existing experience and qualifications.

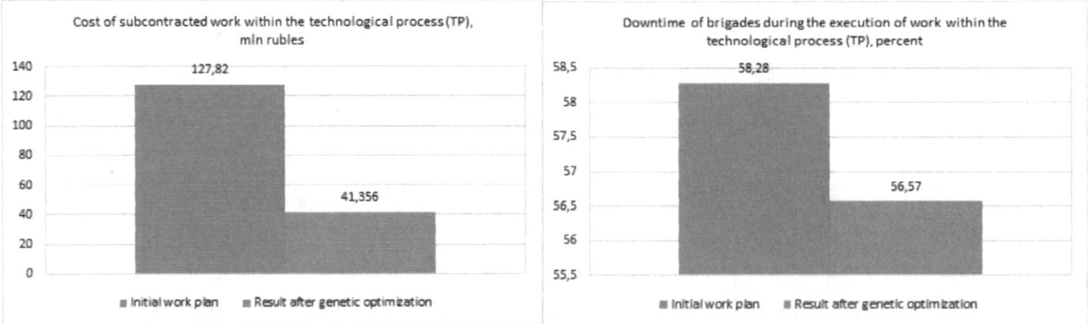

Figure 4. Changing the values of the objective functions of the task depending on the solution method.

Analysis of solutions to the scheduling task found by various methods has shown the effective operation of the genetic optimization method. Its application allowed reducing by 68% (3 times) the cost of work performed by CSO, within TP, by 3% of a simple team within seven months.

4. Conclusion

The paper presents the theoretical results of the study and the possibilities of practical application of evolutionary modeling to solving tasks of construction scheduling in the formation of technological processes of railway construction.

Scheduling of construction works is the task of optimizing the amount of subcontracting involved to ensure maximum utilization of the construction contractor's own resources. The purpose of the task was to determine the sequence of work during the construction of railway facilities. The solution of the task is carried out taking into account the project limitations: resource, technological, and information. The analysis of the specifics of the activities of construction contractors and subcontractors required taking into account the time limit that set the time interval: early and late start of construction work. The set of restrictions and the dimension of the task contributed to the attraction of a genetic algorithm for its solution. A mathematical model was developed for this purpose. The development of the model allowed us to evaluate the fitness functions of individuals (solutions), taking into account the dynamic nature of the distribution of own and subcontracted resources for construction work.

As a result of the study, it was found that the automated solution of construction scheduling tasks using evolutionary modeling significantly increases the efficiency of the solution found. To solve the task, we used a genetic algorithm for directed search in the solution space. At the same time, the dynamic nature of the distribution of own and subcontracted resources for construction work was taken into account.

The introduction of evolutionary modeling tools in the development of technological processes of railway construction contributes to the operational scheduling of construction works. A construction contractor receives a new tool aimed at ensuring the production and economic efficiency of its activities.

References

[1] Zavadskas E K, Antucheviciene J, Vilutiene T and Adeli H 2018 Sustainable decision-making in civil engineering, construction and building technology *Sustainability* **10(1)** 14

[2] Buitragoa M, Adama J M, Alvaradob Y A, Moraguesa J J, Gascha I and Calderóna P A 2016 Designing construction processes in buildings by heuristic optimization *Engineering Structures* **111** 1-10

[3] Arashpour M, Kamat V, Bai Y, Wakefield R and Abbasi B 2018 Optimization modeling of multi-skilled resources in prefabrication: theorizing cost analysis of process integration in off-site construction *Automation in Construction* **95** 1-9

[4] Lin L, Hao X-C, Gen M and Jo J 2012 Network modeling and evolutionary optimization for scheduling in manufacturing *Journal of Intelligent Manufacturing* **23** 2237-2253

[5] Kim H-J and Reinschmidt K 2012 Market Structure and Organizational Performance of Construction Organizations *Journal of Management in Engineering* **28** 212-220

[6] Zavadskas E, Vilutienė T and Tamošaitienė J 2017 Harmonization of cyclical construction processes: a systematic review *Procedia Engineering* **208** 190-202

[7] El-Abbasya M S, Elazounib A and Zayedc T 2020 Finance-based scheduling multi-objective optimization: Benchmarking of evolutionary algorithms *Automation in Construction* **120** 103392

[8] Papadaki I N and Chassiakos A P 2016 Multi-objective Construction Site Layout Planning Using Genetic Algorithms *Procedia Engineering* **164** 20-27

[9] Fung I W H, Huang C and Tam V W Y 2011 Application of GA techniques for solving precedent relationship problem in project scheduling optimization *Procedia Engineering* **14** 2527-2534

[10] Faghihi S V, Reinschmidt K and Kang J 2014 Construction scheduling using genetic algorithm based on building information model *Expert Systems with Applications* **41** 7565–7578
[11] Toklu Y C 2011 Application of genetic algorithms to construction scheduling with or without resource constraints *Canadian Journal of Civil Engineering* **29** 421-429

Considering various conditions during determination of railway curve superelevation

V P Beltiukov and A V Andreev

Emperor Alexander I St. Petersburg State Transport University (PSTU), 9, Moskovsky pr., Saint Petersburg, 190031, Russia

E-mail: Oddman@bk.ru

Abstract. The calculation of the established railway curve superelevation, until recently, was carried out on the basis of two conditions: the condition of equal wear of both rails and the condition of passenger ride comfort. With the introduction of the Interim guideline on determining the railway curve superelevation and speeds limits in curves, this condition was abolished in favor of a more complex condition for ride comfort. Unfortunately, the condition of ride comfort does not take into account many factors that can influence the determination of the optimal railway curve superelevation. Among such factors are economic factors. In the URRAN methodology, for example, among others, the life cycle cost is considered. For track facilities, it is more relevant to consider the concept of overhaul cycle cost, since it is during the overhaul cycle that the curves are operated. The article proposes, along with the Interim guideline, in which safety factors are considered, to consider the definition of the railway curve superelevation, taking into account the life cycle cost.

1. Introduction

Ensuring the same vertical wear and settlement of both rails requires that the sum of the normal pressures on the high side of curve rail (or normal reactions E_h) from all wheels of all trains traveling along a given curve during the accepted time interval be equal to the sum of the normal pressures from the same trains to the inside of curve rail (or its normal reactions E_i). It is customary to consider in this case a design scheme with a carriage without springs (Figure 1).

Additionally, it is assumed that the carriage is symmetrically positioned in the track, and the influence of the wind is neglected. To ensure the stated condition, it is necessary that:

$$\sum E_h = \sum E_i, \tag{1}$$

where $\sum E_h$ - total loads on the high side of curve rail;

$\sum E_i$ - total loads on the inside of curve rail.

Content from this work may be used under the terms of the Creative Commons Attribution 3.0 licence. Any further distribution of this work must maintain attribution to the author(s) and the title of the work, journal citation and DOI.
Published under licence by IOP Publishing Ltd

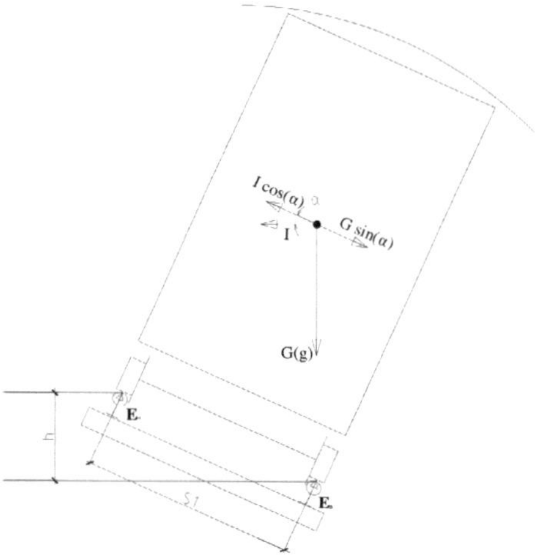

Figure 1. Design scheme for determining the railway curve superelevation

Based on the equality of the total loads (Formula 1), a formula was derived to determine the elevation of the railway curve superelevation based on the condition of equal wear of both rails during the passage of the train traffic (h_{traf}). In the method for calculating the railway curve superelevation [1], the formula looked like this:

$$h_{traf} = 12{,}5\frac{\overline{v_{top}^2}}{R}, mm. \tag{2}$$

In this case $\overline{v_{top}}$ – weighted average speed of trains, obtained on the basis of the top speeds of trains circulating on the study railroad; R – curve radius.

2. Materials and methods

Professor G. M. Shakhunyants and Professor O. P. Ershkov in the 60s proposed adding an additional term (Δh) [2], which would take into account the influence of the bolster part to the wheelset and the work of the carriage springs. In future, the carriage members were taken into account only during determination of railway curve superelevation based on the ride comfort. Formula 2 was abolished with the release of an Interim guideline on determining the railway curve superelevation and speeds limits in curves [3].

Considering the ride comfort during the carriage following the curved section of the track is carried out using the non-compensated acceleration that occurs when the centrifugal force appears.

Years of research experience shows that high values of these accelerations are unpleasant for passengers. The value of the railway curve superelevation, calculated from the weighted average speed, may be insufficient to reduce the non-compensated accelerations that occur when passenger train pass along the curve, the speed of which is higher than the freight train. It is required to set such a railway curve superelevation that the value of the non-compensated acceleration that occurs when the train passes at top speed does not exceed the permissible value for the effect on the passenger.

To simplify the calculation, such a limitation of the non-compensated acceleration of the rolling stock elements is adopted according to the considered scheme, in which the non-compensated acceleration of the passenger does not exceed the permissible medical standards.

Consideration of safety (ride comfort) can conflict with the condition of an equal load on both rails during operation, which leads to a more intensive growth of failures in a curve track section.

In the period from 2016 to 2018, Russian Railways performed an analysis of the non-compensated accelerations and railway curve superelevation in connection with the suspicion of irrational railway curve superelevation on the railway.

The Hexa company, the Railway Vehicle and Track Interaction Testing Center and JSC "VNIKTI" considered the optimal railway curve superelevation for five directions of the railway: Samara – Chelyabinsk, Irkutsk – Chita, Moscow – Smolensk, Aleksandrov – Orekhovo-Zuevo, Yaganovo – Voskresensk. To consider the optimal railway curve superelevation, the values of vertical and lateral loads in the curves were determined using load complexes at various non-compensated accelerations. Studies have shown that minimum loads occur at a non-compensated acceleration of 0 m/s^2. It follows from this that under the most comfortable conditions for passengers there will be an equal distribution of loads on outer and inner rail, which indicates the minimum load on one rail (left or right).

But it should be said that the minimum loads does not mean the minimum costs when operating a railway curve. To determine the minimum cost, it is worth considering the concept of overhaul cycle cost.

3. Considering overhaul cycle cost during determination of railway curve superelevation

The overhaul cycle of the upper track is a part of the life cycle during which the maintenance and repair of the railway track takes place. The overhaul cycle cost is the sum of the costs at the laying of track, maintenance and intermediate repairs of the track, dismantling of track.

Compared to similar stages during the life cycle of an object (implementation, operation and decommissioning and disposal), the overhaul cycle has the difference at the stage of dismantling of track, the elements of the permanent way can be used in the future, that is, after the stage of dismantling of track in another railway track start the laying of track.

When considering the overhaul cycle cost, it was assumed that the amount of costs will be taken without taking into account the further relaying of track in another railway track.

Minimization of the life cycle cost based on the URRAN system [4] will give the optimal maintenance of the railway track.

This principle can also be used to optimize the overhaul cycle. The minimum of the overhaul cycle cost will indicate that the costs at all stages of the cycle will be optimal. If we consider the overhaul cycle cost in relation to the non-compensated acceleration, then we can say that the non-compensated acceleration will be optimal if the minimum the overhaul cycle cost is observed.

When determining the railway curve superelevation, taking into account the overhaul cycle cost, one should not forget about the limiting values of non-compensated accelerations from the condition of the comfort of the passengers. ride.

In general, the railway curve superelevation, taking into account safety (ride comfort), will be determined by the following formula:

$$h_{\max pass} = 12{,}5\frac{v_{\max pass}^2}{R} - a_d \cdot 163, mm. \tag{3}$$

where $v_{\max pass}$ - maximum speed of passenger trains, km / h;
a_d - non-compensated acceleration limit (0,7 m/s^2).

The limit of non-compensated acceleration for passenger traffic can be increased to 1.0 m/s^2 for a number of types of passenger rolling stock with improved dynamic properties.

To take into account the technical and economic component when calculating the railway curve superelevation in the Interim guideline [3], the railway curve superelevation for freight trains are calculated. Taking into account the technical and economic components, in this case, the limiting values of the non-compensated acceleration will be ± 0.3 m/s^2, depending on the considered maximum or minimum speed of freight trains.

To optimize the overhaul cycle, the railway curve superelevation should be considered, taking into account the overhaul cycle cost, on the equal basis with the railway curve superelevation, determined according to the Interim guideline [3], which will determine the optimal railway curve superelevation.

For consideration of the overhaul cycle cost, when determining the railway curve superelevation, the non-compensated acceleration is determined so that the overhaul cycle costs would be minimal.

To solve this problem, the function of the dependence of the overhaul cycle cost on non-compensated acceleration was determined. When calculating the overhaul cycle cost, the main track maintenance-of-way works and track repair works performed during the overhaul cycle were considered.

The cost of track maintenance works was calculated as the sum of the cost of overhauls (track relaying), track raising and ballast cleaning repairs. A set of track repairs for each railroad section was selected based on the class of the railway track according to Specifications [5]. The costs of the works were taken on the basis of aggregated estimates. The calculation of the track maintenance works costs showed that the costs in track with the same overhaul scheme (a set of track repairs) are approximately equal and almost do not depend on the non-compensated accelerations in the curve. Therefore, the determination of the dependence of the overhaul cycle cost on the non-compensated acceleration was reduced to the determination of the dependencies of the costs for the maintenance-of-way works on the non-compensated acceleration.

Before determining the final dependence of the costs for the maintenance-of-way works on the non-compensated acceleration, the dependences of the costs for each individual work on the non-compensated acceleration were determined. During the study, it turned out that the function that would best reflect the dependence is an exponential function of the following form:

$$f(x) = Ae^{Bx} \tag{4}$$

where x – value of non-compensated acceleration for a specific curve, m/s^2.

The study also showed that when considering different costs for the execution of work, the function could be with the opposite sign, which in turn could mean that for different works, costs may be unevenly distributed between the activities carried out on the high side and inside of curve rail.

After determining the coefficients A and B by the least square method [6], the final function was defined to determine the dependence of the cost of the maintenance-of-way works on the non-compensated acceleration. The resulting function was equal to the sum of the functions of the dependence of costs on the non-compensated acceleration. To simplify the determination of the non-compensated acceleration, at which the minimum cost of the maintenance-of-way works will be observed, the exponential function has been replaced by a parabolic function. Figure 2 shows the graphs of the dependence of the costs of the maintenance-of-way works on the non-compensated acceleration and the obtained optimal non-compensated acceleration.

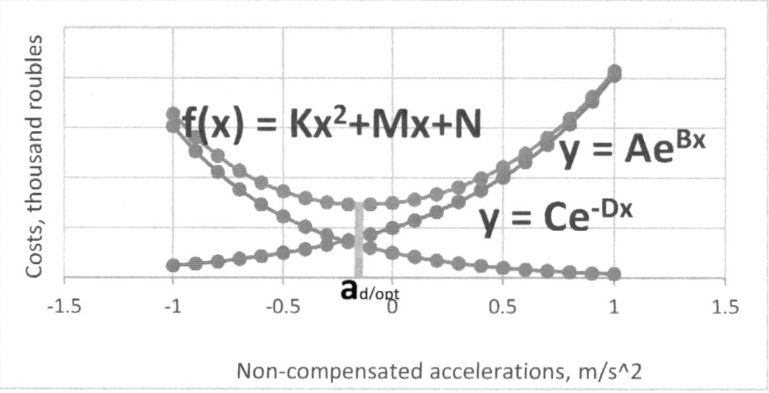

Figure 2. Charts of the dependence of costs on optimal non-compensated acceleration

After determining the optimal non-compensated acceleration from the point of view of the maintenance-of-way works ($a_{d/opt}$), the optimal non-compensated acceleration was determined taking into account the track maintenance.

Based on the determined non-compensated acceleration, the railway curve superelevation were calculated using the following formula:

$$h_{OCC} = 12{,}5\frac{\overline{v_{top}^2}}{R} - a_{\frac{d}{opt}} \cdot 163, mm, \quad (5)$$

where $a_{d/opt}$ – optimal value of non-compensated acceleration for a specific track, determined based on the overhaul cycle cost, m/s^2.

In formula 5, as in formula 2, the weighted average train speed is taken as the speed. The weighted average speed is taken in order to determine such a railway curve superelevation, at which the average conditions of the transportation process will be taken at the optimal non-compensated acceleration.

For example, the railway curve superelevation were determined taking into account the overhaul cycle cost on the curve of the longevity welded track with R65 rails, fastenings of the ARS type, reinforced concrete sleepers on crushed stone ballast.

Table 1 shows the results of calculating the railway curve superelevation, taking into account the overhaul cycle cost:

Table 1. Determination of the railway curve superelevation, taking into account the overhaul cycle cost

Curve No.	h_{act}, mm	$a_{d/opt}$, m/s^2	R, m	$\overline{v_{top}}$, km/h	h_{OCC}, mm
1	104	-0,038	590	51,20	61,73
2	102	-0,038	595	51,85	62,67
3	105	-0,038	618	48,61	53,98
4	114	-0,038	598	49,22	56,84
5	91	-0,038	599	58,01	76,41
6	99	-0,038	585	54,94	70,69
7	82	-0,038	616	42,27	42,44
8	98	-0,038	588	43,63	46,65

To simplify the work with the railway curve superelevations, the railway curve superelevation was rounded in multiples of 5 mm according to the rules of mathematics.

When considering the overhaul cycle cost, one should not forget about the safety of train traffic. To simultaneously consideration safety components and technical and economic components, a calculation was made according to the Interim guideline [3]. The calculation results are shown in Table 2. The railway curve superelevation, determined using the Interim guideline [3], is noted as the value h_{calc}. On the basis of two values of railway curve superelevation, the optimal railway curve superelevation (h_{opt}) was determined, which would simultaneously considerate safety and the overhaul cycle cost.

Table 2. Calculation of the optimal railway curve superelevation

Curve No.	h_{act}, mm	h_{OCC}, mm	h_{calc}, mm	h_{opt}, mm
1	104	60	80	**80**
2	102	65	80	**80**
3	105	55	70	**70**
4	114	55	80	**80**
5	91	75	55	**75**
6	99	70	45	**70**
7	120	40	50	**50**
8	98	45	60	**60**

From Table 2 we can conclude that in most cases the railway curve superelevation determined by the Interim guideline [3] is taken as the optimal railway curve superelevation. But at the same time there are two cases (curves No. 5 and No. 6), in which the optimal railway curve superelevation are the railway curve superelevation, determined based on the overhaul cycle cost. This fact suggests that the use of the concept of the overhaul cycle cost can help in determining the best railway curve superelevation, considering various conditions.

Among other things, we would like to note that the results obtained differ greatly from the actual railway curve superelevation (h_{act}). This fact suggests that an irrational railway curve superelevation has been established in these curves. It is not rational both from the point of view of safety and from the point of view of technical and economic components.

4. Conclusion

The study carried out to account for various conditions of the railway curve superelevation showed that the existing Interim guideline [3] does not take into account all aspects of the permanent way economy. The technical and economic components in the Interim guideline [3] are poorly reflected.

When considering the overhaul cycle cost, it is possible to determine the optimal non-compensated acceleration, which in turn will allow the railway curve superelevation to be determined based on the economic components. It is not rational both from the point of view of safety and from the point of view of technical and economic indicators.

References

[1] Ministry of Railways of the Russian Federation 1997 *On the Determining of Outer Rail Elevation in Curves: Instruction C-333y.* (Moscow: Transport)

[2] Amelin S V and Danovsky L M 1972 *Track and track facilities* (Moscow: Transport)

[3] JSC «Russian Railways» 2009 *Interim Guideline on Determining the Railway Curve Superelevation and Speeds Limits in Curves.* (Moscow: JSC «Russian Railways»)

[4] Russian Railways Standard «STO RZD 02.037–2011». *Resource, Risk and Reliability Management at Life Cycle Stages (URRAN). Management of the Life Cycle Cost of Systems, Devices and Equipment of Facilities* 2012 (Moscow: JSC «Russian Railways»)

[5] JSC «Russian Railways» 2013 *Specifications for the Reconstruction (Modernization) and Repair of Railway Track (TU-75p).* (Moscow: JSC «Russian Railways»)

[6] Linnik Yu V 1962 *The Least Square Method and the Foundations of the Mathematical and Statistical Theory of Observation Processing.* (Moscow: Fizmatgiz).

Stress-strain state of the "wheel-rail" system under different movement conditions

D V Ovchinnikov[1] and D A Koven'kin[2]

[1] Samara State Transport University, 2V, Svobody Str., Samara, 443066, Russia
[2]Irkutsk State Transport University, 15, Chernyshevsky Str., Siberian Federal District, Irkutsk Region, 664074, Irkutsk

E-mail: ovchinnikov@samgups.ru

Abstract. A mathematical model of wheel-rail interaction for unworn profiles in the final-element modeling environment under different movement conditions, i.e. with the addition of additional torque and frictional forces, has been developed and tested in the article. Thus, the movement of rolling stock in the mode of braking (related to all cars and locomotives) and gaining traction (related to locomotives) is simulated. A significant influence of friction forces in the contact zone and additional torque on the stress-strain state during the interaction of the wheel and the rail are shown. The data obtained are additional refinements of one of the important factors in testing the durability of the material from which the rails are made – the loading of the samples and load application frequency. The work also approximated the results to maximize compressing and stretching stresses on the rail surface not only in a few variants of the studied original data. With the help of approximation, it is possible to study the stress state of the "wheel – rail" system at any point in a given range of vertical forces and additional torque.

1. Introduction

The "Russian Railways Holding Development Strategy for the period up to 2030" shows one of the main tasks of the development of the infrastructure business unit of the Russian Railways Holding: the development of infrastructure in order to switch to heavy traffic with train weight standards of nine thousand tons or more. To solve the problems, it is necessary to further modernize the elements of the upper structure of the track and running gears of the rolling stock, to improve the scientific and research base associated with the processes of interaction in the "wheel-rail" system to specify the maximum compressing and stretching stresses arising from the contact of wheels of rolling stock with the railway track.

The increase in axial loads inevitably causes an increase in mechanical stresses in the contact area of the wheel and rail, as well as all elements of the subrail base. That is, one of the main factors limiting the increase in axial loads are the maximum mechanical stresses (equivalent and normal) of the wheel-rail system. The quantitative parameters of the interaction between the rail and the wheel largely depend on the traffic safety and the basic feasibility indicators of the track and rolling stock facilities, including: the intensity of the wear and tear of the rails, the development of contact-fatigue defects, guaranteed handled tonnage, etc. [1-4].

This paper presents a mathematical model of the contact interaction between the wheel and the rail in the environment of the finite-element modeling. Different modes of rolling stock movement were

Content from this work may be used under the terms of the Creative Commons Attribution 3.0 licence. Any further distribution of this work must maintain attribution to the author(s) and the title of the work, journal citation and DOI.
Published under licence by IOP Publishing Ltd

simulated at different values of axial loads, including the addition of auxiliary friction and torque in the contact zone. Additionally, contact stresses for the new wheel and rail profiles were calculated, and a qualitative and quantitative characteristic of their area of impact on the rail steel matrix in the contact area with all sorts of combinations of raw data was given.

As a result of the work, a conclusion on the stress-strain state of the rail and wheel was prepared. In particular, the effect of additional frictional forces in the contact zone was taken into account, and the dependences of the change in normal stresses on the rail surface during interaction with the rolling stock wheels on the values of the axial load and movement mode were obtained for experimental studies of the rail steel durability.

2. Description of the finite-element model

The wheel comes into contact with the head of the rail on a certain platform. The magnitudes and law of stress distribution over the contact platform depend on the dynamic load of the wheel, the ratio of the normal and tangential components of the wheel load vector, the shape of the contacting surfaces, etc.

For the first time, the solution of the main problems about contact stresses and deformations was obtained on the basis of the methods of elasticity theory in 1881—1882 by G. Hertz.

Based on the normal Hertz's problem, the maximum contact stress P_{max} can be calculated according to the formula:

$$P_{max} = \sqrt[3]{\frac{3 \cdot F \cdot E^2}{2 \cdot \pi^3 \cdot r_e^2 \cdot (1-\varepsilon^2)^2}}, \quad (1)$$

where E is the elasticity module; F is the normal loading force of the wheel and rail; r_e is the equivalent radius, depending on the characteristic radii of the wheel and rail interaction at the point of contact; ε is Poisson's coefficient.

It should be kept in mind that Hertz's contact theory is true with the following assumptions:

- the contact surfaces are homogeneous and isotropic;
- friction forces in the contact zone are not active;
- the size of the contact area is small compared to the size of the contact bodies and the characteristic curvature radii of non-deformed surfaces;
- the solution of linear elastic semispace is used for the contact problem.
- contact surfaces are smooth.

When the vehicle moves, the position of the wheel set in relation to the rails changes significantly, resulting in different combinations of the contact zones of the wheel and the rail. Even if the axial load is constant, normal stresses will change due to differences in curvature radii of the contact surfaces of these zones. If there is one curvature radius of the surface in the contact area, one can use Hertz's solution. If there are two or more curvature radii in the contact area, "non-Hertz's" solutions should be used to determine the contact area. This is fundamentally important when solving the problem of the stress-strain state of contact of worn wheels and rails, as well as in the presence of side contact. Classical methods of calculation, despite the increasing accuracy of methods, are able to solve problems only with their considerable idealization, replacing the real structure with its computational scheme.

Modern numerical modeling techniques allow such tasks to be solved with sufficient precision in the finite-element modeling environment, taking into account, among other things, the additional friction forces between the contact surfaces [5-10].

To analyze the distribution of stresses in the contact area of the wheel and the rail, a three-dimensional model of interaction between the new wheel and the new rail with full geometric similarity of full-scale structures – GOST 10791-2011 and GOST R 51685-2013 (Figure 1).

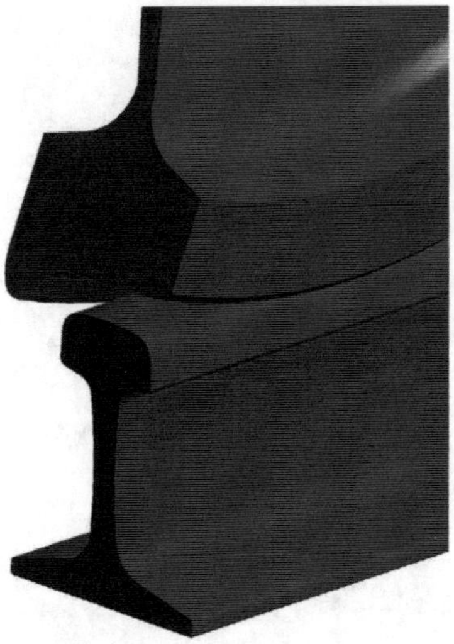

Figure 1. 3D model of the interaction of the new wheel and the new rail

On the basis of three-dimensional geometric models, the finite-element models (FEM) of the interaction between the new wheel and the new rail are built. The power of the estimated models is more than 300.000 nodes and reaches 1 million depending on the area of the contact patch. In the contact area, the element mesh is noticeably thickened (the size of the element is 0.2 - 0.5 mm) for the most accurate display of the analysis results.

It has been experimentally determined that for a particular task, further reduction in the mesh size does not increase the accuracy of output, but only increases the time it takes to solve the system of differential equations.

The following boundary conditions and loads were used in the solution:

- A rigid restraint is modeled along the rail foot area.
- The contact surface of the wheel and axis moves at a single-point contact only in a vertical direction.
- The rest of the elements are not fixed;
- The force is applied with the help of a "remote point", which makes it possible to simulate the impact of the axial load closest to the actual physical process that occurs during contact.

The original data parameters:

- The car wheel is 950 mm in diameter;
- The axial load is 25, 27 and 30 t/axis;
- The wheel axis is in vertical position;
- The rail axis is with a slope of 1/20;
- Contact surfaces are according to specified profiles.

The stress-strain state of the wheel and the rail was also assessed at different modes of rolling stock movement: frictional forces between contacting surfaces and torque attached to the wheel rotation axis were added. The maximum torque is limited by the greatest friction between contacting surfaces, since the wheel slippage occurs further. The wheel and rail cohesion ratio varies from 0.05 to 0.3 under different contact conditions, so it is advisable to calculate the stress-strain state of the wheel-rail system with a smooth increase of tangential forces in the contact area. Torque values range from 20% to 80% of the maximum allowed one depending on the axial load (Figure 2).

Figure 2. The diagram of the application of additional torque, where F_{ax} is the axial load, F_{fr} is the friction force when the wheels and rail are in contact; M is the torque

The maximum value of friction forces depends on the axial load and the friction (cohesion) ratio. The maximum value of torque is limited to the maximum forces of friction, overcoming which, the wheel slippage occurs. Thus:

$$M_{max} = F_{\text{fr}} \cdot l, \tag{2}$$

where l is the force lever and the distance from the wheel's rotation axis to the point of contact. For the vertical contact of the new rail and the new wheel, l is numerically equal to the wheel radius (475 mm for the car wheels). In other types of contact, the force lever is determined depending on the rail and wheel wear profile. In assessing the stress-strain state of side contact, the value of l is defined as the distance from the wheel's rotation axis to the point of ridge contact.

The torque values in modeling linearly increase from 0 (no additional friction force) to 80% of the maximum for further analysis of the results and the ability to approximate the resulting functions.

3. Results

During the simulation, the equivalent stresses in rails and wheels were obtained (Figure 3). But these stresses, which are based on the Mises-Hencky theory, also known as the theory of maximum distortion energy, have no sign, that is, they are always positive and do not show the transition of the material from a compressed state to a stretched and vice versa. Therefore, the drop in normal stresses along the axis (Figure 4), coinciding with the wheel trajectory, is of interest in terms of rail endurance tests, thus creating additional tangential forces before and after the contact patch depending on the mode of movement (traction or braking). Also it is important that due to the railway track elasticity, the wheel always performs extra work when moving upwards, as well as the rolling of the wheel through the elastic reverse half-wave.

Figure 3. Equivalent stresses (MPa) with an axial load of 30 tons in a vertical section

Figure 4. Normal stresses on the rail surface with an axial load of 27 tons, occurring directly on the trajectory of the wheel under an additional effect of a torque of 15.098 kN·m

The general results obtained are graphs of the difference in normal stresses on the rail surface (MPa) at various axial loads under the additional effect of a torque (Figure 5). The X-axis represents the length of the line of influence of contact forces in mm depending on the applied additional torque. The Y-axis represents the magnitude of the normal stresses occurring on the rail surface.

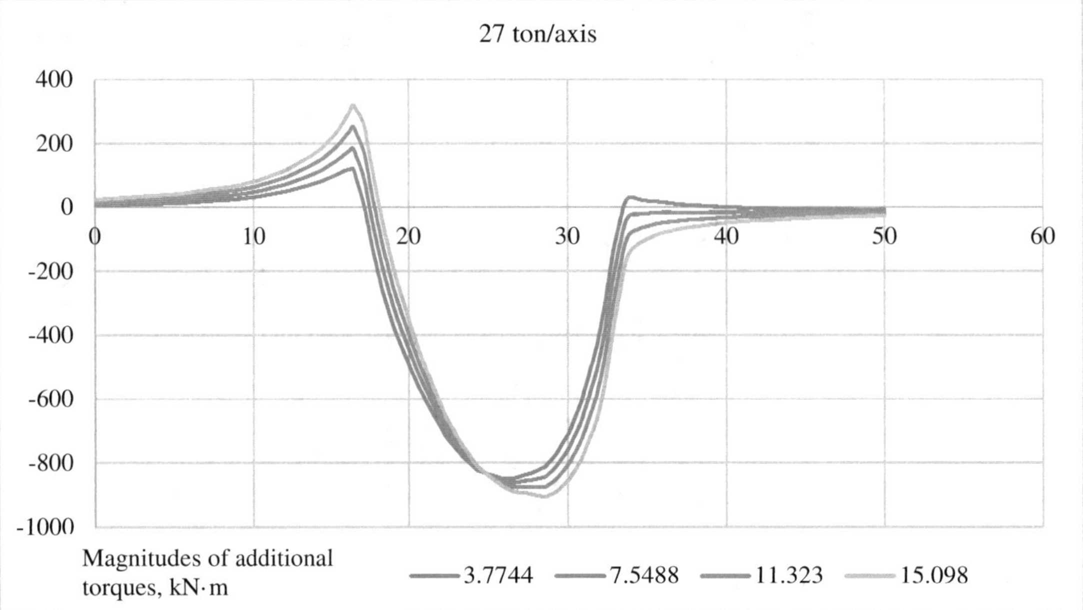

Figure 5. Normal stress difference in the rail surface (MPa) with an axial load of 27 tons

Studies of the graphs show that the maximum and minimum values do not qualitatively change with the addition of torque, that is, the extremum points practically do not depend on the value of the applied torque. The quantitative impact of the addition of the torque is significant, which indicates a substantial influence on the stress-strain state of the "wheel-rail" system of the rolling stock movement mode.

It should be noted that the influence of the braking mode has a more significant effect in connection with the quantitative overbalance of cars in relation to locomotives.

After the performed approximation, a compressive stress curve was obtained (Figure 6).

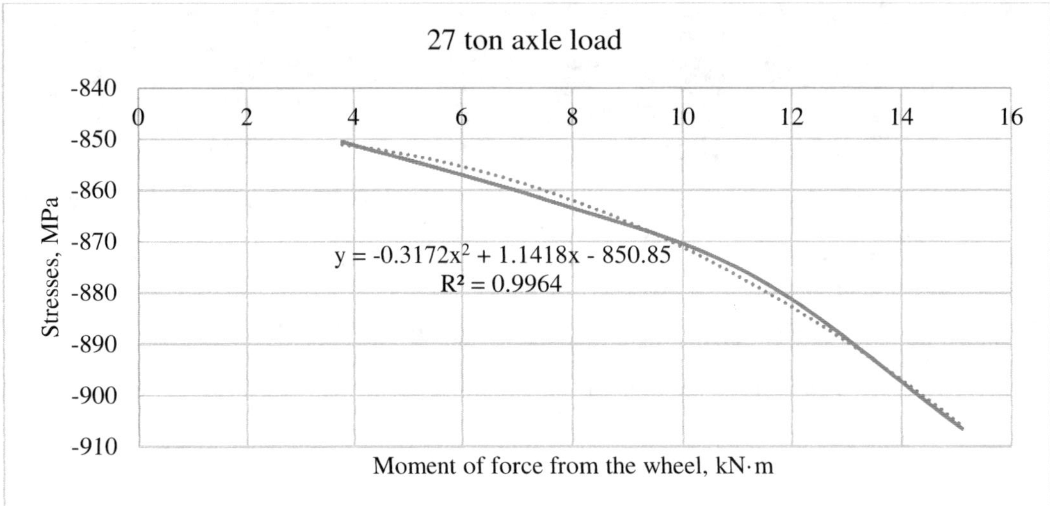

Figure 6. Change in compressive stresses on the rail surface with an axial load of 27 tons depending on the applied torque

Thus, using the approximation curves, it becomes possible to determine the local maximum and minimum stresses, as well as the required frequency of load application to conduct an experimental study of rail steel durability when using the method described earlier. The results are presented below:

Approximation equations with an axle load of 25 tons:
Compression $\sigma = -0.2843 \cdot M^2 + 0.3697 \cdot M - 821.59$
Elongation $\sigma = 16.6 \cdot M + 68.703$

Approximation equations with an axle load of 27 tons:
Compression $\sigma = -0.3172 \cdot M^2 + 1.1418 \cdot M - 850.85$
Elongation $\sigma = 17.689 \cdot M + 51.609$

Approximation equations with an axle load of 30 tons:
Compression $\sigma = -0.2533 \cdot M^2 + 0.1609 \cdot M - 885.27$
Elongation $\sigma = 16.428 \cdot M + 43.629$

Comparative analysis of the largest stress values without additional torque and with its addition is presented in tabular form in Table 1.

Table 1. Magnitudes of normal stresses on the rail surface

Maximum value (the sign is ignored) at the maximum possible torque before slipping	Minimum value (without torque)
-901.94	-821.59
358.77	68.7
-942.27	-850.85
385.44	51.61
-993.27	-885.27
388.11	43.63

4. Results

During the simulation, equivalent stresses in rails and wheels were obtained, taking into account the influence of friction forces in the contact zone, as well as additional torque, on the stress-strain state during the interaction of the wheel and the rail. The results clearly indicate a significant increase in compressive normal stresses – from 5 to 9 times, compared with the Mises-Hencky theory. Thus, the addition of torque creates on the rail surface areas of sharp differences in normal stresses from tension to compression and vice versa. Such differences must be taken into account when testing the material for durability in order to more accurately determine the guaranteed service life of rails in the areas of action of additional tangential forces: prolonged ascents, descents, braking sections.

References

[1] Bin Zhu, Jing Zeng, Dafu Zhang and Yi Wu 2019 A non-Hertzian wheel-rail contact model considering wheelset yaw and its application in wheel wear prediction *Wear* **432–433** 202958

[2] Wen J, Marteau J, Bouvier S, Risbet M, Cristofari F and Secordel P 2020 Comparison of microstructure changes induced in two pearlitic rail steels subjected to a full-scale wheel/rail contact rig test *Wear* **456–457** 203354

[3] Vo K D, Tieu A K, Zhu H T and Kosasih P B 2014 A 3D dynamic model to investigate wheel–rail contact under high and low adhesion *International Journal of Mechanical Sciences* **85** 63-75

[4] Magheri S, Malvezzi M, Meli E and Rindi A 2011 An innovative wheel–rail contact model for multibody applications *Wear* **271 (1–2)** 462-471

[5] Habibollah Molatefi and Hozhabr Mozafari 2013 Analysis of new method for vertical load measurement in the barycenter of the rail web by using FEM *Measurement* **46 (8)** 2313-2323

[6] Samantaray S K, Mittal S K, Mahapatra P, Sanjeev K 2019 Assessing the flexion behavior of bolted rail joints using finite element analysis *Engineering Failure Analysis* **104** 1002-1013

[7] Qinglin Lian, Guanyu Deng, A. Kiet Tieu, Huijun Li and Hongtao Zhu 2020 Thermo-mechanical coupled finite element analysis of rolling contact fatigue and wear properties of a rail steel under different slip ratios *Tribology International* **141** 105943.
[8] Daves W, Kubin W, Scheriau S, Pletz M 2016 A finite element model to simulate the physical mechanisms of wear and crack initiation in wheel/rail contact *Wear* **366–367** 78-83.
[9] Lei Wu, Zefeng Wen, Wei Li, Xuesong Jin 2011 Thermo-elastic–plastic finite element analysis of wheel/rail sliding contact *Wear* **271 (1–2)** 437-443
[10] Wu Cai, Zefeng Wen, Xuesong Jin, Wanming Zhai 2007 Dynamic stress analysis of rail joint with height difference defect using finite element method *Engineering Failure Analysis* **14 (8)** 1488-1499

Data analysis in the fire safety system of rail transportation and dumping of municipal solid waste

L A Koroleva[1], A G Khaidarov[2], G K Ivakhnuk[2] and D V Brusyanin[1]

[1] Saint Petersburg University of State Fire Service of Emercom of Russia, 149, Moskovskiy ave., Saint-Petersburg, 196105, Russia
[2] Saint-Petersburg State Technological Institute (Technical University), 26, Moskovskiy ave., Saint-Petersburg, 190013, Russia

E-mail: lyudamil@mail.ru

Abstract. The problems concerning the municipal solid waste (MSW) management have been discussed. A particular attention has been payed to the fire safety solutions. An original system of the MSW flammability classification that is based on the exergetic coefficient was introduced. The advantages of the application of this system to the MSW railway transportation and burial have been discussed. To implement the exergy approach, a data mining system has been developed. The input data includes the following characteristics: separate waste collection, the waste morphological composition; the location, capacity, inclination, a load level, and the number of the waste containers; the MSW temperature, the explosive gases concentrations, and other parameters. The analysis of data obtained with the waste management system allows us to optimize processes for the waste collection, transportation, and burial. This approach is proved to optimize the waste transportation and stocking, and improve the fire safety. This article provides the best practices for landfilling and burying MSW for Siberia.

1. Introduction

At present, the waste management becomes an extremely important challenge for industrialized countries. However, none of the countries could completely resolve the waste recycling and burial problems [1, 2].

The most frequently used scenario for managing municipal solid waste (MSW) is the waste burial in landfills. At the same time, increasing the number of illegal landfills, and the legal landfills situated near human settlements, results in the environmental degradation and increases the risk of emergencies.

The disposition and recycling MSW is a very serious problem for big cities [3-5]. Accounts Chamber of Russia reported that 65 million tons (or 450 kg per a person) of MSW had been produced in 2019. The major part of MSW (about 90%) is sent to landfills. The treatment plants can process only 7% of waste. This situation is caused by a poor infrastructure for transportation, collection, safe saving, and processing MSW. In Siberia, the MSW landfills are situated near settlements that results in the environmental degradation. This situation is considered to be critical.

The disposal of garbage from big cities becomes more and more important challenge [6]. From this point of view, the railroad transport is considered to be the safest and perspective for MSW transportation.

Content from this work may be used under the terms of the Creative Commons Attribution 3.0 licence. Any further distribution of this work must maintain attribution to the author(s) and the title of the work, journal citation and DOI.
Published under licence by IOP Publishing Ltd

The technologies for MSW management used in Russia have a number of weaknesses. For example, a system of collecting and analyzing statistical information about MSW does not take into account all sources of MSW. There are no reliable techniques for accounting the quantity of the produced MSW and their properties. It is necessary to pay more attention to the fire safety during transportation and processing MSW.

A lot of fires often take place in the Siberian landfills. For example, in last years, the following important fires occurred: a fire in the landfill situated in Ivanovka settlement, Chita region (January 2019); the fire in the MSW landfills situated near Ust-Abakan settlement, Khakasiya, and Yarovoe, Altay region (July 2020). A fire occurred in a waste landfill in Novosibirsk (August 2020); an important smoldering fire occurred in Surgut (October 2020).

Thus, the observation of the fire safety procedures for handling, burial, and transportation of MSW is an actual challenge.

2. Methods and scenarios of use

The Waste management system is aimed to decrease the waste impacts on the human health and environment. In this scenario, the logistic support considerably influences the process efficiency. The railroad transportation of MSW requires a comprehensive approach to the use of the Waste management system (Figure 1).

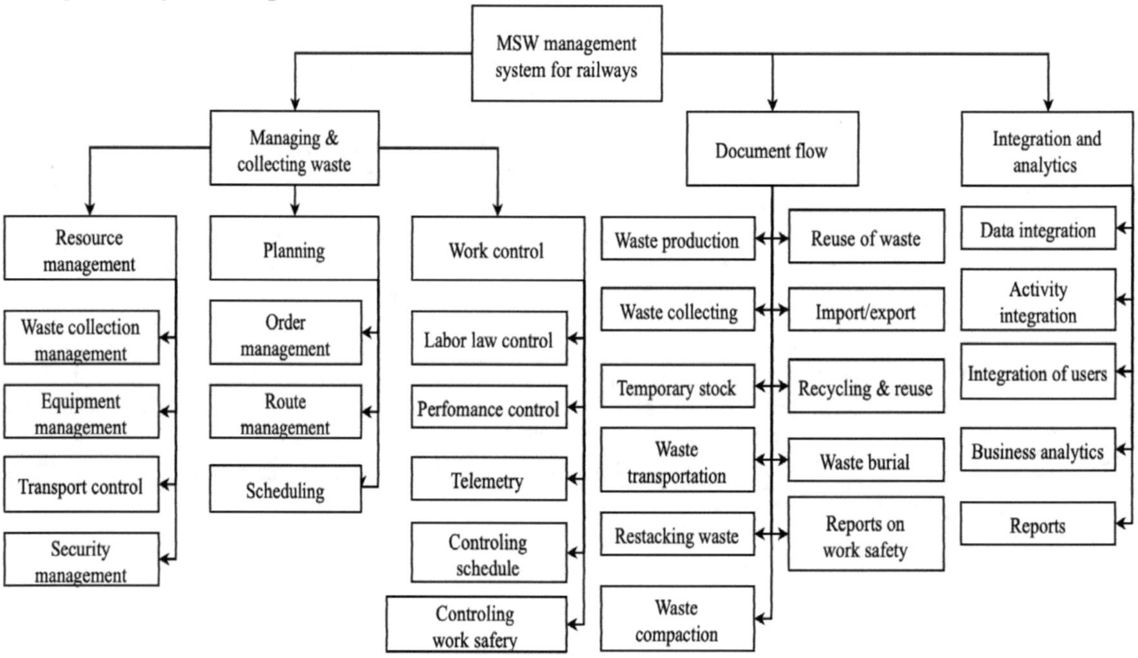

Figure 1. The use of the MSW management system for organization of the railroad transportation.

The work [7] proves the use of the exergy approach for studying the fire risks in the MSW railroad transportation. The use of the exergetic coefficient lets us standardize an estimation of the MSW flammability evolution during their life cycle. This approach also lets us estimate the role of the environment parameters, such as the temperature, pressure, and the MSW composition. With this method, it is possible to derive real conditions for the MSW inflammation, improve estimations of fire risks, and forecast the fire development.

Due to studying the chemical, physical, and the flammable properties of MSW, we could systemize MSW by their exergetic coefficient (Table 1) and prove advantages of this systematization for organization of the MSW railroad transportation and burial.

Table 1. MSW systemized by the exergetic coefficient.

Flammability group	Exergetic coefficient	Flammability rating
Non-flammable	$\Pi_e<0,1$	E0 – non-flammable
Low-flammable	$0,1 \leq \Pi_e < 0,5$	E1 – low value of the exergetic coefficient
Flammable	$0,5 \leq \Pi_e < 1,0$	E2 - middle value of the exergetic coefficient
	$\Pi_e \geq 1,0$	E3 - high value of the exergetic coefficient

$$\Pi_e = \frac{e}{30}$$

e is the exergetic ratio of the transported MSW (in MJ/kg) that is calculated as a sum of the physical and chemical exergetic contributions;

30 [MJ/kg] is a minimum exergy of MSW with the flammability rating E3 (high value of the exergetic coefficient).

The proposed method for calculating exergy, which is used to estimate the fire risks in the MSW transportation, includes two steps:

1) Determining an initial (zero) level of exergy (it depends on the environment parameters).

2) Forecasting changes of exergy caused by physical and chemical processes.

This exergy method allows us to estimate the possible fire risks, investigate negative impacts of MSW to the environment, and also estimate the exergy efficiency of possible techniques for the waste disposal [7, 8].

For practical implementation of this method, it is necessary to gather and analyze information about MSW during their collection, triage, transportation, and burial. This information helps us optimize the processes of the MSW stocking and transportation, and also select optimum methods for their processing and improving the fire safety.

The following is a list of the tasks to perform for analyzing data used by the system for fire protection in the MSW railroad transportation and burial:

- Estimating the temporal changes of quantities of MSW to be transported and buried.
- Planning the waste transportation, investigate the MSW morphological composition and take into account their changes.
- Preventing the fires, explosions, and to avoid the work downtime, optimize the procedures for MSW stocking and transportation including the following tasks: collecting, landfilling and burying the waste; a temporary stocking the waste; loading and unloading the waste.
- Monitoring the MSW parameters (such as temperature, pressure, humidity, etc.).
- Taking into account the environment parameters (temperature and pressure).
- Analyzing changes of concentration of the biomass and explosive substances, and early detect their critical values.
- Calculating the car filling level during the operations of the waste load, unload, or transportation.
- Forecasting the waste flammable properties, and the fire probability.

The obtained information is useful for developing the fire safety regulations for the waste management that takes into account the waste composition, their properties, and the environment parameters.

3. Results and discussion

3.1. Studies of the MSW morphological composition and their changes: an application for planning the waste transportation and burial

The data on the MSW morphological composition is a general information that can be used to estimate and forecast potential fire risks [5, 9]. An analysis of available data shows generals trends in the MSW composition changes, namely a rise in the proportion of polymers, food waste, paper, cardboard, and textile. The average data on morphological composition of MSW in big Siberian cities is shown in Figure 2.

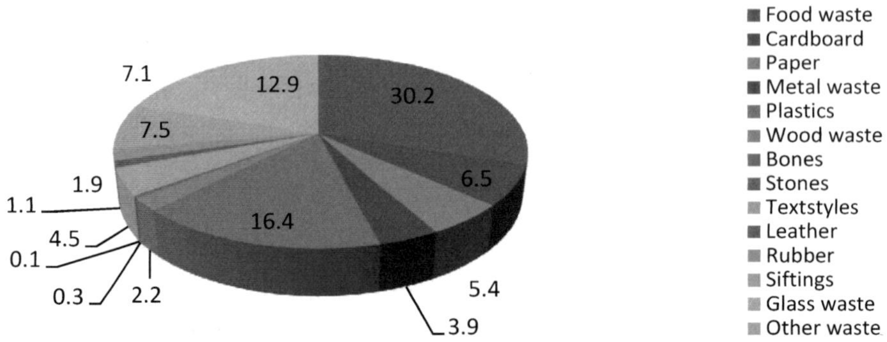

Figure 2. An average MSW morphological composition in big Siberian cities (in percentages).

The following factors have significant effects on the self-ignition, flammability, and self-heating of MSW: a variety of the waste composition, the complex biological and chemical processes, changes in the oxygen concentration and in the decomposition products, catalysts, humidity, and other parameters. However, it is difficult to monitor and control those factors because of their temporal variability, local specificities, and the MSW inhomogeneity [9].

The most comprehensive information on variations of the MSW morphological composition has been collected in Moscow and Saint-Petersburg (former Leningrad). This information was used for calculating the exergy (Figure 3) with the use of the technique reported in [7, 10].

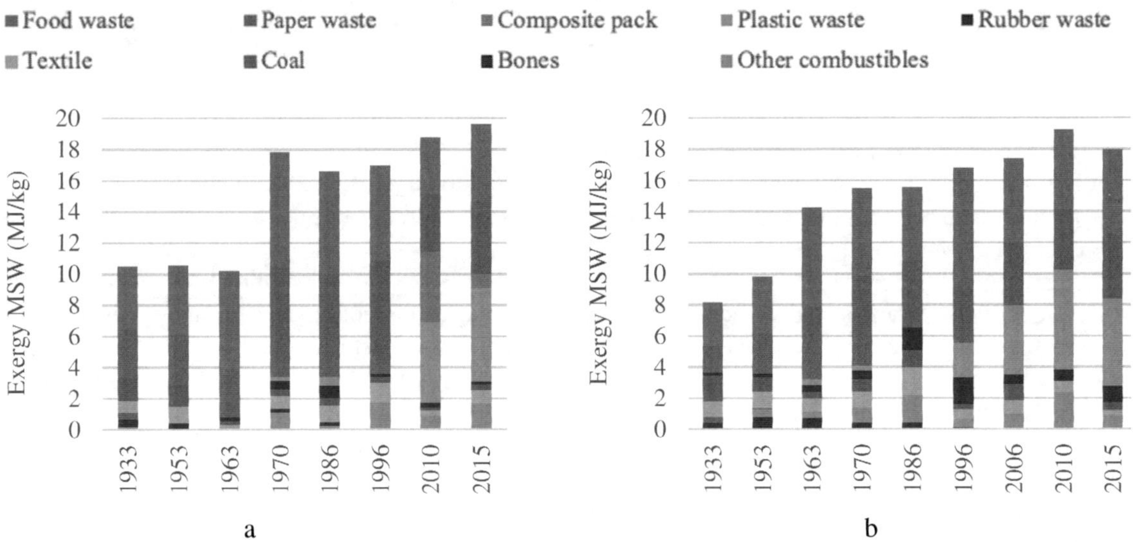

Figure 3. Variations of the MSW exergy from 1933 to 2015: a) Moscow; b) Saint-Petersburg (former Leningrad).

The plastics waste is found to have a maximum exergy. In big cities, the MSW has high values of exergy and those values continue to rise. On the one hand, the high values of the MSW exergy favor using MSW for biogenesis of energy. On the other hand, the high value of the exergy is an indicator of a high-fire risk. It is necessary to take into account both factors for organization of the safe transportation and burial of MSW.

3.2. Monitoring the MSW parameters

An increase of temperature is an indicator of the thermal, chemical, and microbiological processes giving off heat. The temperature rise depends on a ratio between the heat emission and heat removal. In some scenarios, the temperature rise might result in a self-ignition of MSW or their decomposition products, and in extension of the combustion zone. The physical exergy rises due to the increase of the difference between temperatures of the MSW substance and environment.

An increase of pressure does not change the ratio between a combustible and oxidizer, but could increase the volume concentration of the reagents. For the most of substances, this scenario results in the increase of the combustion reaction rate and even in an explosion. The physical exergy rises due to the increase of the difference between pressures inside the waste stock-pile and environment.

3.3. Changes in the biomass concentration, an analysis of the explosive substances concentrations, and a timely detection of their dangerous values

The simulations of the kinetics of the aerobic and anaerobic biological processes, which take place during a transportation and burial of MSW, allow us to forecast some of dangerous scenarios. For example, it is possible to simulate the temporary generation rate of the explosive gases, and estimate the effects of different parameters on the current processes. These studies help us suggest timely the measures to decrease the fire risks related to the MSW processing.

The changes in the concentration of the aerobic and anaerobic biomasses, methane, carbon monoxide, and oxygen define the temporal dependence of the MSW inflammability, and are initial data for calculating chemical exergy.

3.4. Using a data visualization system to estimate the fire risks in the MSW railroad transportation and burial

The following tools are required to collect and analyze data related to the MSW railroad transportation and their burial: the sensors for monitoring the container load level and temperature, a network equipment for data transmission, a software for data processing, and others.

The used cloud-based platform supports a daily information monitoring and the real-time management of the MSW data. This system allows the users to do the following: optimize the MSW collection, transportation, stocking, disposal, and burial; estimate the fire risks, and implement the fire protection measures. Due to the use of built-in GPS systems, it is possible to track the train routes and schedule.

This system uses the free analytics database management system (DBMS) for big data ClickHouse. The sensors send information through network to a server, and write data in the fact tables of the data model. MS Power Bi is used to automatically generate reports and visualize data. The data mining functionality is implemented using the Python algorithmic language. When the monitored parameters take critical values, the management system sends the mail or push notification messages.

Examples of the data visualization for a combined transportation scheme (railroad and truck transport) are shown in Figure 4.

The process of the data analysis starts at the MSW collection stage. The waste is collected into special containers installed in settlements (Figure 4a). The processing system considers the following parameters: a separate waste collection, the waste morphological composition, the location, capacity, and number of the waste containers. The software plots a map that displays the container positions, their load levels, and the MSW parameters (such as temperature). This information is used to optimize the traffic of the garbage trucks, estimate the MSW volume to be landfilled or transported, and minimize the vehicle idle time.

The proposed system for data mining can be used to monitor the container load levels and their inclination, the MSW temperature, and the explosive gas concentration. This monitoring can be performed during the MSW transportation or when loading or unloading MSW. Figure 4b displays reports generated using the obtained data (the table and visual formats are available).

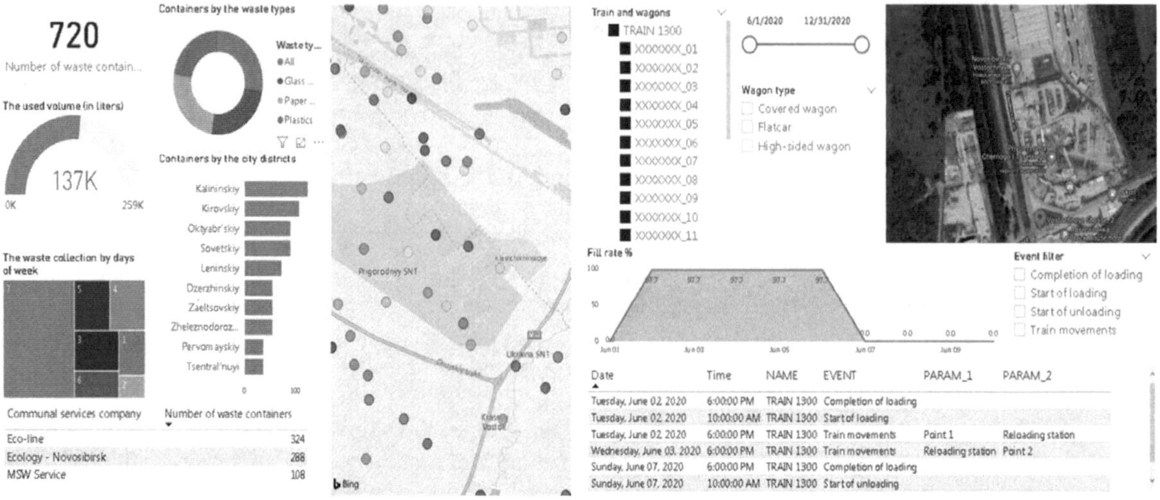

a b

Figure 4. The dashboards created with Power BI: a – collection of MSW into the waste containers installed in settlements; b – railroad transportation of MSW.

The obtained information allows the users to calculate and forecast the MSW exergy changes. The data mining lets you handle the situation and timely take appropriate measures for minimizing the fire risks and environment pollution that could arise during the MSW collection, transportation or burial. This information is used to choose an optimal method for disposing the waste.

4. Conclusion

The implementation of the exergetic approach allows us to perform a complex energetic and environment estimation and forecast the fire risks related to the MSW handling. The indicators of the MSW fire risks, which have been obtained using the exergetic method, improve an objectivity of the procedure for classification of the hazardous materials for the railroad transportation. Our method lets the users take into account the conditions that favor the MSW inflammation during their transportation and burial in Siberia.

To implement the exergetic approach, it is necessary to gather and process information on the quantity, composition, and other properties of the MSW, and their temporal evolution during the waste collection, triage, transportation, and burial. The proposed method provides a possibility to choose a strategic approach to the waste handling procedures.

The use of the proposed system of data mining allows the users to optimize the waste stocking procedures and find the best solutions for ensuring the fire safety in the MSW railroad transportation and burial in Siberia.

References

[1] Bovea M D, Ibáñez-Forés V, Gallardo A and Colomer-Mendoza F J 2010 Environmental assessment of alternative municipal solid waste management strategies. A Spanish case study *Waste Management* **30** 2383-95

[2] Hong J, Li X, Zhaojie C 2010 Life cycle assessment of four municipal solid waste management scenarios in China *Waste Management* **30** 2362-69

[3] Duan N, Li D, Wang P, Ma W, Wenga T, Zhong L and Chen G 2020 Comparative study of municipal solid waste disposal in three Chinese representative cities *Journal of Cleaner Production* **254** 120134

[4] Lou C X, Shuai J, Luo L and Li H 2020 Optimal transportation planning of classified domestic garbage based on map distance *Journal of Environmental Management* **254** 109781

[5] Di Foggia G, Beccarello M 2018 Improving efficiency in the MSW collection and disposal service combining price cap and yardstick regulation: The Italian case *Waste Management* **79** 223-31
[6] Peri G, Ferrante P, Gennusa M La, Pianello C and Rizzo G 2018 Greening MSW management systems by saving footprint: The contribution of the waste *transportation Journal of Environmental Management* **219** 74-83
[7] Khaydarov A G, Koroleva L A and Ivakhnyuk G K 2018 Exergetic assessment of fire hazards of cargo transportation on railway transport *Fire and Explosion Safety* **27(10)** 26-37
[8] Laner D, Rechberger H, De Soete W, De Meester S and Astrup T F 2015 Resource recovery from residual household waste: An application of exergy flow analysis and exergetic life cycle assessment *Waste Management.* **46** 653-67
[9] Moody C.M, Townsend T G 2017 A comparison of landfill leachates based on waste composition *Waste Management* **63** 267-74
[10] Eboh F C, Ahlström P and Richards T 2016 Estimating the specific chemical exergy of municipal solid waste *Energy Science & Engineering* **4** 217-31

Protection of critical information in external data stores using the dissection-separation method

George N Shurkhovetsky

Irkutsk State Transport University, Chernyshevsky str., 15, Irkutsk, 664074, Irkutsk region, Siberian Federal district, Russia

E-mail: gshn5@yandex.ru

Abstract. Today, due to the large spread of cloud technologies, more and more individuals and businesses use their services. It is common practice in the field of cloud technologies to store in-formation (data) from one service provider in one physical and geographical location. At the same time, we must remember that the information passes through third parties whose intentions are not always known. Therefore, valuable critical information is usually encrypted before it is sent to the cloud on the user's side. To do this, you can use both simple tools (for example, creating password-protected archives) and more advanced tools (software for creating encrypted partitions). However, in addition to normal encryption, you can use the fact that there are many unrelated repositories. Such methods include, for example, the dissection-separation method, which is discussed in this paper.

1. Introduction

It is a common practice in the information technology and communications industry to store information (data) at the same service provider, in the same physical and geographical location. Often, the service provider also provides redundancy and re-storage of data in remote locations, but this is only to ensure the performance and availability of data in the event of an accident or other failure. Information (data) is still entrusted to this single service provider, which must implement a variety of security measures to ensure the protection of information, namely its confidentiality, integrity, and availability. These measures may be preventive, investigative or corrective in nature and include, but are not limited to, physical and logical access control, data encryption at rest and during transmission, backup, monitoring, notification, logging, antivirus, firewalls, intrusion detection and prevention systems, physical and environmental protection measures, security policies, procedures and standards, data verification, security awareness programs, audit and certification plans, incident response, breach notification, risk assessment, etc.

Encryption is an important element. However, it may be poorly implemented, contain backdoors unknown to the user or data owner, and the keys themselves may be transferred to third parties without the knowledge or consent of the user or data owner. This makes it an urgent problem to develop methods that are additional or even alternative to encryption. For example, you can use the fact that there are many unrelated repositories. Such methods, in particular, include, for example, the dissection-separation method, let's consider its effectiveness and reliability in more detail.

Content from this work may be used under the terms of the Creative Commons Attribution 3.0 licence. Any further distribution of this work must maintain attribution to the author(s) and the title of the work, journal citation and DOI.
Published under licence by IOP Publishing Ltd

IOP Conf. Series: Materials Science and Engineering 1151 (2021) 012022 doi:10.1088/1757-899X/1151/1/012022

2. The general principle of operation and the essence of the dissection-separation method

Dissection (separation) or dissection-separation of information is a cryptographic method of information protection, which consists in the fact that the array of protected data is divided into parts, each of which separately does not allow revealing the content of the protected information. These fragments can be transmitted across multiple sources, distributed over time and at the recording location on a floppy disk or any other recording device. The dissection-separation method is divided into semantic and mechanical according to its implementation [1-3].

Let us consider the following example. We take the text as the source message:

'Who_controls_the_past,'_ran_the_Party_slogan,_'controls_the_future:_who_controls_the_present_controls_the_past.' (George Orwell. 1984) and let us try to evaluate the effectiveness of the method using this example. The total number of characters in the message is 110.

To split a block of plain text, we use the technique described in [2]. We cut the text into 12 parts and write the plain text in the following format (table 1)

3 rows and 4 columns are selected to split the text into 12 parts. Let **sj** columns be selected in the sequence {4, 1, 3, 2}, and the **ri** strings are in the sequence {2, 3, 1} [4].

Text characters are placed in blocks that are identified by numbers. Number **k** of the block **F(k)**, where the next character of the plain text is written, is determined by the formula:

$$k = (ri - 1) \cdot n + sj$$

where **n** is the number of columns [5]. Then the first character "**W**" will be recorded in the block number (ri = = 2, sj = 4): k = (2 – 1) · 4 + 4 = 8; the second character "**h**" – block number (ri = 2, sj = 1): k = (2 – 1) · 4 + 1 = = 5; the third symbol "**o**" in block number (ri = 2, sj = 3): k = (2 - 1) · 4 + 3 = 7; etc. (table 2).

Table 1. Dissecting an open text

	*	4	1	3	2	4	1	3	2	4	1	3	2	4	1	3	2	4	1	3	2	4	1	3	2	4	1	3	2
**		1	2	3	4	1	2	3	4	1	2	3	4	1	2	3	4	1	2	3	4	1	2	3	4	1	2	3	4
2	1	**W**	**h**	**o**	**_**	**_**	**t**	**h**	**e**	**r**	**a**	**n**	**_**	**y**	**_**	**s**	**l**	**o**	**n**	**t**	**r**	**f**	**u**	**t**	**u**	**c**	**o**	**n**	**t**
3	2	**c**	**o**	**n**	**t**	**_**	**p**	**a**	**s**	**t**	**h**	**e**	**_**	**o**	**g**	**a**	**n**	**o**	**l**	**s**	**_**	**r**	**e**	**:**	**_**	**r**	**o**	**l**	**s**
1	3	**r**	**o**	**l**	**s**	**t**	**,**	**‘**	**_**	**P**	**a**	**r**	**t**	**,**	**_**	**‘**	**c**	**t**	**h**	**e**	**_**	**w**	**h**	**o**	**_**	**_**	**t**	**h**	**e**

Continuation of table 1

2	1	**_**	**p**	**r**	**e**	**t**	**r**	**o**	**l**	**s**	**t**	-	-
3	2	**s**	**e**	**n**	**t**	**s**	**_**	**t**	**h**	-	-	-	-
1	3	**_**	**c**	**o**	**n**	**e**	**_**	**p**	**a**	-	-	-	-

* - **sj** columns are selected in the sequence {4, 1, 3, 2}
** - **ri** strings - in the sequence { 2, 3 ,1 }

Table 2. Splitting plain text (which was previously split)

	1	2	3	4	1	2	3	4	1	2	3	4	1	2	3	4
1	8	5	7	6	8	5	7	6	8	5	7	6	8	5	7	6
2	12	9	11	10	12	9	11	10	12	9	11	10	12	9	11	10
3	4	1	3	2	4	1	3	2	4	1	3	2	4	1	3	2

Continuation of table 2

1	8	5	7	6	8	5	7	6	8	5	7	6	8	5	7	6
2	12	9	11	10	12	9	11	10	12	9	11	10	12	9	11	10
3	4	1	3	2	4	1	3	2	4	1	3	2	4	1	3	2

Continuation of table 2

1	8	5	7	6	8	5	-	-
2	12	9	11	10	-	-	-	-
3	4	1	3	2	-	-	-	-

As a result, blocks F(k) written in number order will contain the following characters:
F1 = {**o,a_hhtc_...**}
F2 = {**s_tc_ _ena...**}
F3 = {**l'r'eohop...**}
F4 = {**rtp,tw_ _e...**}
F5 = {**hta_nuoprt...**}
F6 = {**_e_lrutel...**}
F7 = {**ohnsttnro...**}
F8 = {**W_ryofc_ts...**}
F9 = {**ophgleoe_...**}
F10 = {**ts_n_ _sth...**}
F11 = **{naeas:lnt...**}
F12 = {**c_toorrss...**}.

Thus, one block of plain text is replaced by twelve blocks, which together give the length of the source text. Then the resulting blocks are distributed to different storage locations, and all this is a reversible process. This implementation of the split-diversity method is called mechanical. Let us consider its persistence.

3. Evaluation of cryptograhic strength

The cryptanalysis method dissection-separation will be performed for the case by the brute force method in 3 directions:

1. when an attacker knows all 12 blocks of source code that they need to collect;
2. when one knows how to implement the split-split method and 1 block of 12 source messages;
3. when one knows only 1 block.

Case 1. If the attacker has all the blocks F1-F12, he must arrange the available characters in the correct order. Then you can calculate the number of combinations of possible solutions using the iteration method:

$$\boldsymbol{P_m = m!} = 110! = 1.5882455415 \text{ x } 10^{178}$$

where m is the total number of characters [6]. It is easy to see that the brute force method for the first case, even on such a short message, gives an excessive amount of calculation. You can reduce it a little by noting that the letters "**W**" and "**P**" are capital letters, which means that one of them should go 1st.

$$\boldsymbol{P_m = m!} = (110\text{-}1)! = 109! + 109! = 2.8877191664 \text{ x } 10^{176}.$$

Instead of the brute force method, you can use frequency analysis if the message is long enough (in this case, it is 110 characters). It allows you to determine whether encryption was used or whether the original message was simply divided into blocks. If there was encryption, the frequency of characters in the message did not match their frequency in English, and this could also indicate the use of some other language.

Seeing that we have 17 spaces (" _ ") suggests that we should have 18 words. Accordingly, if we have 18 words in the message, the number of combinations will be significantly reduced. In this case, you can rely, for example, on dictionaries:

- Cambridge Advanced Learner's Dictionary-English dictionary edition for advanced users. Suitable for students with B2 Upper Intermediate - C2 High Advanced proficiency levels. Contains 140.000 words and phrases.

- Longman Collocations Dictionary and Thesaurus – English dictionary of phraseology, phrases and synonyms for users with intermediate – Advanced proficiency levels. The dictionary includes 700.000 phrases and synonyms to help improve your spoken and written British English skills.

- Oxford Advanced Learner's Dictionary-English dictionary for students and adults with B2 Upper-Intermediate – C1 Advanced proficiency. Contains 185.000 words and expressions.

- Collins English Dictionary – one of the most complete dictionaries of the English language. The dictionary covers 722.000 words and phrases.

– and other.

To do this, you can build combinations of these words and our known letters by filtering through these dictionaries. Moreover, when using statistical laws of the English language, to improve the search, you can pay attention to the most frequent bigrams and trigrams, as well as pay attention to pairs of repeated letters. In English, the most commonly repeated letters are **ss, h, tt, ff, 11, mm** and **oo**. If there are any duplicate characters in the ciphertext, we can assume that they represent one of these pairs.

If there are spaces between words in the ciphertext, we will try to identify words consisting of one, two or three letters. The only words in English that consist of a single letter are **a** and **I**. The most common two-letter words are **of, to, in, it, is, be, as, at, so, we, he, by, or, on, do, if, me, my, up, an, go, no, us, am**. The most commonly used three-letter words are **the** and **and** [7].

If possible, prepare a table of letter frequencies for the message we are trying to decrypt. For example, military reports tend to omit pronouns and articles, and the absence of words such as **I, he, a** and **the** will reduce the frequency of some of the most common letters [7].

If we know that we are working with a military report, we should use the letter frequency table created from other military reports.

One of the most useful skills for a cryptanalyst is the ability to recognize words or even entire phrases through their own experience or purely intuitively. Al-Khalil, one of the first Arab cryptanalysts, demonstrated his abilities when he cracked the Greek ciphertext. He suggested that the ciphertext begins with the greeting "**In the name of God**". Having established that these letters correspond to a certain fragment of the ciphertext, he was able to use them as a scrap and reveal the rest of the ciphertext. This was called the crib [8].

By skipping our options through these bigrams and trigrams, we can reduce the amount of iteration. There is a fairly significant probability that the original message will satisfy this option, but here, in order to better work out all the options, it is advisable to use a search for climbing to the top or an algorithm for simulating annealing. In the desired version, some of the letter combinations will be put down at the expense of bigrams and trigrams, so you can continue to deliver the missing letters based on the meaning of words, phrases, morphological rules, the approximate subject of the message, etc. If the original message is able to restore at least 5-10% of the words, and in all other words at least some letter combinations, then it is already quite possible to restore the entire message.

Case 2. In this case, when the attacker has one of the blocks (for example, F11 = {**naeas:lnt**...}) and its number, and also knows the method by which the split-spacing method was implemented, it can be concluded that there are 12 blocks in total, since the text is separated into 3 rows and 4 columns (Table 3 and 4).

And since our 11 block has 9 characters instead of 10, we can conclude that some blocks have either 9 or 10 characters.

You can immediately put it in the table 4 after the dash in block 11, there are also dashes in subsequent blocks.

3 rows and 4 columns are selected to split the text into 12 parts. Let **sj** columns be selected in the sequence {4, 1, 3, 2}, and the **ri** strings are in the sequence {2, 3, 1}.

Then the number k of the block **F(k)**, where the next character of the plain text is written, is determined by the formula:

$$\mathbf{k = (ri\text{-}1) \cdot n + sj}$$

where **n** is the number of columns.

In principle, we can restore a table with blocks (Table 5).

All letters of the alphabet are in upper and lower case 52 + 1 (space) + 10 numbers (0-9) + 12 punctuation marks (– -, . ! ? : ; " " ( ) ' ...). A total of 75 is possible options per 1 cell. There are 114 cells in total (minus the known 6 empty cells in table 4).

Table 3. Dissecting an open text

	1	2	3	4	1	2	3	4	1	2	3	4	1	2	3	4	1	2	3	4	1	2	3	4	1	2	3	4
1																												
2			**n**				**a**				**e**				**a**				**s**				**:**				**l**	
3																												

Continuation of table 3

1												
2			**n**				**t**				-	-
3									-	-	-	-

Table 4. Splitting plain text (which was previously split)

	1	2	3	4	1	2	3	4	1	2	3	4	1	2	3	4
1																
2			11				11				11				11	
3																

Continuation of table 4

1																
2			11				11				11				11	
3																

Continuation of table 4

1								
2			11				-	-
3					-	-	-	-

Let's use the placement formula with repetitions.

$\overline{\mathbf{A}}_{\mathbf{n}}^{\mathbf{k}} = \mathbf{n}^{\mathbf{k}} = 75^{114} = 5{,}7145761808 \times 10^{213}$, k – total cells where elements with repetitions are placed, n-all types of placed elements [9].

It is necessary to exclude all variants with repeated letters and punctuation marks 2 or more times in a row, this does not happen in English with the exception of the double spelling of the most frequently repeated letters: **ss, ee, tt, ff, 11, mm** and **oo.**

It seems that this will significantly reduce the search. $n^k{}_1\,(65) = 65^{114} = 4{,}7002683899 \times 10^{206}$, this is a combination of punctuation marks and letters that should not be repeated or can be repeated, but with rare exceptions compared to the total number of variants. It can be assumed that letters and punctuation marks can be repeated once or twice or more. Then you can search for the original message in the same way as in the previous version, starting with bigrams and trigrams.

Case 3. The attacker has one of the blocks, for example F11 = {**naeas:lnt...**}. In this case, a favorable outcome is very unlikely. The first possible option, if there are only 2 blocks of equal length, then we get:

$$\overline{C}\frac{m}{n} = C\frac{m}{n+m-1} = \frac{(n+m-1)!}{m!\,(n-1)!}$$

n – all known elements, **m** - variants of the arrangement [10].

$P_1 = (n!\cdot \overline{C}\frac{m}{n})! = (9!\cdot \frac{83!}{75!*8!}) = 1{,}4313158201 \text{ x } 10^{16}$

$P_2 = 1{,}4313158201 \text{ x } 10^{16} \cdot \frac{83!}{75!*8!} = 5{,}6455714754 \text{ x x } 10^{26}$

etc.

Then you need to take into account that there may be more than 2 blocks. And in 2, 3 and further blocks, the number of characters can be more or less, which will constantly increase the number of options many times. Exponentially.

Moreover, the number of meaningful words that can mean our original search message will constantly increase, also exponentially. For successful implementation of the search, you will still need additional information, such as the approximate meaning of the message or what preceded this message (user correspondence), etc.

Table 5. Splitting plain text (which was previously split)

	1	2	3	4	1	2	3	4	1	2	3	4	1	2	3	4
1	8	5	7	6	8	5	7	6	8	5	7	6	8	5	7	6
2	12	9	11	10	12	9	11	10	12	9	11	10	12	9	11	10
3	4	1	3	2	4	1	3	2	4	1	3	2	4	1	3	2

Continuation of table 5

1	8	5	7	6	8	5	7	6	8	5	7	6	8	5	7	6
2	12	9	11	10	12	9	11	10	12	9	11	10	12	9	11	10
3	4	1	3	2	4	1	3	2	4	1	3	2	4	1	3	2

Continuation of table 5

1	8	5	7	6	8	5	7/-	6/-
2	12	9	11	10	12/-	9/-	-	-
3	4	1	3	2	-	-	-	-

4. Discussion

In addition to this split-spread method, there are such "classical" encryption methods as substitution, permutation, and XOR cipher [11-13]. They are inherently linear in the sense that the length of the encrypted message is equal to the length of the source text. Non-linear transformation of the substitution type is possible instead of the original characters (or whole words, phrases, sentences) of pre-selected combinations of characters of different lengths [12- 14]. Also recently, the so – called computer steganography (from the Greek words steganos – secret, secret and graphy - record), which is hiding a message or file in another message or file, has become popular [15, 16]. For example, you can hide an encrypted audio or video file in a large information or image file.

But in this article, first of all, we consider the possibility of using different storage locations in the "clouds" or any other media. The only difference is that the above numerical methods allow you to store the modified source file in one storage location, and backup copies in another. And if the attacker can get this file and hack it, then the goal of the crypto attack will be achieved. The dissection-separation method considered in this article allows you to split the source file into several parts and store each in different storage locations. if an attacker can get one part and successfully hack it, then they will still not be able to achieve the goals of the crypto attack. This will only be possible under favorable circumstances, for example, if it can extract all parts of the source file or most of them.

From the above, we can conclude that this method can be effectively hacked under certain conditions.

If we turn to the well-known cryptographer Auguste Kerkhoff, who is the author of the fundamental work "Military cryptography" ("La Cryptographie Militaire"), in which he formulated General requirements for the cryptosystem, and also approved cryptanalysis as the only true way to test ciphers [17].

Six general criteria are for a cryptographic system.

1. The system must be physically, if not mathematically, undetectable.

2. It should not be necessary to keep the system secret; getting the system into the hands of the enemy should not cause inconvenience ("the Kerkhoff Principle").

3. Key storage and transfer should be possible without the use of paper records; correspondents should be able to change the key at their discretion.

4. The system must be suitable for Telegraph communication.

5. The system should be easily portable, and working with it should not require the participation of several people at the same time.

6. Finally, the system is required, given the possible circumstances of its application, to be easy to use, not require significant mental stress or compliance with a large number of rules [18, 19].

For points 1 and 2, this split-split method cannot be considered reliable. This method can be considered according to these six criteria, despite the fact that it does not belong to the cryptosystem, but only a method. Because it can be included in the cryptosystem itself, which can increase the risks of a successful cryptographic attack, in view of its vulnerability, and increase the cost of improving its reliability.

If we turn to the doctrine of the famous French cryptologist Antoine Rossignol, according to which the strength of the military field cipher must be such as to ensure the secrecy of the cipher message until it is received by an army unit and the order is executed. A diplomatic cipher must be such that it takes several tens or even hundreds of years to reveal it, since in diplomacy, an encrypted document must often remain secret for a very long time after it is encrypted [20]. Today, this doctrine can be paraphrased as follows: a prerequisite for today's ciphers is non-disclosure not only in diplomacy, but also in military Affairs and in other areas of human and managerial activity, in particular when working with critical information infrastructures.

5. Conclusion

From the work done, we can conclude that this cryptographic method of protecting information in external data warehouses is very interesting and meets the "spirit of the time", but the dissection-separation method itself is significantly vulnerable. This is confirmed by the identified two directions in this article: knowledge of the method of implementation of this method, knowledge of all the threads into which the split source message was distributed. The topic and direction need to be further explored.

6. References

[1] Bezbokov A A, Yakovlev A V and Simkin V N 2006 *Methods and means of computer information protection: a Training manual,* available at: https://tstu.ru/book/elib/pdf/2006/shamkin2.pdf

[2] 2010 *Cryptogram. A practical guide to manual encryption.* - 4th ed. (Saratov: Novy Veter publishing house)

[3] Kondratenko S V and Novikov Yu V 2005 *Basics of local networks. National research nuclear* (University "MEPhI")
[4] Partyka T L and Popov I I 2002 *Information security. Training a textbook for students of institutions of secondary vocational education* (Moscow: FORUM: INFRA-M)
[5] Lygin E A 2010 *L 88, Secret Writing. Practical guide to manual encryption* (Saratov: Novy Veter publishing house)
[6] Vilenkin N Ya 1969 *Combinatorics* (Moscow: Nauka)
[7] *Cryptographic methods and data protection tools,* available at: http://develab.narod.ru/crypt/crypt04.htm
[8] *Other encryption methods,* available at: http://www.univer.omsk.su/omsk/Edu/infpro/1/kodir/drmetsh.html
[9] Ryser H J 1963 *Combinatorial Mathematics, The Carus Mathematical Monographs (#14)* (The Mathematical Association of America)
[10] Riordan J 1958 *An Introduction to Combinatorial Analysis* (New York, Wiley)
[11] Ravi Gupta and Rahul Singh 2015 SSCC - An Improved Substitution Method for Data Encryption Using DNA Sequence and CDMB. *Communications in Computer and Information Science*, pp. 197-206 DOI http://dx.doi.org/10.1007/978-3-319-22915-7_19
[12] khlas Abbas Albahrani, Tayseer Karam Alshekly 2017 New Chaotic Substation and Permutation Method for Image Encryption *International Journal of Applied Information Systems* **12(4)** 33-39 DOI http://dx.doi.org/10.5120/ijais2017451698
[13] Fei Huo, Guang Gong 2015 XOR Encryption Versus Phase Encryption, an In-Depth Analysis *IEEE Transactions on Electromagnetic Compatibility,* **57(4)** 903-911 DOI http://dx.doi.org/10.1109/temc.2015.2390229
[14] Muhammed Jassem Al-Muhammed 2018 Light but Effective Encryption Technique based on Dynamic Substitution and Effective Masking *International Journal of Advanced Computer Science and Applications* **9(9)** DOI https://dx.doi.org/10.14569/IJACSA.2018.090909
[15] Sefa Tuncer 2016 Information Encryption and Hiding into an Image by Steganography Methods to Improve Data Security *Journal of new results in science* **5(12)** 170-177
[16] Gotfried C Prasetyadi, Achmad Benny Mutiara, Rina Refianti 2017 File encryption and hiding application based on advanced encryption standard (AES) and append insertion steganography method *Second International Conference on Informatics and Computing (ICIC),* Nov 1, 2017. DOI http://dx.doi.org/10.1109/iac.2017.8280584
[17] Kahn D 1996 *The Codebreakers.*
[18] *Kerckhoffs' principles,* Available at: https://www.petitcolas.net/kerckhoffs/la_cryptographie_militaire_i.htm#desiderata
[19] Kerckhoffs A 1883 La cryptographie militaire. *Journal des sciences militaires*
[20] Perrault Ch 1697 *Les Hommes illustres (*Paris: Dezaillier)

Environmental and floristic analysis of undesired plants on the railway formation canvas

S A Dontsov[1] and L F Drozdova[2]

[1] Russian University of Transport, 9 build. 9, Obraztsova Str., 127994, Moscow, Russia
[2] Baltic State Technical University, 1, 1st Krasnoarmeyskaya Str., 190013, Saint-Petersburg, Russia

E-mail: dontsovgroup@mail.ru

Abstract. Negative effects caused by the vegetation of unwanted plants on the railway track skeleton have been studied. The ballast prism is shown to be a potentially areal for a vegetation of specific plants under particular effects of biotic and abiotic factors. The phytoindication technique was used to estimate the plants vegetation in the railroads with different volumes of traffic. The typical plants species and the average degree of the manifestation in the photoautotroph areal have been determined. To improve efficiency of the weed control measures, various forms of herbicides have been tested. The paper reports techniques for the environmental and floristic analysis of unwanted plants that grow in the tie cribs and provides recommendations on how to avoid an appearance of those plants. These measures ensure a stable operation of the railway infrastructure.

1. Introduction

The photoautotrophs are both undesired and natural elements of the biocenos process that has the track skeleton of the railway formation. The photoautotrophs can negatively affect the technical parameters of the railway tracks and the railway operation.

For example, the following negative effects are possible: distorting the results of diagnostics and monitoring of the railway tracks, and GPS sensors; overlapping the train markers and sidelights; decreasing the special equipment efficiency; degradation of the drainage and carrying capacities of the railway canvas; accelerating the corrosion and bio-corrosion of the track elements, and others.

Any linear technospheric object is a potential areal of biota that is typical of soils and climate of current region.

The railway track consists of a sand sub-ballast, broken-stone ballast, spring-actuated rail fastening, sleepers and rails.

The broken-stone ballast contains no plant food required for producers. However, the railway tracks receive various organic and mineral substances due to the train traffic.

A lack of the water availability might limit the growth of photoautotrophs on the railway track. However, a set of the local plants includes the drought-resistant crops that can form producers for local biocoenosis.

The syntaxonomical approach is the important tool for bioenvironmental estimations.

For environmental and floristics classification, Russian phytocenology school widely uses Braun-

Content from this work may be used under the terms of the Creative Commons Attribution 3.0 licence. Any further distribution of this work must maintain attribution to the author(s) and the title of the work, journal citation and DOI.
Published under licence by IOP Publishing Ltd

IOP Conf. Series: Materials Science and Engineering 1151 (2021) 012023 doi:10.1088/1757-899X/1151/1/012023

Blanquet method [1, 2]. Some authors [3, 4] criticize this method. However, we support and use Braun-Blanquet method.

The phytoindication-based method is a universally valid technique that can be used for studying the phytocenosis state.

2. Materials and methods

Presently, some physical and chemical methods are used for eliminating unwanted plants. These methods are expensive and environmentally destructive.

The following questions are actual: populating the railway track skeleton with photoautotrophs, studying their species composition, and methods for preventing their reproduction.

For studying the species composition of photoautotrophs and estimating the zones of presence of their Biodestructor-producers on the tie cribs, we selected three test sites: constantly operated railroads, rarely operated railroads, and abandoned railroads (see Table 1).

The area of each site is about 100 m^2. In our paper, we use the species composition as described in [5, 6], and Ellenberg indicator [6, 7].

Table 1. Biology characteristics of investigated phytocenosis

Species	Braun-Blanquet cover-abundance score	Values of Ellenberg indicators				Areal type according to Meusel et al., *(TA)*
		Requirements for illumination *(L)*	Requirements for soil moisture *(F)*	Soil reaction *(R)*	Mineral nitrogen availability *(N)*	
		Site No. 1 Constantly operated railroads				
Achillea millefolium	2	8	4	x	5	Ews
Geranium robertianum	2	4	x	x	7	Ea
		Site No. 2 Rarely operated railroads				
Bromopsis inermis	3	8	4	8	5	Cir
Calamagrostis epigeios	3	7	x	x	6	Ea
		Site No. 3 Abandoned railroads				
Rubus idaeus	3	7	5	6	4	Ea
Fragaria vesca	4	7	5	x	6	Cir

3. Results and discussion

3.1. Studies of phytocenosis on the sleeper box for railroads with different operation levels

The maximum number of species 24 (twenty-four) was found in Site No. 3 (abandoned railroads). These results might be caused by intensive successional processes, an accumulation of organic substances by photoautotrophs, and absence of the weed control measures. The minimum number of species 7 (seven) was found in Site No. 1 (constantly operated railroads). These results are because of the wide use of herbicides, and intense impact from trains and special equipment.

The Braun-Blanquet cover-abundance score shows the following most abundant photoautotrophs: Achillea millefolium and Geranium robertianum for Site No.1; Bromopsis inermis and Calamagrostis epigeios for Site No. 2; Rubus idaeus and Fragaria vesca for Site No.3.

The average degree of the manifestation in the areal of Biodestructor-producers is calculated using the following expression:

$$X = \frac{B_1 X_1 + ... + B_n X_n}{B_1 + ... + B_n}, \text{ where}$$

X – average value of degree of the manifestation;
B_1-B_n – Braun-Blanquet cover-abundance scores.
Table 2 illustrates the results of calculations.

Table 2 – Average degrees of manifestation of the environment factors according to Ellenberg indicators

Experimental site	∑X			∑KX			Average value of degree of manifestation X		
	F	*R*	*N*	*F*	*R*	*N*	*F*	*R*	*N*
Site No.1	7	2	7	33	14	54	4,7	7,0	7,7
Site No. 2	13	8	13	59	53	70	4,5	6,6	5,4
Site No. 3	32	21	35	171	140	211	5,3	6,7	6,0

The following results have been obtained for each site:

- Site No. 1: high moisture (F=4.7). Weak acid medium (R=7.0) with nitrogen enriched (N=7.7);
- Site No. 2: semi-dry (F=4.5). Middle acid medium (R=6.6). Poor in nitrogen (N=5.4);
- Site No. 3: optimum moisture content (F=5.3). Middle acid medium (R=6.7) with optimum nitrogen content (N=6.0).

3.2. *Estimation of the phytocides efficiency*

To estimate an efficiency of the weed control measures conducted in the operated railroads (Site No. 1), we selected four equivalent test areas (S=100 m^2) that have been exposed to the herbicide impact. Table 3 provides more details on this experiment.

Table 3 Description of experiments on treating test areas with various herbicides

Test area	Herbicide tradename	Active ingredient of herbicide, Preparation form	Usage rate
I	Reference area	-	-
II	Tornado 500	Glyphosate isopropylamine salt 500 g/l, water solution	4.0 l/ha
III	Magnum	Metsulfuron-methyl, 600 g/kg, aqua-dispersed granules	0.3 l/ha
IV	Grayder	Imazapyr, 250 g/l, water glycol solution	2.5 l/ha

Figure 1 illustrates the experimental results that describe killing of weeds in the tie crabs.

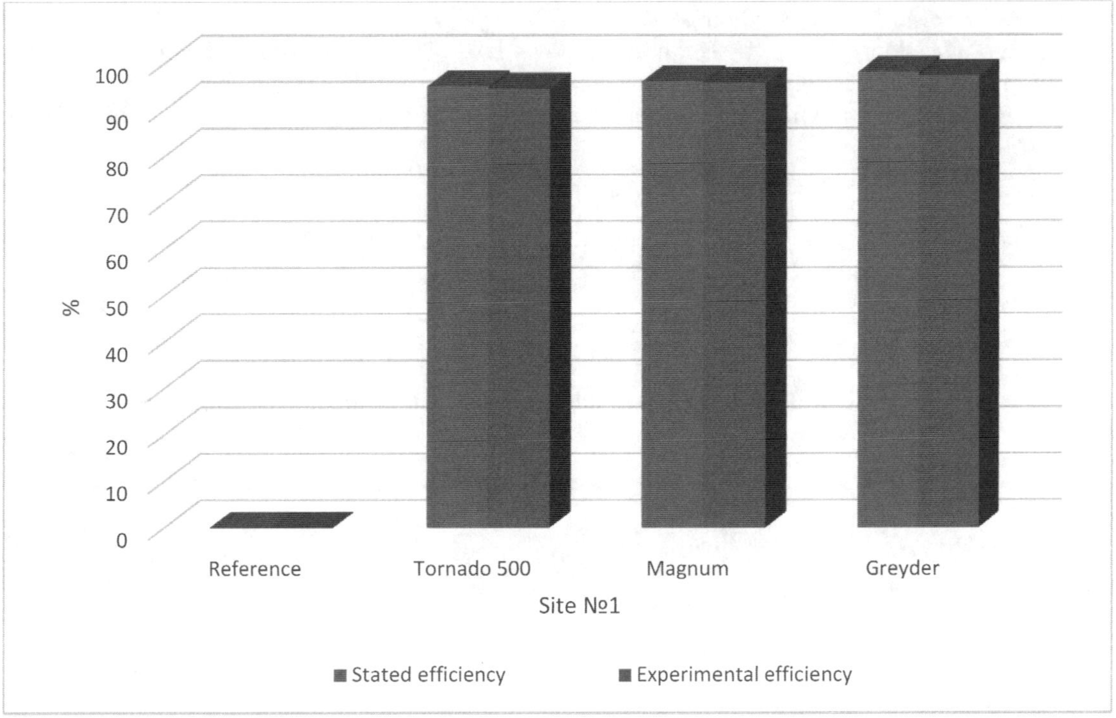

Figure 1 - Comparative analysis of the efficiency of various herbicides used in the test areas

Analysis of experimental results proved a high efficiency of all investigated herbicides. The "Greyder" herbicide manifested a maximum efficiency while the "Tornado 500" has a minimum efficiency.

4. Conclusion

1. Probably, the high concentration of mineral nitrogen in the permanent way is caused by an inflow of the organic and mineral substances from several sources: blanket sand and tree leaves; components of trailing loads; wastewater from passenger trains. Saprophytic microorganisms turn those substances into the forms that are easily absorbable for plants.

2. Variations of the water availability of the ballast prism reflect local environmental and geographic features, and accumulation of the organic substances. Different values of the hydrogen index (pH) can be caused by morphological properties of metamorphic rocks that form the broken-stone ballast, and specifics of the railway track pollutants.

3. The environmental estimations of the photoautotrophs species help us develop the efficient phytocide measures. Those measures can be divided into two groups: environment-friendly measures and environment hazards.

Environment-friendly measures include the activities that suppress the growth and propagation of the photoautotrophs. For example: hermetic sealing the equipment and wagons to avoid the spillage of trailing load, and the leaking of the lubricants, process liquids, and wastewater.

The following measures might be hazardous for environment: the use herbicides, and removing the weeds with mechanical tools. In addition, those measures might be expensive and labor-consuming.

This paper describes techniques for the environmental and floristic analysis of unwanted plants that grow on the railway canvas. The methods for removing unwanted plants as a way to ensure the stable operation of the railroad transport have been reported.

References

[1] Mirkin B M and Naumova L G 2009 Method of vegetation classification by Brown-Blank in

Russia *Russian Journal of General Biology* **70(1)** 66-77
[2] Mirkin B M, Martynenko V B and Naumova L G 2004 The importance of vegetation classification for modern ecology *Russian Journal of General Biology* **65 (2)** 167-177
[3] Gilarov A M 2003 Darwinism as a means of limiting environmental *Russian Journal of General Biology* **64(5)** 439-448
[4] Gilarov A M 2005 Restructuring in ecology: from the description of the visible to the understanding of the hidden *Bulletin. Russian Academy of Sciences* **75 (3)** 214-223
[5] Kiselyova K V, Mayorov S R and Novikov V S 2019 F*lora of the Middle Stripe of Russia: Atlas Determinant* (Moscow: "Fiton XXI")
[6] Ellenberg H 1988 *Vegetation ecology of Central Europe* (Cambridge: Cambridge University Press)
[7] Wamelink G W W, van Dobben H F and Berendse F 2003 Apparently we do need phytosociological classes to calibrate Ellenberg indicator values *Journal of Vegetation Science* **14 (4)** 619-620

IOP Conf. Series: Materials Science and Engineering 1151 (2021) 012024 doi:10.1088/1757-899X/1151/1/012024

Intellectual diagnosis components of electrical equipment in the sphere of transportation

M P Dunaev[1], A M Dunaev[1], Y O Dunaeva[2], L V Arshinskiy[3]

[1] Irkutsk National Research Technical University, 83, Lermotova st., Irkutsk, 664074, Russia
[2] Irkutsk State University, 1, Karla Marksa st., Irkutsk, 664003, Russia
[3] Irkutsk State Transport University, 15, Chernishevskogo st., Irkutsk, 664074, Russia

E-mail: mdunaev10@mail.ru

Abstract. Ensuring the reliability of the electrical equipment in transport means a necessity of its inspection and diagnosis. Frequency convertors are part of this equipment. The article is focused on creation of the optimal logical algorithm for frequency convertors diagnosis. The algorithm determines the most effective knowledge base structure of the diagnostic expert system. The algorithm is developed, the knowledge base structure is constructed in accordance with the proposed algorithm, and the diagnostic expert system used is implemented. Correctness and effectiveness of the expert system functioning are confirmed in practice.

1. Introduction

Ensuring the reliability of the electrical equipment (EE) components in transport is especially relevant today. The EE components in transport include a sophisticated EE of electric drives (ED) [1]. Examples of EE components are frequency-regulated electric drives of pumps and fans as well as the asynchronous traction drive and auxiliary electric drive of locomotive [2].

Industrial frequency convertors (FC) are parts of the EE mentioned above, and their diagnosis is more effective being supported by the diagnostic expert system (ES) [3-5]. Thus, developing a knowledge base (KB) structure for the expert system for diagnosis of frequency convertors is an important and actual task. This paper is dedicated to the development of the optimal logical algorithm for diagnosis of the industrial frequency converters, which can determine the most effective KB structure for the ES intended for diagnosis of this equipment.

2. Structure and characteristics of the object for diagnosis

The functional scheme of industrial FC connection to the electrical network is represented in Figure 1. The scheme shows the basic functional blocks of the system and signals existing in it and uses the following abbreviations: AS – automatic switch; IF – inlet filter; UR – unguided rectifier; RF – ripple filter; AI – autonomous inverter; L – electric load of converter; PU – power unit of control system; CS – control system of converter; CP – control panel; Up – voltage of the power line of alternating current; U_i – converter circuit input voltage; U_o – output voltage of autonomous inverter; U_b – output voltage of power unit; U_s – output frequency task signal; I_l – electric current of the load.

Content from this work may be used under the terms of the Creative Commons Attribution 3.0 licence. Any further distribution of this work must maintain attribution to the author(s) and the title of the work, journal citation and DOI.
Published under licence by IOP Publishing Ltd

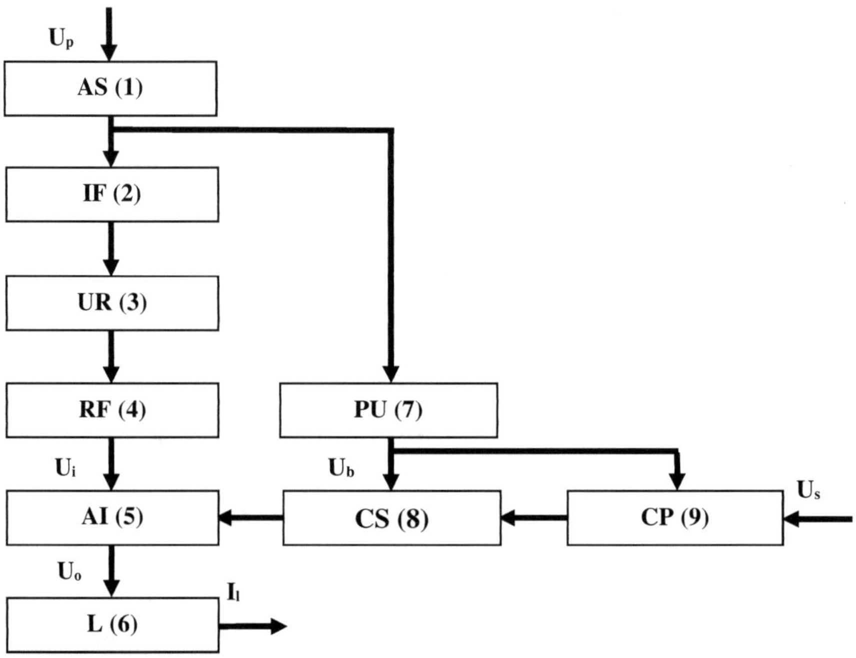

Figure 1. The functional scheme of industrial FC connection to the electrical network.

3. Development of the optimal diagnostic algorithm

To develop the optimal logical algorithm for EE diagnosis a special procedure is proposed. It includes the following stages:

1. FC functional scheme analysis in order to determine whether the functional blocks checks are expediential in different situations;
2. selection of the first check block (top of the algorithm tree);
3. construction of the block nomenclatures for the branches of the algorithm tree;
4. forming the complete arrays of check sequences for the branches;
5. finding the sequences satisfying the optimality criterion;
6. synthesis of the algorithm tree from the obtained components.

The fourth stage is the most complex and time-consuming one of the listed above, it forms a full array of checks sequences. This process can be illustrated by the block diagram shown in Figure 2. The diagram contains the following legend: LACE0 – logical array of the checks execution formed in the case of obtaining the defective output signals; LACE1 – logical array of the checks execution formed in the case of obtaining the proper output signals; SBNB – set of the block nomenclature for the branch; C1 – condition which means that the candidates of the low-order level have been exhausted and the remaining candidates have become unable to form the sequence which would appear to be the equivalent to the nomenclature of the branch without the block corresponding to the component SBNB [i].

IOP Conf. Series: Materials Science and Engineering 1151 (2021) 012024 doi:10.1088/1757-899X/1151/1/012024

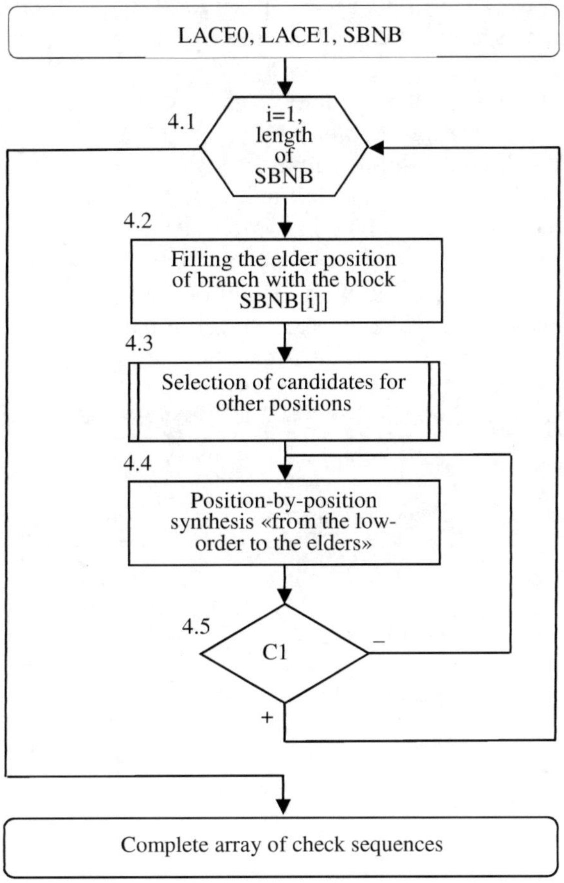

Figure 2. The block diagram of algorithm.

In order to implement the procedure represented above for the industrial FC diagnosis, the authors have developed the specialized application *FC_OPTIM*. The structure of its working prototype is shown in Figure 3 that contains the following legend: TC – table of the FC functional blocks characteristics; LACE – logical arrays of the check execution; FACS – full array of check sequences; OA – tree of the optimal logical algorithm for diagnosis of the industrial frequency converters (algorithm output).

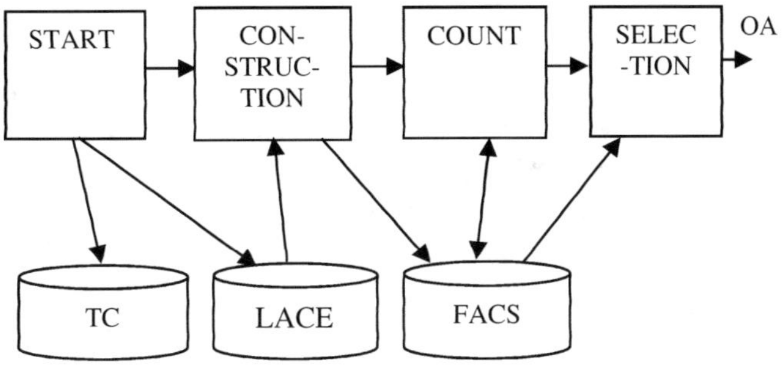

Figure 3. Structure of the application for the optimal diagnostic algorithm development.

IOP Conf. Series: Materials Science and Engineering 1151 (2021) 012024 doi:10.1088/1757-899X/1151/1/012024

Procedure *Start* implements the first stage of the algorithm development. Procedure *Construction* executes the second, third and fourth stages. Procedures *Count* and *Selection* realize the fifth and sixth stages.

Figure 4 shows application FC_OPTIM output. The procedure *Start* implements the first stage of the algorithm development. The representation of the algorithm tree in the *FC_OPTIM* output has the following form: the table cell containing the number of block that should be checked when a proper output signal is obtained (left child position) located in the following table row directly under the cell with a number of just checked block (parental position). The cell containing the number of the block that should be checked when a defective signal is obtained (right child position) located in 1, 2, 2^2, … , 2^{n-k} cells respectively to the right of the left child position. Here n is a total number of levels of blocks seen as candidates, k is an index of level of parental position under consideration, and the level of the first check block child positions is taken equal to 0).

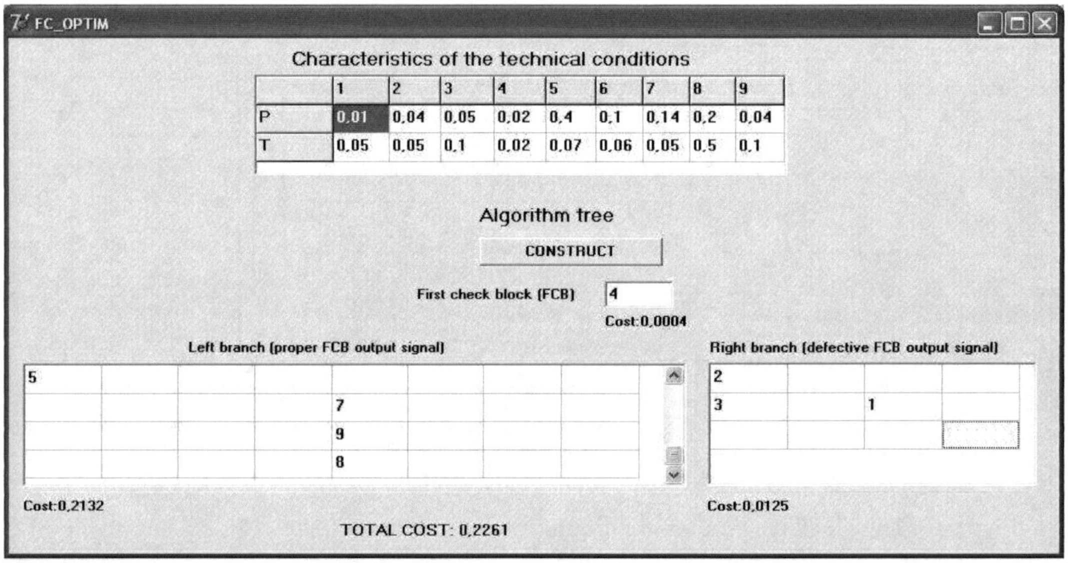

Figure 4. Results of *FC_OPTIM* work.

The obtained configuration of the optimal logical algorithm for industrial FCs diagnosis corresponds to the tree of developed by the authors MPT algorithm for FC diagnosis shown in Figure 5.

Figure 5 contains the following legend: circles with the numbers inside mean the elementary checks of functional blocks with corresponding numbers (Figure 1); numbers 1 and 0 represent the results of the block checks (proper and defective output signals respectively); rectangles are the results of diagnosis (number inside the rectangle indicates the number of the defective block).

The MPT algorithm uses the criterion of the first check minimal time and takes into account the technical state characteristics of blocks. In accordance with this algorithm, the first elementary check should have a minimal check time $t(e_i)$. In case of equality of the check times, a check that controls a functional block located nearer to the central section of the diagnosed object functional scheme should be the first. The subsequent checks are selected based on the characteristics of technical states in a descending order of $p(e_i)/t(e_i)$ value.

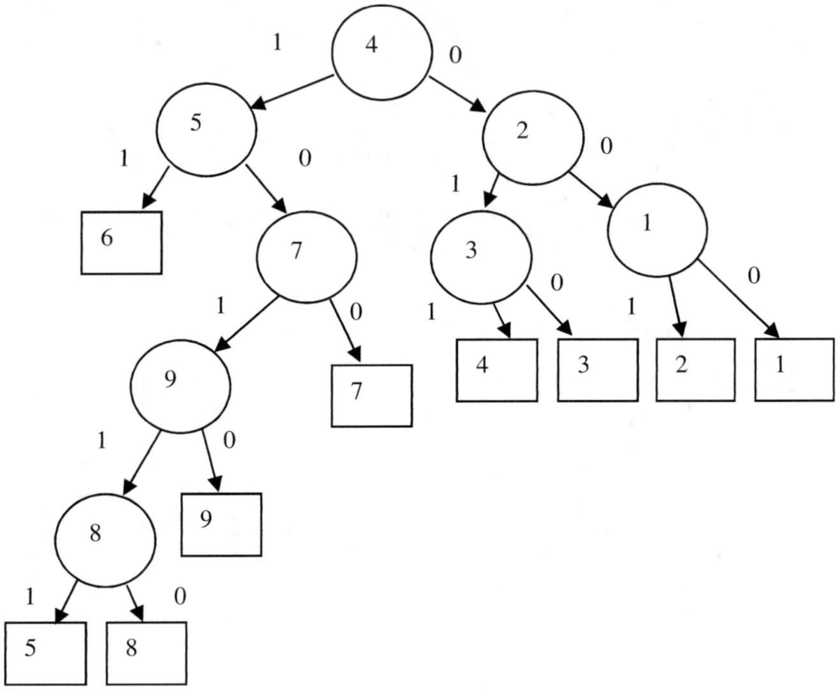

Figure 5. Tree of the optimal logical algorithm for diagnosis of the industrial frequency convertors.

4. Conclusion

As a result of a detailed analysis of industrial FC structure made in this paper, the configuration of the optimal logical algorithm for FC diagnosis was created. Based on this algorithm and experts' heuristics, a knowledge base for diagnosis of industrial frequency convertors that are components of frequency-regulated electric drives of pumps, fans, and locomotives is developed. The expert system *FC3* used the proposed knowledge base structure is implemented. Correctness and effectiveness of the expert system functioning are confirmed in practice.

References

[1] Kryukov A V, et al. 2017 *Far East Journal of Electronics and Communications (FJEC)* **17(5)** 1127-1140

[2] Kladiev S N, Solozhenko A V and Filipas A A 2016 *MATEC Web of Conferences*, **91** article 01040

[3] Dunaev M P, et al. 2017 *Far East Journal of Electronics and Communications (FJEC).* **17(5)** 1011-1019

[4] Arshinskiy L V et al. 2017 *Far East Journal of Electronics and Communications (FJEC).* **17(5)** 1119-1125

[5] Arshinskiy L V 2007 *Automation and Remote Control.* **68(1)** 139-148

Requirements for a digital twin of the bulldozer work tool

I V Leskovets

Belarusian-Russian University, 43, Mira str., Mogilev, 212000, Belarus

E-mail: le@bru.by

Abstract. The paper presents a brief description of the bulldozer as a widely used machine for earthworks. The method for determining resistive forces to digging is analyzed based on the theory of layer-by-layer excavation works. The relationship for determining the area of the longitudinal section of the dragging prism is presented, and on its basis the productivity is further calculated. The characteristics of forces resistive to digging are constructed with the accepted parameters of the blade and soil for different angles of blade setting. The criterion for determining the efficiency of earthmoving operations is presented, which was developed on the basis of unit costs. It is stated that excavation works are performed effectively at a certain angle of blade setting. Requirements for a digital twin of the bulldozer work tool, which is designed to improve the efficiency of its operation, have been developed.

1. Introduction

Bulldozers are earth-moving equipment widely used in the construction industry. The simplicity of their design and operation accounts for their widespread use in earthwork construction. The bulldozer working tool can use pivoting or non-pivoting blades. Depending on the blade type, a universal blade frame or the one with push bars is used. The operation of bulldozers performing different construction works or conducting mining and stripping operations has led to the use of various types of blades [1]. The long-term use has led to the optimization of the parameters of their operating equipment and, in particular, blades. Today, the designs of bulldozer blades have reached a high degree of perfection and provide high productivity at the lowest possible costs in earthmoving operations and extraction of mineral resources.

Conditions for carrying out earthwork operations are very changeable and highly dependent on a number of factors, such as soil type, humidity, ambient air temperature, operator qualification, etc. The productivity of the machine, therefore, depends not only on the excellence of its work tool, but also on the conditions of its operation. The operation of the bulldozer work tool during carrying out earthworks is almost completely entrusted to the operator, who is guided only by his experience and perception of ongoing events. While performing earthworks, he does not have access to the information on the degree of the effect of blade position, penetration depth and travel speed on the efficiency and costs during each pass of the machine.

2. Methods and techniques

The theory of the bulldozer design offers a significant amount of scientific investigations [2] which allows choosing the optimal parameters for the blade and the working tool of the bulldozer depending on the soil type and its condition (Figure 1).

Content from this work may be used under the terms of the Creative Commons Attribution 3.0 licence. Any further distribution of this work must maintain attribution to the author(s) and the title of the work, journal citation and DOI.
Published under licence by IOP Publishing Ltd

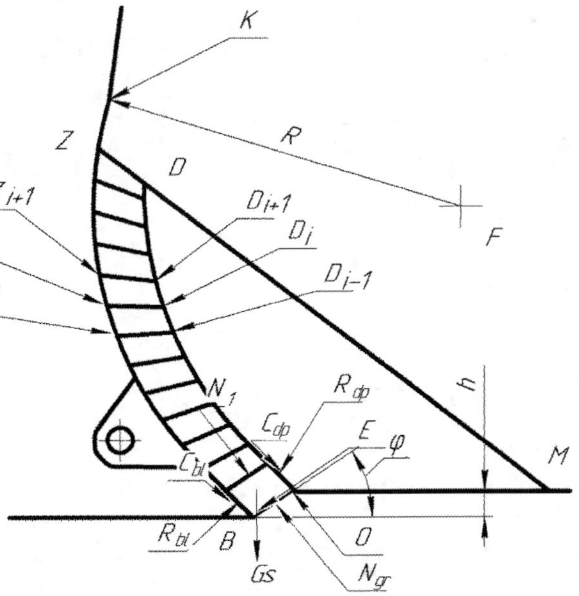

Figure 1. Diagram of blade-soil-dragging prism interaction.

The blade parameters are determined based on the criterion for minimizing resistive forces to digging. At the initial stage of digging, the blade penetrates the soil by the value *h*. While the blade moves forward, a compacted zone is formed in the soil under the action of compressive forces, and normal and tangential stresses originate [3]. These stresses are increasing, and a force arises that acts on the blade and resists its movement. When the stresses reach their limit values, the soil is shifted along the area BO (Figure 1). At this time the force acting on the blade reaches its maximum value, as this corresponds to the moment of shear area formation. The magnitude of forces acting on the blade at this stage depends on the cohesion forces of the soil and the slope angle of the shear area. Upon further movement of the blade, a dragging prism is formed from the soil separated from the massif. The soil moves up the blade and forms the dragging prism from the segments Z_{i-1}, Z, D, D_{i-1}. To perform calculations, it was assumed that the dragging prism consists of a conditionally stationary part OMZ and a conditionally mobile part BODZ, the dimensions of its elements being dependent on the size of the shear area and the radius of the blade curvature R. The elements of the mobile part are acted upon by the gravity forces, the normal reaction N from the lower and upper soil segments, the normal forces from the blade R_{bl} and the stationary dragging prism R_{dp}, the friction forces between the blade C_{bl} and the stationary dragging prism C_{dp}. The interaction of these forces forms the resultant resistive force to digging that is applied to the bulldozer blade. In the process of digging, the dragging prism is growing. Its gravity force affects the magnitude of ultimate stresses in the soil, which is in the undisturbed condition and additionally loaded by the dragging prism. Due to the presence of the additional load from the force of gravity of the soil and the increase in the ultimate stresses, the inclination angle of the shear area BO changes and the resistive forces to digging increase.

During the design process, the designers choose the blade parameters that provide the optimal magnitude of digging resistive forces. But when the work is performed on a specific work site, the resistive forces to digging can be far from minimal, because during the digging process, the operator measures their magnitude by the engine loading, i.e. by indirect indicators not having any numerical expression [5]. In addition, the size of the dragging prism and the productivity depend on the angle of blade setting, the value of which can vary in modern bulldozer.

The volume of the dragging prism depends on the blade width and the triangle, which is limited by the segments DM and MO and the curve OD. The dimensions l_{DM} and l_{MO} depend on the height of the dragging prism H and the angle of repose γ. (Figure 2)

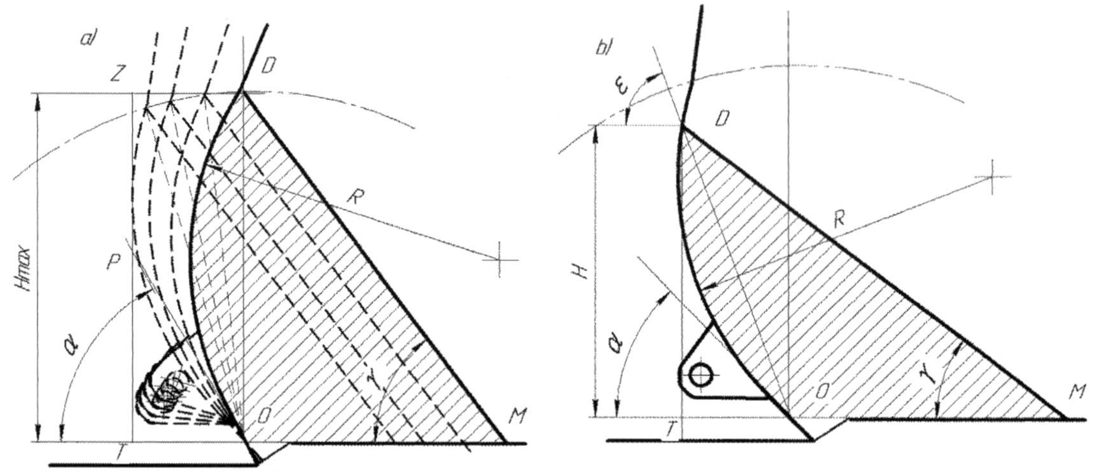

Figure 2. Parameters of the dragging prism.

As a rule, when digging the ground, the blade is mounted at some angle ε, other than 90° (Figure 2b). This angle is optimal for soils of different types which have different properties. But, when working on a specific work site and under specific ambient conditions, there can be found such a blade setting angle ε, at which the productivity will be maximum, and the resistive forces are minimum.

The determining factor for calculating productivity is the volume of the dragging prism, which depends on the area of the shaded region in Figure 2 and can take different values depending on the blade angle (Figure 2a).

The area of the dragging prism as a function of ε can be found by the formula:

$$S = \left[\left[H_{\max}(1 - \tan(\varepsilon)) \right]^2 \cdot \tan(\gamma) + R^2 \left[acr\sin\left[\frac{1}{\cos(\varepsilon)\cdot R} \right] - \frac{1}{\cos(\varepsilon)\cdot R} \right] \right] \quad (1)$$

where H_{max} is the maximum blade height corresponding to the case when $\varepsilon = \pi/2$ (Figure 2b), γ is the angle of repose, R is the radius of the blade curvature.

Formula 1 and Figure 2 show that the dragging prism has the maximum area in the blade position, in which $\varepsilon = \pi/2$. Thus, in the bulldozer operation this is the position of blade that needs to be maintained in order to achieve maximum productivity.

In addition to productivity, an important component of the operation process is the cost of the work performed. The costs directly depend on the forces resistive to digging because the operation speed of the bulldozer is 1 ... 1.4 m/s, and the resistive forces to digging can vary within wide limits.

With the help of the software developed on the basis of the investigations [4] forces of resistance to digging by the bulldozer blade were calculated for the soil parameters presented in table 1 and the blade.

Table 1. Soil parameters

Parameter name	Measurement unit	Magnitude
Internal friction angle	deg	28
Soil-metal friction angle	deg	26
Slope angle of the segment DM to the horizon	deg	35
Specific cohesion of soil (undisturbed)	kPa	25
Specific resistance to cutting	Pa	70
Soil density in the dragging prism	t/m^3	1.8

Table 2. Blade parameters

Parameter name	Measurement unit	Magnitude
Blade height	m	1.2
Height of blade's top spill plate	m	0.2
Length of blade knife	m	0.2
Width of blade	m	3
Blade curvature radius	m	1.2
Blade setting angle (ε)	deg	85, 75, 65
Cutting angle (α)	deg	changes based on the blade position

Based on the calculations, plots of the resistive forces to digging were obtained for the blade moving through the soil and the dragging prism being formed (Figure 3). The diagrams show that the blade setting angle can have a significant effect on the magnitudes of the resistive forces to digging, which, in turn, will affect the value of the unit costs.

The analysis of Figure 3 shows that the resistive forces to digging mostly depend on the height of the dragging prism. The effect of the blade angle is also significant. By increasing the blade angle by 20 degrees, a significant reduction in resistive forces to digging can be achieved.

An important task is to choose such an angle of blade setting, at which the productivity will be maximum, and the costs involved in performing earthworks will be minimal.

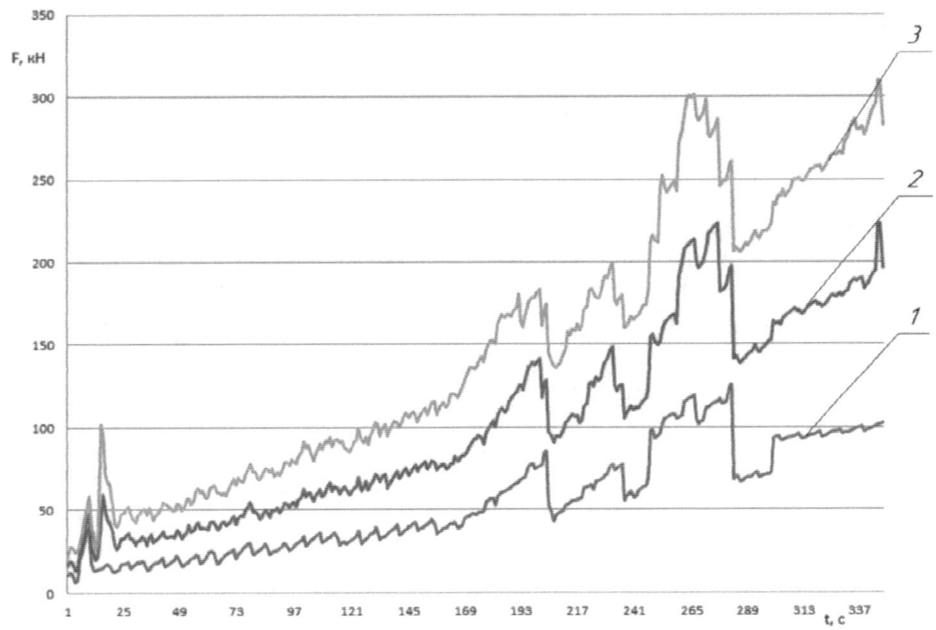

1. Blade setting angle ε = 65°; 2. blade setting angle ε = 75°; 3. blade setting angle ε = 85°.
Figure 3. Magnitudes of forces resistive to digging depending on the blade setting angle.

Calculating the productivity by the formula:

$$P = \frac{3600 V_{dp}}{t_w + t_{rm}} \tag{2}$$

where V_{dp} is the volume of the dragging prism, t_w is the time of the operation pass, t_{rm} is the time of the reverse travel, and considering that the bulldozer operates at the rated power corresponding to some constant value, it is possible to establish the criterion for the optimal ratio of productivity and power as the relationship:

$$K = N / P \quad (3)$$

where N is the accepted constant power value.

Let us assume that the length of the work area is 130 m, the reverse travel speed is 2.5 m/s, then the speed of the operation pass is calculated as:

$$v_{rm} = N / F_i$$

where F_i is the average magnitude of the resistive forces to digging at a fixed angle of the blade. After determining the productivity under the accepted conditions, the value of the criterion K can be defined. The calculation results for the accepted values are presented in Figure 4.

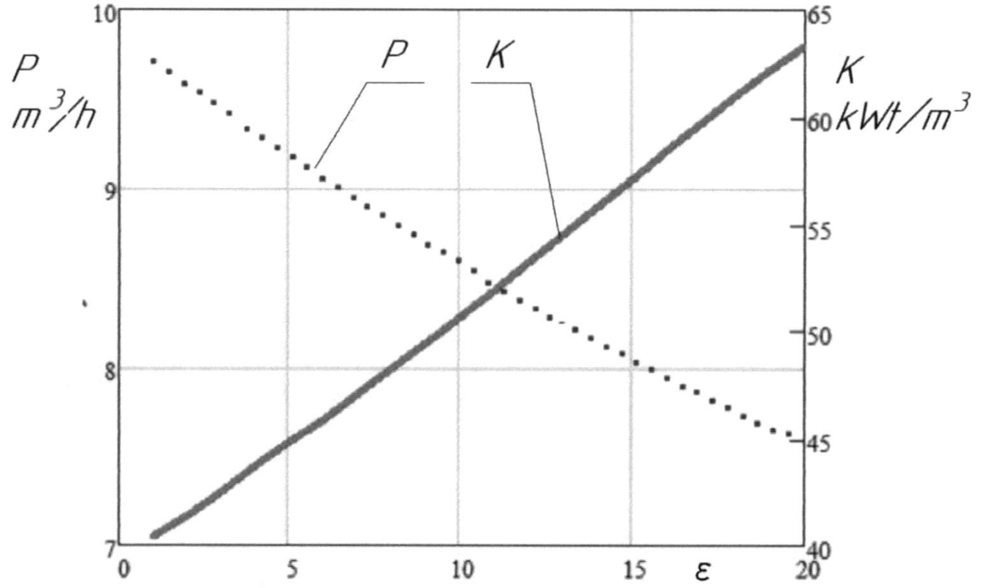

Figure 4. Relationships between productivity and the efficiency criterion of the bulldozer, and the angle of blade setting.

The analysis of Figure 4 shows that the blade angle of 11 ... 13 degrees is the most effective under the accepted conditions and the maximum dragging prism. The calculations show that the efficiency of earthmoving works is affected not only by the angle of blade installation, but also by the size of the work area, bulldozer speed, gear selected, maximum engine power, soil parameters and weather conditions prevailing at the time of the work.

3. Conclusion

Thus, the digital twin of the bulldozer work tool should receive, as initial data, the height of the dragging prism or its volume, the magnitude of resistive forces to digging, and the bulldozer speed. The software package of the digital twin must calculate bulldozer performance during one pass and throughout a number of passes; determine the energy required for digging the soil and the specific fuel consumption per 1 m^3. The information storage unit should contain information on several passes, and, based on the statistical data processing, either set automatically the optimal position of the blade or provide the operator with the information about it.

References

[1] Dovgyalo V A and Bochkarev D I 2010 *Road Construction Machines part I: Earthmoving Machines* (Gomel: BelGut).

[2] Berestov E I and Leskovets I V 2011 Effect of soil properties on resistance to digging by bulldozer's work tool *J. Polotsk St. Univ. Bulletin* **3** 45 – 52.

[3] Berestov E I 2017 Nauchny`e osnovy` vzaimodejstviya rabochego oborudovaniya zemlerojno-transportny`x mashin s gruntom (Mogilev: Belorus.-Ros. un-t).

[4] Berestov E I and Leskovets I V 2010 Effect of soil properties on forces of resistance to digging with a bulldozer work tool *J. Polotsk St. Univ. Bulletin* **3** 45 – 52.

[5] Kutuzov V V 2012 Povyshenie effektivnosti ekspluatacii stroitel'nyh i dorozhnyh mashin s uchetom ih tekhnicheskogo sostoyaniya (Mogilev: Belorus.-Ros. un-t).

IOP Conf. Series: Materials Science and Engineering 1151 (2021) 012026 doi:10.1088/1757-899X/1151/1/012026

Development of mathematical models of the relationship between the main parameters of a container depot

D I Ilesaliev, F K Azimov, G R Ibragimova, N F Svetasheva, SH R Abduvaxitov and Z SH Tursunov

Tashkent State Transport University, 1, Odilxojayev str., Tashkent, 100167, Uzbekistan

E-mail: ibragimova.gulshana@mail.ru

Abstract. The purpose of the article is to develop a method for reliably determining the most rational values of container depot parameters, thereby improving and increasing the level of customer service for transport in general. Research methods are based on a system analysis and generalization of existing methods and methods for calculating the capacity of container terminals. The results of the study are the technique of determining the capacity of the warehouses in railway transport, ensuring more qualified calculation and selection of parameters of the container depot, as well as reducing erroneous design decisions and improving the reliability of development projects and improvement of the transport network as a whole.

1. Introduction

Today, there is a trend of development of container transportation all over the world, but this can not be said about the level of containerization in Uzbekistan. This is mostly due to the shortage of containers in the freight transport market, as well as the lack of logistics facilities, and a small share of transit traffic.

For example, the park of containers of the country's main operator, Uztemiryulcontainer JSC, is less than 1000 units in twenty-foot equivalent (TEU). In this regard, the task is to solve this problem by organizing a container depot, which will increase the level of customer service for railway transport and the transport network as a whole.

2. The description of the object

Modern cargo terminal complexes are complex technical objects. In this study, a system analysis of a technical object is used as the main methodology for studying a container depot. According to which any object under study can be considered as a system that has certain elements (technological sections of a container depot), the relationship between elements (information and container flows), and are united by a common goal of functioning (rational transformation of container flow parameters).

The structure of a container depot in accordance with the main provisions of the system analysis is a variety of relationships between the main elements: technological sections and loading and unloading machines. Figure 1 shows the technological structure of a container depot. A container depot, unlike a container terminal, consists of only three elements: anarea for storing empty containers, and a railway and automobile loading and unloading area (LUA). Because of this, the container depot has a simpler structure.

Content from this work may be used under the terms of the Creative Commons Attribution 3.0 licence. Any further distribution of this work must maintain attribution to the author(s) and the title of the work, journal citation and DOI.
Published under licence by IOP Publishing Ltd

The task of parametric description of container depots is reduced to their meaningful description, mathematical expression, and output of formulas for their calculation. To describe a container depot, we use its representations as a set of interrelated elements and take into account their parameters.

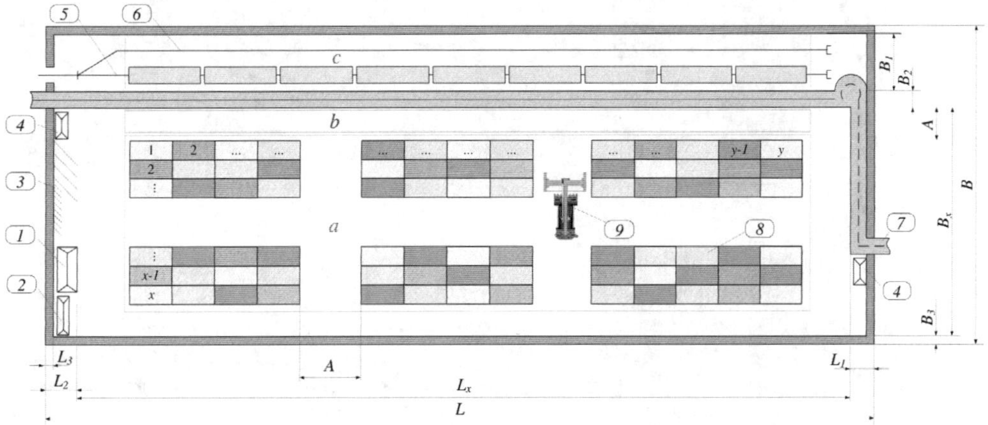

Figure 1. Diagram of a container depot by elements: *a* – empty container storage area; *b*– automobile LUA; *c* – railway LUA

Designations in figure 1: 1 – administrative and household complex; 2 – repair shop; 3 – parking for cars and more heavy vehicles; 4 – checkpoint; 5 – railway loading and unloading area; 6 – exhibition way; 7 – two-lane roadway; 8 – stack of empty containers; 9 – loading and unloading machine.

Consequently, the total number of initial and calculated parameters of a container depot, as well as their mutual influence on each other, is quite large, which also confirms the complexity of the warehouse structure as a system object. Correctly calculated capacity will allow you to determine with greater accuracy all other parameters necessary to improve the functioning of the container depot.

3. Comparative analysis of the main parameters of loading and unloading machines

The advantage of forklift trucks compared to the traditional use of gantry cranes is the following: simplicity and short time to enter the container depot, the absence of crane tracks, the ability to stack containers above three tiers, etc. All these advantages have become crucial for the recommendation of these types of loading and unloading machines in a container depot. Types of loading and unloading machines are related to the layout of the general plan of the container depot and depend on the type of containers being processed. Figure 2 show the types of container loaders designed for processing and stacking various types of containers.

The most widespread in Western container terminals are forklifts with an extendable crane boom (reach stacker) (see figure 2, a). A container forklift with a side spreader grip (double stacker) has the ability to stack two empty containers at once (figure 2, b).In this regard, the technical performance of this loading and unloading machine is approximately twice as high as that of other types of forklifts. Container forklifts with top spreader grip are also used at Western terminals (see figure 2, b). The container forklift with a fork grip is a universal loading and unloading machine (see figure 2, d). Special load-handling devices for forklift trucks are allowed to perform loading and unloading operations directly and inside the container. The table 1 shows the main parameters of individual models of container loaders.

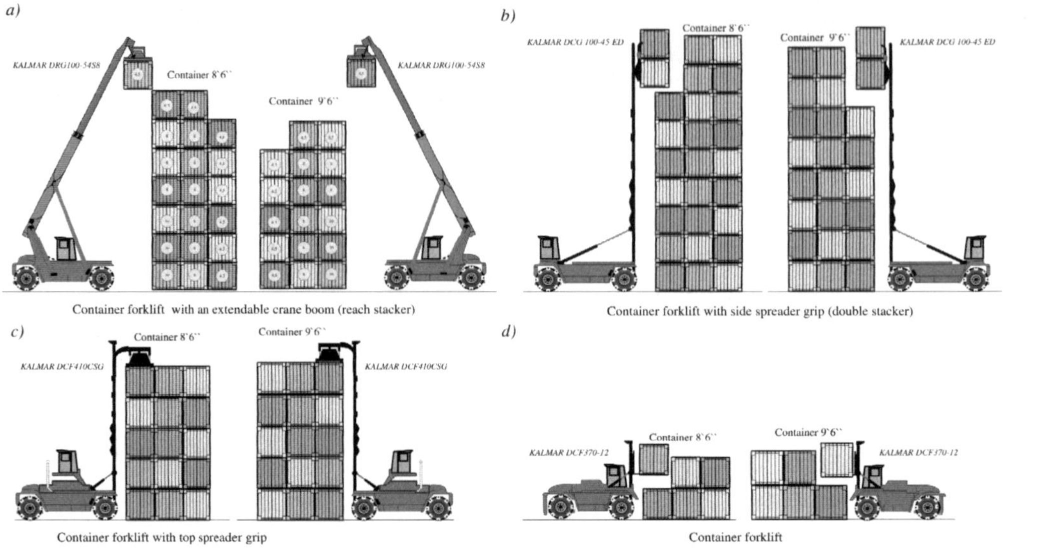

a b c

Figure 2. Container forklifts: *a* – reach stacker; *b* – double stacker; *c* – forklift with top spreader grip; *d* – with fork grip.

Table 1. Characteristics of container handling trucks

Parameters	Types of container forklifts			
	Reach stacker DRG100-54S8	*Double reach stacker DCG 100-45 ED*	with top spreader grip *DCF410CSG*	with fork grip*DCF370-12*
Lifting capacity, in container row, t				
1 row	10.0	22.0	41.0	37.0
2 row	8.0	-	-	-
3 row	4.5	-	-	-
Stacking height, no. of containers				
Container 8`6``	8	9	5	2
Container 9`6``	7	8	5	2
Aisle width, with 20 ft – 40 ft container				
Container 20 ft	11200	10846	11990	10795
Container40 ft	14200	14326	14680	10795
Service weight, standard truck, kgs	40300	44497	69750	50100
Lifting speed, m/s				
Unloaded	0.50	0.65	0.42	0.35
At rated load	0.42	0.59	0.25	0.25
Lowering speed, m/s				
Unloaded	0.42	0.54	0.31	0.25
At rated load	0.48	0.59	0.36	0.40
Travelling speed, forward – reverse, unloaded – at rated load, m/s				
speed / forward	8.3/8.3	6.9/6.9	7.2/7.2	7.2/7.2
Unloaded / At rated load	8.3/8.3	6.9/6.9	6.4/6.4	6.6/6.6

* – number of tiers for stacking empty containers

Taking into account the parameters of container loaders will allow a more qualified calculation.

4. Development of mathematical models of relationships between container depot parameters

The following are mathematical models that show the dependencies between individual parameters of a container depot. The total capacity of a container depot is determined by the formula:

$$R_i = x \cdot y \cdot z \rightarrow \max \quad (1)$$

where x– the number of empty containers across the width of the storage area, cont; y – number of empty containers along the length of the storage area, cont; z –the number of stacking of empty containers in height.

To begin with, the useful width of the container depot is determined (the width of the technological area for storing empty containers):

$$B_x = B - B_1 - B_2 - B_3, \text{ m} \quad (2)$$

where B –width of the container depot, m; B_1 – part container depot allotted for railway loading-unloading station and also the exhibition path, m; B_2 – part container depot allotted for two-lane roadway, m; B_3 –width of lawns along the fence, m.

The number of containers to be placed by width x of the empty container storage area:

$$x = \varepsilon\left\{\frac{B_x - n_2 \cdot A}{b + \omega}\right\}, \text{ cont.} \quad (3)$$

where $\varepsilon\{...\}$ – the designation of the integer part, rounded down; n_2 – number of longitudinal passages, which depends on the number of containers installed at a depth of stacks $x1$; A is the width of the passage for the loading and unloading machine, m; b is the width of the container, m; ω is the technological gap between containers, located along the width of the storage area, m.

$$n_2 = \varepsilon\left\{\frac{B_x - (b+\omega)\cdot x_1 - A}{2\cdot(b+\omega)\cdot x_1 + A}\right\} + 1, \text{ passage} \quad (4)$$

where 1 is the longitudinal passage along the railway track.

$$L_x = L - L_1 - L_2 - L_3, \text{ m} \quad (5)$$

where L is the length of a container depot, m; L_x – the length of the storage area empty containers, m; L_1 – length part of the container depot, occupied by cross passages for vehicles, m; L_2 – length part of the container depot, occupied office building, repairing shop , m; L_3 is the width of the lawn along the fence, m;

The number of containers placed along the length y of the storage area is determined as follows:

$$y = \varepsilon\left\{\frac{L_x - n_1 \cdot A}{l + \lambda}\right\}, \text{ cont.} \quad (6)$$

whereλ is the technological gap between containers located along the length of the storage area, m; l is the length of the container, m;

The number of cross passages (n_1) is calculated as follows:

$$n_1 = \varepsilon\left\{\frac{L_x}{80}\right\}, \text{ passage} \quad (7)$$

where 80 is the approximate length through which cross passages are installed for forklifts.

The advantage of the proposed mathematical models is that they describe the relationship of the container depot parameters to each other and to the depot operation.

Based on the developed mathematical models of the relationship between parameters, calculations were made to determine the capacity of the storage area for empty containers equipped with various types of forklifts. Figures 7-10 show graphs of the capacity of container depots depending on the depth of stacking of empty containers and the length of the storage area.

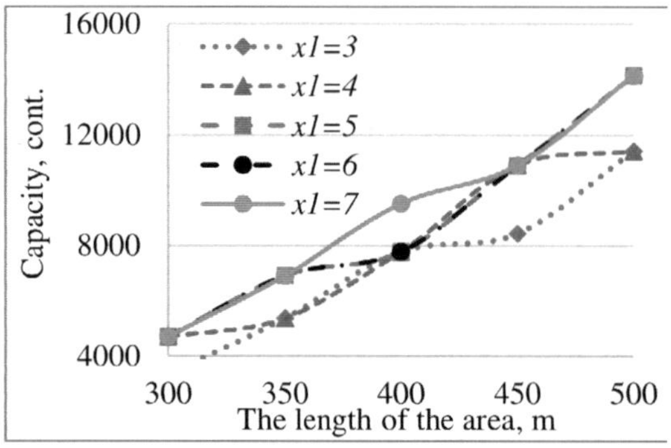

Figure 3. Container depot is equipped with forklift with an extendable crane boom

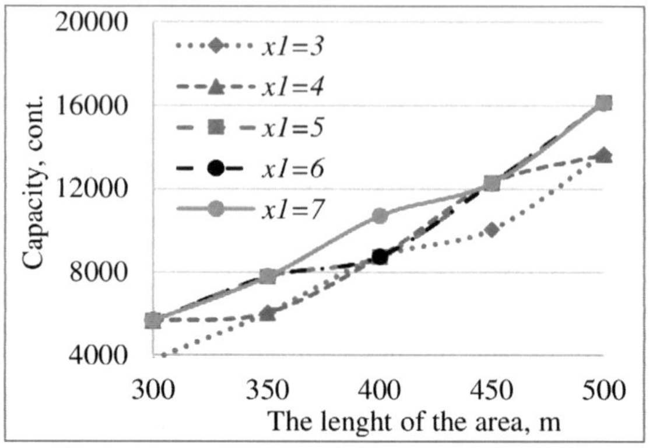

Figure 4. Capacity of a container depot equipped with container forklift with a side spreader grip

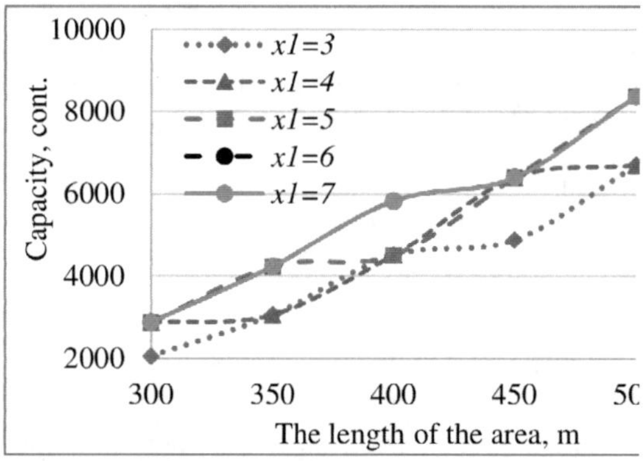

Figure 5. Capacity of a container depot equipped with container forklift with the top spreader grip

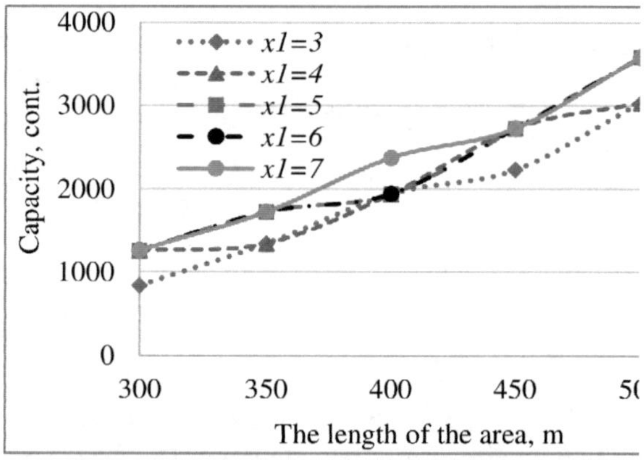

Figure 6. Capacity of a container depot equipped with container forklift with the fork grip

Figures 3-6 show that for different values of x1, the capacity of a container depot can be the same. Therefore, it is recommended to choose the lowest value x1, which will ensure a high processing capacity of the depot.

5. Conclusion

In this paper, the issue of reliable determination of the most rational values of container depot parameters is considered. During the research, the following tasks were solved and the results were obtained:

- As a rational technological equipment for container depot areas, it was proposed to use container loaders.
- The study developed mathematical models that show the relationship between the parameters of a container depot, they can be used to determine and optimize the main parameters of a container depot, as well as other types of container terminals.

- Using the developed mathematical models, it becomes possible to solve various problems, such as determining the capacity of container depots, based on taking into account all possible parameters of the technological storage area, containers and forklifts.
- The organization of a container depot on the railway transport can solve the problem of deficit of containers.

References

[1] Lubentsova V S and Efremov A V 2005 Solution of the problem of determining the capacity of a container terminal using the "death and birth" model. *VestnikSamGTU* **38** 155-158

[2] Ilesaliev D I Abduvakhitov S R Ismatullaev A F and Makhmatkulov S G2019Research of the main warehouse of the container terminal.*International Journal of Mechanical Engineering and Advanced Technologies***1** 4625-4630.doi: 10.35940/ijeat.A2923.109119

[3] Rakhmangulov A N and Muravyev D S2016Assessment of the development of the systems "seaport -" dry "port by the method of simulation.*Vestnik UrGUPS* **3 (31)**54-72

[4] Kuznetsov A L 2009Genesis of agent simulation in the course of development of methods for technological design of ports and terminals.*Ekspluatatsiyamorskogotransporta***4 (58)** 3-7

[5] Mittal N, Boile M, Baveja A and Theofanis S 2013 Determining optimal inland-empty-container depot locations under stochastic demand. *Research in Transportation Economics.***42(1)** 50–60. doi:10.1016/j.retrec.2012.11.007

[6] Pascual J, Aranda D, Hidalgo F, Smith A E, Karakaya E and Gonzalez-Ramirez R G 2016 Empty container stacking operations: Case study of an Empty Container Depot in Valparaiso Chile. *2016 Winter Simulation Conference (WSC).* doi:10.1109/wsc.2016.7822419

[7] Boile M, Theofanis S, Baveja A and Mittal N 2008 Regional Repositioning of Empty Containers. *Transportation Research Record: Journal of the Transportation Research Board.* **2066(1)** 31–40. doi:10.3141/2066-04

[8] Petering M E H 2011Decision support for yard capacity, fleet composition, truck substitutability, and scalability issues at seaport container terminals. *Transportation Research Part E: Logistics and Transportation Review.***47(1)** 85–103. doi:10.1016/j.tre.2010.07.007

[9] Chu C and Huang W 2005 Determining container terminal capacity on the basis of an adopted yard handling system. *Transport Reviews.***25(2)** 181–199. doi:10.1080/0144164042000244608

[10] Guan C & Liu R 2009 Container terminal gate appointment system optimization.*Maritime Economics & Logistics.***11(4)** 378–398. doi:10.1057/mel.2009.13

[11] Kozan E 2006 Optimum Capacity for Intermodal Container Terminals. *Transportation Planning and Technology.***29(6)** 471–482. doi:10.1080/03081060601075716

IOP Conf. Series: Materials Science and Engineering 1151 (2021) 012027 doi:10.1088/1757-899X/1151/1/012027

Influence of anomalous truth values on logical inference in V^{TF}-logics as a basis for verification of rule-based systems

L V Arshinskiy[1], **V L Arshinskiy**[2], **M P Dunaev**[2], **M S Nitezhuk**[1]

[1] Irkutsk State Transport University, 15, Chernyshevskogo str., Irkutsk, 664074, Russia
[2] Irkutsk National Research Technical University, 83, Lermontova str., Irkutsk, 664074, Russia

E-mail: larsh@mail.ru

Abstract. The paper is devoted to the problem of expert systems knowledge bases verification. The methodological basis of verification is logics with vector semantics in the V^{TF}-logics form. The knowledge model is a rule-based system. The issues of algorithmization of contradictions and other problems detection are considered in the paper. Algorithmization is based on the characteristic features of in V^{TF}-logics inference. It is shown that in the formalism under consideration, the anomalous truth of a small premise generates the same conclusion (a large premise is considered strictly true). Anomalies such as falsity, uncertainty, and contradiction are considered. The problems of reducing the computational complexity of algorithms are considered.

1. Introduction

Despite well-known criticism, expert systems (ES) are often important components of managing for complex technical, industrial, and other systems (see, for example, [1-4]). This is mainly due to the weakly structured and poorly formalized complex subject areas (SA). Knowledge models used in ES are often an alternative approach to modeling. The advantage of knowledge models, compared, for example, with neural networks, which are also used in this area [5-8], is an explicit and editable representation of knowledge. In artificial neural networks, knowledge is "buried" in the network architecture and its changing is possible only through retraining, or conceptual rethinking of the network. In this regard, modeling by means of knowledge will be in demand at least in the nearest future. In this meaning, knowledge models are also of interest when managing critical infrastructures, where timely response to a situation and its meaningful understanding is an important element of safety.

Like any other model, the knowledge model may contain errors and must be verified. A large number of studies have been devoted to verification of knowledge models (see, for example, [9-19]). The main emphasis is placed on rule-based models as the most popular in practice [14-17, 19]. Many authors consider such classes of anomalies of rule-based knowledge bases (KB) as a violation of consistency (inconsistency, cyclicality, redundancy, intersections) and a violation of integrity (incompleteness in a narrow sense, lack of links, incorrect attribute values) [14]. Static and dynamic anomalies are also considered [15, 16]. The first ones are those that are identified by analyzing the KB structure. The others are anomalies that appear during the inference engine operation. Static anomalies

Content from this work may be used under the terms of the Creative Commons Attribution 3.0 licence. Any further distribution of this work must maintain attribution to the author(s) and the title of the work, journal citation and DOI.
Published under licence by IOP Publishing Ltd

include, for example, incompleteness, redundancy, and the presence of facts that are not used anywhere. The dynamic anomalies include inconsistency of the knowledge system. Thus, incompleteness may be manifested by the presence of unattainable, missed, terminating rules, and redundancy – by duplicate rules and rules that do not lead to any goal [16]. The interest in anomalies of rule-based models is linked with that fact; they are easily modified unlike network models. Moreover, we can modify the KB without coordinating the next rule with the existing ones. This external ease is the reason for the discussed problems (modularity is also typical of logical models, but the situation there is similar, so we are talking only about rules).

Despite the relatively large number of possible KB artefacts, the contradiction in KB is the most interesting case for researchers. This is, for instance, [17]:

- contradiction within the rules when parts of the rules exclude each other (for example, $a \to \neg a$, here «$\to$» and «$\neg$», respectively, implication and negation);
- contradictions between rules that lead to mutually exclusive conclusions, such as a and $\neg a$;
- internal inconsistency of reasoning chains, when facts that contradict known facts are introduced into the system;
- contradictory chain of inference when the same data set creates a mutually exclusive situation.

In addition, the contradiction may not be known immediately. For example, if it is hidden in synonymous words or SA features [16, 18, 19].

Static anomalies are a simple case. They can be identified at the stage of creating the knowledge base. Dynamic ones require computational verification by starting the solver. Authors would like to:

- automatize this procedure;
- reduce the number of solver launches to identify artefacts.

There are other anomalies also covered in this paper.

2. Logics with vector semantics as an artefacts detecting tool

To solve this problem, we will use the capabilities of logics with vector semantics in the form of V^{TF}-logics [20-23]. Their distinctive feature is the ability to perform inference in conditions of data deficit and/or inconsistency. The classic inference is impossible in this situation. At the same time, we consider only artefacts of the type of uncertainty and/or contradictions in the formal sense without reference to the content of the SA. Errors in the form of logical circles, non-use facts in rules, and other static anomalies can be detected without using the logical inference procedure.

In V^{TF}-logics, uncertainty are formalized by the truth vector $\langle 0;0\rangle$, a contradiction by any $\|a\| = \langle a^+;a^-\rangle$, where $a^+,a^- \in [0,1]$. Here «a^+» is a measure of confidence that the proposition a is true, and «a^-» that a is false. In classical inference, uncertainty in this sense occurs if there is no inference step that results in the truth of some proposition. In other words, a does not participate in the inference or it is terminated. The conclusion may fail if the antecedent of the rule is always false. The problem arises in case of conflict when one branch of the inference is conclusion a, and the other is $\neg a$. Application of the V^{TF}-logics allows you to bring the inference to the end. If there is uncertainty or contradiction at any of the stages, they «stretch» to the end. Moreover, in certain cases, we can stretch the false premise influence to the end of inference. It generates a false result at the output. Detection of such a situation is a signal of the presence of an artifact.

The core of the product is the implication $i = a \to b$. The inference rule is «modus ponens» in the form:

$$a,\ a \to b \vdash b\colon \|b\| = \|a\&i\| \div \|\neg a \vee i\|. \qquad (1)$$

After the colon, the method of calculation of the truthfulness of the conclusion is specified. It is based on truth of small and large premise: a and $i = a \to b$. The «&» is a symbol of conjunction, «$\vee$» is the symbol of disjunction; «$\div$» is the range of values of the vector of the truth. In this case it is the rectangle in the square $[0,1]\times[0,1]$. It is defined by its upper left and lower right corners: $\|a\&i\|$ and $\|\neg a \vee i\|$. The value of $\|b\|$ in this case is equal to:

IOP Conf. Series: Materials Science and Engineering 1151 (2021) 012027 doi:10.1088/1757-899X/1151/1/012027

$$\|b\| = \langle [a^{+}\bullet i^{+}, a^{-}\oplus i^{+}];[a^{+}\bullet i^{-}, a^{-}\oplus i^{-}]\rangle. \quad (2)$$

Here • is a triangular norm, ⊕ is a triangular co-norm with an additional axiom:

$$x\bullet y = 1-(1-x)\oplus(1-y);$$

or

$$x\oplus y = 1-(1-x)\bullet(1-y).$$

Several situations are considered:

(i). One of the premise is false or strictly false: $\langle 0;1\rangle$. For a strictly false small premise, the area of possible values of the conclusion truth vector is the square $[0,1]\times[0,1]$. It makes the value $\|b\|$ as blurry and inaccurate as possible. In this sense, the «entropy of truth» becomes maximal.

With a strictly false large premise, the value of the truth of the conclusion falls into the area close to the strict lie:

$$\|b\| = \langle [a^{+}\bullet 0,\ a^{-}\oplus 0];[a^{+}\bullet 1,\ a^{-}\oplus 1]\rangle = \langle [0,\ a^{-}];[a^{+},1]\rangle.$$

This allows us to conclude that the falsity of a small premise makes the truth of the conclusion unknown; the falsity of a large one generates a lie in the conclusion.

(ii). One of the premises is uncertain: $\langle 0;0\rangle$. For a small premise, it turns out:

$$\|b\| = \langle [0\bullet i^{+}, 0\oplus i^{+}];[0\bullet i^{-}, 0\oplus i^{-}]\rangle = \langle [0,i^{+}];[0,i^{-}]\rangle.$$

The uncertainty of a small premise placed the truth vector of the conclusion (somewhere) in the region $[0,i^{+}]\times[0,i^{-}]$. The borders of this area are set by the vector $\|b\|$.

If the implication is uncertainty, we get:

$$\|b\| = \langle [a^{+}\bullet 0, a^{-}\oplus 0];[a^{+}\bullet 0, a^{-}\oplus 0]\rangle = \langle [0,a^{-}];[0,a^{-}]\rangle.$$

In other words, the area of truth of the vector $\|b\|$ is a square with sides $[0,a^{-}]\times[0,a^{-}]$.

If the small premise is strictly true or close to it, the uncertainty of the large premise generates the uncertainty of the conclusion (the vector $\|b\|$ is located near the point $\langle 0;0\rangle$). If it is strictly false, we come to the already known square $[0,1]\times[0,1]$.

Thus, the uncertainty of both small and large premises leads to uncertainty in the conclusion. At the same time, with a high degree of likelihood of a small premise, the uncertainty generated by a large premise is significant.

(iii). One of the premises is contradictory: $\langle 1;1\rangle$. If it is a small premise then:

$$\|b\| = \langle [1\bullet i^{+}, 1\oplus i^{+}];[1\bullet i^{-}, 1\oplus i^{-}]\rangle = \langle [i^{+},1];[i^{-},1]\rangle.$$

is the area $[i^{+},1]\times[i^{-},1]$ – relatively close to $\langle 1;1\rangle$ (more precisely, to $[1,1]\times[0,1]$). If a large premise is inconsistent, that:

$$\|b\| = \langle [a^{+},1];[a^{+},1]\rangle.$$

That is, with a sufficiently reliable small premise, the region close to $\langle 1;1\rangle$ is obtained again. If it is strictly false or close to it – the entire or almost the entire square $[0,1]\times[0,1]$.

The inconsistency of both small and large premises leads to a contradictory conclusion.

The points (i)-(iii) allow automatizing the search for anomalies. Automatizing looks like as a special logical inference at the end of which an artifact appears. The actual location of the artifact is detected by a back tracing of the reasoning chain.

3. Algorithmization of artefact detection

We assume that the knowledge model artefacts are a consequence of the content of the KB. The logical inference shows up the artefacts. Moreover, artefact exists if there is a combination of input data from a valid set that leads to undesirable consequences. These include:

- non-participation of a knowledge unit in the inference for all valid truth values of input assumptions;
- uncertainty of the conclusion for some of the valid values truth of the input facts and the presence of not used anywhere facts;
- inconsistency of the conclusion for some of the valid values truth of the input assumptions.

Verification makes it possible to establish that such situations do not occur.

"Silent" rules (the first case) are detected by counting the number of calls to each rule for all valid input facts. If there were no calls, the product may be excluded from the KB.

The second case is related to the break in the chain of reasoning. For example, if antecedent is false. This leads to the uncertainty of the consequent, which will affect the terminal facts (hypotheses). Or, if you do not break the chain by the condition of insufficient truth of the small premise, to the maximum entropy of the hypothesis's truth: $[0,1]\times[0,1]$.

Finally, a contradiction occurs when one chain produces a statement and the other a denial of the same fact. Because of evidence combining (it is a necessary element of plausible inference) according to the rule:

$$\|b_1 \vee_2 b_2\| = \langle b_1^+ \oplus b_2^+;\ b_1^- \oplus b_2^- \rangle, \tag{3}$$

where $\|b_1\| = \langle b_1^+;b_1^-\rangle$, $\|b_2\| = \langle b_2^+;b_2^-\rangle$ we also get a contradiction in the output. Here $\vee_2$ is the second form of disjunction [20-22] calculated by the formula:

$$\|a \vee_2 b\| = \langle a^+ \oplus b^+;\ a^- \oplus b^- \rangle.$$

If $\|b\| = \langle 1;0\rangle$, the results b and $\neg b$ according to (3) lead to $\|b\| = \langle 1;1\rangle$. It is a complete contradiction. It will also appear in the terminal fact named hypothesis.

The verification algorithm may look like this [24]:

1. Declare that rules are strictly true (by memorizing, if necessary, the truth-values set by the experts).
2. Establish the values truth of starting premises into one of valid combination of the strictly true and strictly false: $\langle 1;0\rangle$ or $\langle 0;1\rangle$.
3. Establish the truth of intermediate and terminal facts in $\langle 0;0\rangle$.
4. Perform a direct inference and obtain the truth of the conclusions (hypotheses).
5. Check the truth of hypotheses for admissibility.
6. If the truth of conclusion is inadmissible (for instance the truth vector is equal to $\langle 0;0\rangle$ or $\langle 1;1\rangle$) then:
   6.1. Perform a back tracing of the inference to find out the source of the problem – the KB artifact.
   6.2. Eliminate the artifact (or add it to the appropriate list).
   6.3. Return to step 2
7. If not all possible input values of premises are checked then:
8. Return to step 2.
9. Return the original truth values of the rules.
10. STOP!

There is a question about the procedure for setting the truth values of the start facts. In the limiting case, N start premises generate $2N$ truth combinations. Even for a small KB containing a few tens or hundreds input facts, it leads to an explosive increase of the number of checks. To solve this problem, it is proposed to use the fact that each hypothesis is based on a relatively small number of start assumptions. Then for each hypothesis, you can select only the facts related to it. Since in real problems n is usually much smaller than N, where n is the number of start facts of the hypothesis, we get a significant reduction in the computational load. The general scheme of the algorithm becomes the following:

1. Declare that rules are strictly true (by memorizing, if necessary, the truth values set by the experts).
2. Highlight a terminal fact (hypothesis).
3. Identify all the associated with hypothesis startup facts.
4. Establish the values of truth of starting facts are into one of valid combination of the strictly true and strictly false values: ⟨1;0⟩ or ⟨0;1⟩.
5. Establish the truth of intermediate and terminal facts in ⟨0;0⟩.
6. Perform a direct inference and obtain the truth of the conclusion (hypothesis).
7. Check the truth of hypothesis for admissibility.
8. If the truth of conclusion is inadmissible (for instance the truth vector is equal to ⟨0;0⟩ or ⟨1;1⟩) then:
   8.1. Perform a back tracing of the inference to find out the source of the problem – the KB artifact.
   8.2. Eliminate the artifact (or add it to the appropriate list).
   8.3. Return to step 4.
9. If not all possible input values of premises are checked, then return to step 4
10. If not all possible hypotheses are checked, then return to step 2.
11. Return the original truth values of the rules.
12. STOP!

4. Conclusion

In this paper, we consider the problem of checking the rule-based KB for internal anomalies in the form of contradictions and breaking of the inference by means of V^{TF}-logic. The advantage of the approach is the ability of such logics to remain operational in conditions of low veracity (uncertainty), falsity and inconsistency of assumptions. At the same time, the uncertainty or contradiction arising during the inference is projected onto a terminal fact (hypothesis), which can serve as a signal of trouble in the KB. This allows you to detect anomalies using the direct inference procedure for all valid combinations of the truth of input premises. To reduce the number of combinations, which may be exponentially large, it is proposed to consider sets of facts for each individual hypothesis, which significantly reduces the amount of calculations.

References

[1] Jiarratano J C and Riley G D 2004 Expert Systems: Principles and Programming, 4th edn. (Course Technology)
[2] Gavrilova T A, Kudryavtsev D V and Muromtsev D I 2016 Knowledge engineering. Models and methods (SpB: Izdatelstvo «Lan'») (in Russian)
[3] Gegechkori E T and Chaadayev A V 2017 Modern problems of developing expert systems in the banking sector when reengineering business processes. *International Journal of Applied and Fundamental Research* **8(2)** 195-197 (in Russian).
[4] *Priyali Sharma. Expert system: meaning, classes and advantages.* Available at: http://www.yourarticlelibrary.com/management/mis-management/expert-system-meaning-classes-and-advantages-mis/70413/
[5] Davies T 1990 Knowledge Bases and Neural Network, in Synthesis in Artificial Intelligence in the Pacific Rim. *Proceeding of the Pacific Rim International Conference on Artificial Intelligence* (Nagoya) 717-722
[6] Golovko V A et al. 2018 Integration of artificial neural network and knowledge bases, *Ontology of designing* **8(3)** 366-386 DOI: 10.18287/2223-9537-2018-8-3-366-386 (in Russian)
[7] Nguyen D Q et al. *A Convolutional Neural Network-based Model for Knowledge Base Completion and its Application to Search Personalization.* Available at: https://pdfs.semanticscholar.org/5b7b/ be73153f383fa96de1218b66602d2ef3dc564.pdf
[8] Socher R, et al. *Reasoning With Neural Tensor Networksfor Knowledge Base Completion.*

Available at: https://nlp.stanford.edu/pubs/SocherChenManningNg_NIPS2013.pdf
[9] Nguen T A at al. 1987 Knowledge Base Verification, *AI Magazine* **8(2)** 69-75
[10] Meseguer P and Preece A 1995 Verification and validation of knowledge-based systems with formal specifications, *The Knowledge Engineering Review* **10(4)** 331-343 DOI: 10.1017/S0269888900007542
[11] Vermesan A A and Coenen F (eds.) 1999 *Validation and Verification of Knowledge Based System* (NY: Springer Science + Business Media) DOI: 10/1007/978-1-4757-6916-6
[12] Bindilatti A and da Silva A 2010 Verification and Validation of Knowledge Bases Using Test Cases Generated by Restriction Rules, *International Journal of Artificial Intelligence and Expert Systems (IJAE)* **3(4)** 117-125
[13] Racunas S A, Shah N H and Fedoroff N V 2006 A case study in pathway knowledgebase verification *BMC Bioinformatics* **7** DOI: 10.1186/1471-2105-7-196
[14] Smirnov V V 2006 Methods and tools for knowledge bases verifying in integrated expert systems *Thesis of the candidate of technical sciences* (Moscow) (in Russian)
[15] Rybina G V and Smirnov V V 2005 Features of static anomalies detecting in knowledge bases containing *certain* types of NON-factors *3rd Int. Sci. and Pract. Seminar Integrated Models and Soft Computing in AI* (Moscow: Fizmatlit) 358-363 (in Russian)
[16] Dolinina O N 2010 Classification of errors in knowledge bases of expert systems *Vestnik of Saratov State Technical University* **2** 125-130 (in Russian)
[17] Pospelova L Ya and Chukanova O V Poisk protivorechiy v produktsionnykh bazakh znaniy (Search for contradictions in rule-based knowledge bases). Available at: *http://www.vmest.ru/nuda/poisk-protivorechij-v-produkcionnih-bazah-znanij/main.html*. (in Russian).
[18] Proskuryakov D P 2014 Search for contradictions using a rule management strategy based on the domain ontology *Proc. Baikal all-Russian conference "Information and mathematical technologies in science and management". Part III.* (Irkutsk: ESI SB RAS) 166–170 (in Russian)
[19] Proskuryakov D P 2015 Conflict resolution managing in rule-based expert systems *Vestnik of Irkutsk State Technical University* **8** 47-51 (in Russian)
[20] Arshinskiy L V 1998 *Methods of processing of nonstrict propositions* (Irkutsk: East-Siberian Institute of MIA of Russia) 40 (in Russian)
[21] Arshinskii L V 2007 Substantial and Formal Deductions in Logics with Vector Semantics, *Automation and Remote Control* **68(1)** 139-148
[22] Arshinskiy L V et al. 2017 Modeling of Complex Subject Areas Based on the Logics Vector Semantics *Far East Journal of Mathematical Sciences (FJMS)* **101(4)** 813-823
[23] Arshinskiy L V, Ermakov A A and Nitezhuk M S 2019 Logic with Vector Semantic as a Means of Knowledge Bases Verification *Ontology of designing* **9(4)** 510-521 DOI: 10.18287/2223-9537-2019-9-4-510-521 (in Russian)

IOP Conf. Series: Materials Science and Engineering 1151 (2021) 012028 doi:10.1088/1757-899X/1151/1/012028

Potential of the railways of the multimodal transport network of the Eastern Siberia and Far East of Russia

V A Anisimov[1], L A Bogdanova[1], O S Morozova[1], S V Shkurnikov[1] and N S Nesterova[2]

[1] Emperor Alexander I St. Petersburg State Transport University, Moskovsky pr., 9, St. Petersburg, 190031, Russia
[2]Far Eastern State Transport University, Serysheva st., 47, Khabarovsk, 680021, Russia

E-mail:Anisvl@mail.ru

Abstract. The article notes the urgency of the problem of the formation and development of the regional multimodal transport network of Eastern Siberia and the Far East of Russia as a set of multimodal transport corridors. The enlarged functional structure of the design methodology for the integrated development of a multimodal transport network is presented. Using this methodology, the authors formulated the problem of designing the development of a regional multimodal transport network in Eastern Siberia and the Far East of Russia, gave a description of the proposed version of its appearance and the possible throughput of existing and new railways in the region under study. The options for the possible carrying capacity of the railway links of the network presented in the article represent an assessment of the potential of multimodal transport corridors, which is used to determine the options for the required traffic volumes along the network facilities.

1. Introduction

The Transport Strategy of the Russian Federation for the period up to 2030 (enacted by Russian Federation government decree No. 1734-r on November 22, 2008) (hereinafter Strategy) defined the main task of the state in the transport industry, which should provide access to safe and high-quality transport services and turn the geographical features of Russia into its competitive advantage to increase the competitiveness of the national economy and the quality of life of the population.

A special place in the Strategy is occupied by Eastern Siberia and Far East of Russia.

These regions are located in the Asian part of Russia. From the west they are bounded by the Yenisei, from the north by the Arctic Ocean, from the east by the Pacific Ocean, in the south by the state border of Russia with China, Mongolia and North Korea. The territory of Eastern Siberia and the Far East has the richest reserves of natural resources of global importance and makes up more than half of Russia. The main transport routes of Russia pass through the following regions: the Trans-Siberian mainline (hereinafter – the Transsib), the Baikal-Amur Mainline (hereinafter - the BAM) and the Northern Sea Route (hereinafter – the NSR).

At the same time, Eastern Siberia and the Far East have an insufficient level of accessibility of transport services for the economy and the population, which is one of the main reasons hindering the development of these regions. Therefore, the Strategy sets the task of the priority development of the

Content from this work may be used under the terms of the Creative Commons Attribution 3.0 licence. Any further distribution of this work must maintain attribution to the author(s) and the title of the work, journal citation and DOI.
Published under licence by IOP Publishing Ltd

IOP Conf. Series: Materials Science and Engineering 1151 (2021) 012028 doi:10.1088/1757-899X/1151/1/012028

transport infrastructure of Eastern Siberia and the Far East of Russia, primarily the Trans-Siberian and Baikal-Amur Mainlines, the Northern Sea Route, seaports, border crossings and approaches to them, which should ensure the development of production for production, processing and transportation of natural resources and implementation of the country's transit potential.

2. Materials and methods

The analysis of the works of scientists and specialists in the transport industry [1-20 and others], given in [21-25], showed the particular relevance of the problem of developing transport infrastructure to ensure multimodal transportation along international transport corridors and confirmed the need to develop a design methodology for the integrated development of a multimodal transport network (MTN) [21, 22] (hereinafter the methodology).

The methodology is a scientifically grounded tool for solving the strategic tasks of transport development in Russia, set in the Strategy. These tasks are very difficult due to their large size. Their solution is also significantly complicated by the unstable economic and international geopolitical situation.

In this regard, it can be concluded that there is a need for scientific and methodological support for the implementation of the tasks of the Strategy.

To reduce the dimension of Strategy tasks, the methodology proposes to form a multimodal transport network from objects of a single transport network (ETN) as a set of multimodal transport corridors (MTK).

The methodology was developed and outlined in [21-25]. Its enlarged functional structure is represented by the following stages:

1. The formulation problem of design the development of the MTN.
2. Formation of the MTN shape variants.
3. Formation a set of possible strategies for stage-by-stage changes in the appearance and power of the MTN objects.
4. Formation of an area of effective strategies for stage-by-stage changes in the appearance and power of the MTN.
5. Choosing the optimal solution from the field of effective strategies, taking into account the accepted restrictions.
6. Support for the implementation of the adopted strategy.

The authors applied the methodology to solve the problem of the formation and development of a regional multimodal transport network in Eastern Siberia and Far East of Russia (hereinafter - RMTN). The authors' attention to this region is due to the fact that:

– Siberia and Far East of Russia is considered in the Strategy as a territory of large-scale projects focused on the production, processing and transportation of the richest reserves of natural resources of global importance, the main share of which is hydrocarbons;

– main competitive advantages of the Russian transport system are: i) the shortest land routes between the countries of the Asia-Pacific region and Europe; ii) the presence of an extended sea coast.

These factors determined the formulation of the problem of designing the development of the RMTN (the first stage of the methodology):

We form an RMTN and define effective strategies for a stage-by-stage changes in its appearance and power, ensuring:

1. advance development of transport infrastructure in the areas of export deliveries of goods, primarily the development of seaports and approaches to them;
2. realization of the country's transit potential.

At the second stage of the methodology, the RMTN shape variant was formed according to the set-theoretic representation in [21]:

$$MTN = \{MTK, E, C, H\}, \quad (1)$$

where MTK is the set of multimodal transport corridors;

$$MTK = \{MTK_k\},\ k = 1,2,\ldots,n_{mtk}, \tag{2}$$

where E is the set of technological processes and events that occur on the MTN and affect its technical condition; C is the set of criteria that characterize the goals of MTN development in accordance with the scenario of socio-economic development of a country or region; H is the set of relations between MTK, E and C.

The model of the multimodal transport corridor in [21] is represented by the set of subsets:

$$MTK_k = \{MTH_k, ZV_k\},\ k = 1,2,\ldots,n_{mtk}, \tag{3}$$

where MTH_k is the subset of multimodal transport hub included in MTK_k, $MTH_k \in MTH$; MTH is the set of multimodal transport hub MTN; ZV_k is the subset of transport links included in MTK_k, $ZV_k \in ZV$; ZV is the set of unimodal and multimodal transport links MTN.

In [21, 27-34], the possible directions of MTK of the studied region were considered, the points of arose and terminated traffic flow and the territorial location of MTK were indicated, and the analysis of external and internal factors affecting the formation of RMTN shape variants was given. Using this information and the methods of forming the MTN shape [21], the authors obtained a variant of the RMTN shape as a set of the following MTK (Figure 1).

Multimodal transport corridors with access to the Trans-Siberian mainline:

1. MTK_1: Vladivostok multimodal transport hub (VMTH) – Transsib – Tayshet.
2. MTK_2: Ulan-Bator (Mongolia) – Naushki – Ulan-Ude – Transsib.
3. MTK_3: Manchuria (China) – Zabaikalsk– Karymskaya – Transsib.
4. MTK_4: Heihe (China) – Blagoveshchensk – Belogorsk – Transsib.
5. MTK_5: Tongjiang (China) – Nizhneleninskoe – Birobidzhan – Transsib.
6. MTK_6: Tumangan (NorthKorea) – Hasan (Russia) – Baranovsky – Transsib.

Multimodal transport corridors with access to the Baikal – Amur Mainline:

7. MTK_7: Vanino-Sovgavan multimodal transport hub (VSMTH) – Komsomolsk – BAM – Tayshet.
8. MTK_8: Hokkaido (Japan) – Sakhalin multimodal transport hub (SMTH) – Nysh – Selikhin – Komsomolsk – BAM – Tayshet.
9. MTK_9: VMTH – Selikhin – Komsomolsk – BAM – Tayshet.

Multimodal transport corridors from China with access to the seaports of the Far East and the Arctic zone of Russia:

10. MTK_{10}: Suifenhe (China) – Grodekovo – VMTH (MTK «Primorye-1»).
11. MTK_{11}: Hunchun (China) – Kamyshovaya – VMTH (MTK «Primorye 2»)
12. MTK_{12}: Mohe (China) – Reinovo – Skovorodino – Tynda – Yakutsk – seaports Tiksi – NSW.
13. MTK_{13}: Fuyuan (China) – Khabarovsk multimodal transport hub (HMTH) – seaports of Samarga.
14. MTK_{14}: Fuyuan (China) – HMTH – seaports of Nikolaevsk on Amur.
15. MTK_{15}: Mishan (China) – Tugiy Rog – Sibirtsevo – Novochuguevka – seaports Olga.

This version of the RMTN shape of the Russian Eastern Siberia and Far East allows us to combine the growth of production, processing and transportation of natural resources of global significance to seaports in the studied region and the realization of the country's transit potential.

According to the methodology, the next stage of designing an integrated RMTN development is the formation of a set of possible strategies for step-by-step changes in the shape and capacity of RMTN's objects [21-23]. To do this, for the three scenario variants of the Strategy, it is necessary to assign estimated cases in accordance with [21]. In [24], a mathematical model for the formation of estimated cases is proposed. It allows you to take into account the influence of factors that generate uncertainty in the initial information, and determine possible variants for the required volumes of traffic on multimodal transport hubs of MTY_k and links ZV_k that are part of MTK_k.

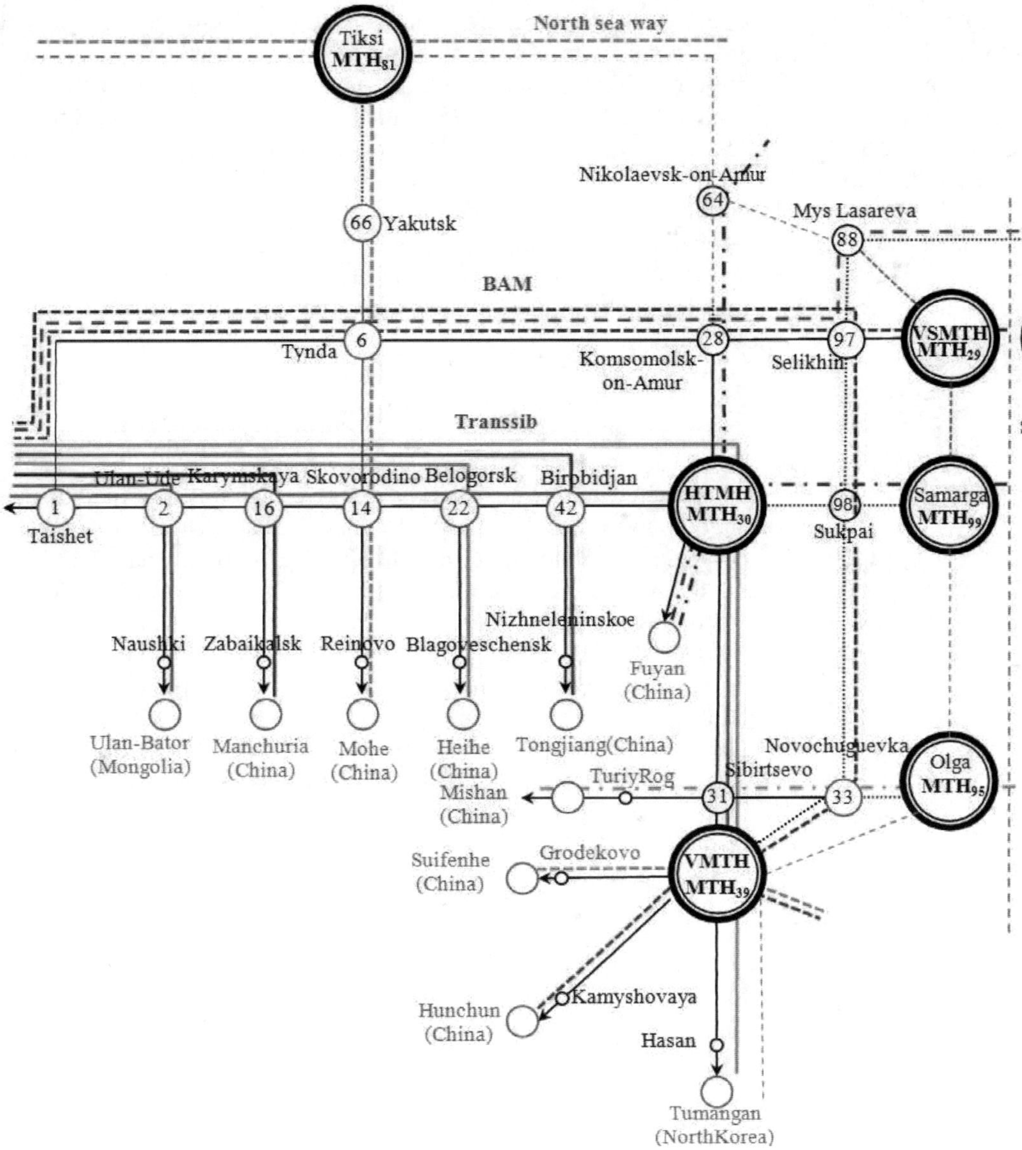

Figure 1. The design scheme of the shape of RMTN of the Far East and the Arctic zone of Russia

Each estimated case includes:

1. RMTN shape scheme that combines MTY_k and ZV_k of various modes of transport into MTK_k.
2. Existing and prospective variant for loading RMTN's objects by years of the estimated period and directions of movement without taking into account traffic flows through the multimodal transport corridors.
3. Variant of required traffic volumes for each MTK_k by years of the estimated period and directions of movement.
4. Possible volume of attracted investments and tariffs for transport services by year of the estimated period.

Variants of the required traffic volumes through the MTK_k are dependent on scenarios of socio-economic development of the country, studied region and bordering countries, load of RMTN's objects and potential of RMTN's multimodal transport corridors.

The Strategy has developed three scenarios: inertia, energy and innovation, which take into account the probabilistic scenarios of socio-economic development of the Federal districts of Russia, the development of interregional economic relations and possible changes in the conjuncture of commodity and transport markets. In all scenarios, the tasks of advancing the development of transport infrastructure in the areas of export deliveries of goods and the implementation of the country's transit potential are set.

The Institute for Transport Economics and Development, based on three scenarios, evaluated the prospective volumes of cargo transportation on the Railways of the Siberian and far Eastern Federal districts of Russia and determined the load of objects of RMTN that the authors are studied.

According to the mathematical model for generating estimated cases for determining the variants of required volumes of traffic through MTK_k, the potential of RMTN's multimodal transport corridors should be evaluated.

3. Results

The potential of *MTK* of RMTN shape variant, proposed by the authors, is limited by the possible capacity of the: Railways of Russuan Eastern Siberia and Far East (the two main railway mainlines – Transsib and BAM), railways to seaports and to land border crossings. The capacity of railways is characterized by capacity and carrying capacity and is determined depending on their main technical parameters: the number of main paths; the type of traction; the type of train schedule; the useful length of the receiving-sending station tracks; the weight of the train.

Railway RMTN's links can be divided into four groups: 1) sections of the Transsib; 1) sections of the BAM and sections connecting Transsib and BAM; 3) sections of railway to seaports and to land border crossings; 4) new railway sections.

The links of the first group ($ZV_{1\text{-}2}$, $ZV_{2\text{-}16}$, $ZV_{16\text{-}14}$, $ZV_{14\text{-}22}$, $ZV_{22\text{-}42}$, $ZV_{42\text{-}30}$, $ZV_{30\text{-}31}$, $ZV_{31\text{-}39}$) are double-track electrified sections with a possible capacity of 132 pairs of trains per day.

The second group includes the link $ZV_{1\text{-}6}$ that includes a double-track electrified section and single-track sections with electric and diesel traction. The remaining links of the second group ($ZV_{6\text{-}28}$, $ZV_{28\text{-}97}$, $ZV_{97\text{-}29}$, $ZV_{14\text{-}6}$, $ZV_{30\text{-}28}$) are single-track sections with diesel traction. In the future, all links of the second group can be strengthened to double-track electrified railway lines with a possible capacity of up to 132 pairs of trains per day.

In the third group, links ($ZV_{2\text{-Na}}$, $ZV_{16\text{-Za}}$, $ZV_{14\text{-Re}}$, $ZV_{22\text{-Bl}}$, $ZV_{42\text{-Ni}}$,$ZV_{31\text{-TR}}$, $ZV_{39\text{-Ha}}$) connect land border crossings to the Transsib, links $ZV_{39\text{-Gr}}$ and $ZV_{39\text{-Ka}}$ link land border crossings to the *VMTH*, and link $ZV_{31\text{-}33}$ is a continuation of the MTK_{15}. The technical condition of links $ZV_{16\text{-Za}}$ and $ZV_{39\text{-Ka}}$ can be strengthened to a double-track electrified section with a possible capacity of up to 132 pairs of trains per day. The remaining links can be strengthened to single-track electrified sections with a possible capacity depending on the railway line length: 10 km – up to 50 pairs of trains per day; 20 km – up to 40 pairs of trains per day. Link $ZV_{6\text{-}66}$ is a single-track section on autonomous traction with a possible capacity depending on the railway line length: 20 km – up to 28 pairs of trains per day; 30 km – up to 21 pairs of trains per day. Link $ZV_{6\text{-}66}$ is an element of corridor MTK_{12}, the continuation of which will provide access from the Mohe-Reinovo land border crossing to the Tiksi seaport of the Northern sea route.

The new links of the fourth group ($ZV_{66\text{-}81}$, $ZV_{97\text{-}88}$, $ZV_{88\text{-Ny}}$, $ZV_{30\text{-}98}$, $ZV_{98\text{-}99}$,$ZV_{98\text{-}97}$, $ZV_{39\text{-}33}$, $ZV_{33\text{-}98}$, $ZV_{33\text{-}95}$) will mainly pass through undeveloped territory, so it is advisable to consider them as single-track railway sections on autonomous traction with a possible capacity for the railway line length of the stages up to: 20 km – 28 pairs of trains per day; 30 km – 21 pairs of trains per day.

The options considered by the authors for the possible capacity of RMTN railways are summarized in table 1. They represent an assessment of the potential of the RMTN. Without this assessment, it is impossible to determine the variants of required traffic volumes through MTK_k for different scenarios of socio-economic development and load of RMTN's objects. Also without this assessment, it is impossible to assign the estimated cases for the formation of a set of possible strategies for stage-by-stage changes of shape and capacity of the RMTN's objects.

Table 1. Potential of the railways of the multimodal transport network of the Eastern Siberia and Far East of Russia

The links	The main technical parameters	Length of the limiting stages, km	A possible capacity, pairs of trains per day
$ZV_{1\text{-}2}$, $ZV_{2\text{-}16}$, $ZV_{16\text{-}14}$, $ZV_{14\text{-}22}$, $ZV_{22\text{-}42}$, $ZV_{42\text{-}30}$, $ZV_{30\text{-}31}$, $ZV_{31\text{-}39}$, $ZV_{1\text{-}6}$, $ZV_{6\text{-}28}$, $ZV_{28\text{-}97}$, $ZV_{97\text{-}29}$, $ZV_{14\text{-}6}$, $ZV_{30\text{-}28}$, $ZV_{16\text{-}Za}$, $ZV_{39\text{-}Ka}$	2 main paths; electric traction; inter-train interval of 8 min	-	132
$ZV_{2\text{-}Na}$, $ZV_{14\text{-}Re}$, $ZV_{22\text{-}Bl}$, $ZV_{42\text{-}Ni}$, $ZV_{31\text{-}TR}$, $ZV_{39\text{-}Ha}$, $ZV_{39\text{-}Gr}$	1 main path; electric traction; inter-train interval of 8 min	10 20	50 40
$ZV_{6\text{-}66}$, $ZV_{66\text{-}81}$, $ZV_{97\text{-}88}$, $ZV_{88\text{-}Ny}$, $ZV_{30\text{-}98}$, $ZV_{98\text{-}99}$, $ZV_{98\text{-}97}$, $ZV_{39\text{-}33}$, $ZV_{33\text{-}98}$, $ZV_{33\text{-}95}$	1 main path; autonomous traction; inter-train interval of 10 min	20 30	28 21

4. Conclusion

To form a set of possible strategies for staged changes in the appearance and capacity of objects of a multimodal transport network, it is necessary to know their required capacity by years of the billing period [23]. For this purpose, based on the analysis of the scenarios of the socio-economic development of the country, the region under study and the neighboring states, the load of MTN facilities and the *MTK* potential for each estimated scheme of the appearance MTN, the options for the required freight and passenger traffic according to MTK_k are determined and estimated cases are assigned [24].

Further, using the mathematical model of estimated cases, the required traffic volumes for multimodal transport hubs MTY_k and links ZV_k and the required capacity of MTN objects (railway stations, sections, port terminals, transshipment points from one type of transport to another, etc.) are calculated by the years of the calculation period [21, 24]. Comparing the required capacity of objects with the possible capacity, we assign a set of possible measures to change their technical condition and form possible strategies for stage-by-stage changes in their appearance and capacity [22, 23].

References

[1] Sazonov S, Kudryavtsev E, Wu Zi 2012 *Far Eastern Affairs* **2** 47-58
[2] Korolyova E 2000 *Organizational-Economic Bases of Functioning of Transport Corridors Dr Sc Thesis*
[3] Morozov V 2010 *Methodology of organization of functioning of international transportation corridors on the basis of the cluster approach with the use of multimodal logistics centers Dr Sc Thesis*
[4] Droździel P, Buková B, Brumerčíková E 2015 *Transport Problems* **10 (4)** 5-13
[5] Goncharenko E 2015 *The Russian Sections of International Transportation Corridors as an object of economic research PhD Thesis*
[6] Tchumlyakov K, Tchumlyakova D 2015 *Bulletin of Transport Information* **11 (245)** 8-13
[7] Macheret D 2015 Soiskatel – *Supplement to the World of Transport Journal* **1 (9)** 58-61
[8] Kovalev S 2015 Soiskatel – *Supplement to the World of Transport Journal* **1 (9)** 62-67
[9] Shapoval D 2016 *Technique and Technology of Construction* **1 (5)** 8
[10] Bykov Y, Vasilyev A, Fadeeva V 2015 *Transport of the Urals* **1 (44)** 53-56
[11] Zhang J, Liao F, Arentze T, Timmermansa H 2011 *Procedia Computer Science* **5** 912-919

[12] Zhang J, Liao F, Arentze T, Timmermansa H 2011 *ProcediaSocial and Behavioral Sciences* **20** pp 313–322
[13] Regmi M, Hanaoka S 2012 *Research in Transportation Business & Management* **5** 27-37
[14] Reis V, Meier J, Pace G, Palacin R 2013 *Research in Transportation Economics* **41 (1)** 17-30
[15] Droździel P, Buková B, Brumerčíková E 2015 *Transport Problems* **10 (4)** 5-13
[16] Wang Y, Yeo G-T 2016 *The Asian Journal of Shipping and Logistics* **32 (1)** 41-47
[17] Dib O, Manier M-A, Moalic L, Caminada A 2017 *Transportation Research Procedia* **20** 134143
[18] Oberg M, Nilsson K, Johansson C 2017 *Transportation Research Procedia* **25** 3694-3702
[19] Grigoriev Y 2019 *IOP Conf. Ser.: Earth Environ. Sci.* **403** 012229
[20] Stoilova S, Munier N, Kendra M, Skrúcaný T 2020 *Sustainability*, **12(4)** 1482
[21] Nesterova N 2017 *Multimodal Transportation Network Designing Methodology DrSc Thesis*
[22] Nesterova N, Goncharuk S, Anisimov V, Anisimov A, Shvartcfeld V 2016 *Procedia Engineering* **165** 1547-1555
[23] Nesterova N, Anisimov V 2019 *Izvestiya PGUPS* **16 (3)** 329-338
[24] Nesterova N, Anisimov V, Goncharuk S, Anisimov A 2018 *Advances in Intelligent Systems and Computing* **692** 177-183
[25] Nesterova N, Anisimov V 2019 *IOP Conf. Ser.: Earth Environ. Sci.* **403** 012204
[26] Nesterova N, Anisimov V 2019 *IOP Conf. Ser.: Earth Environ. Sci.* **403** 012235
[27] Cvetkov V, Zoidov K, Medkov A 2013 *On the possibility and expediency of organizing cargo transit through Russia between the countries of the Pacific region and Europe* (Deponent Socionet)
[28] Cvetkov V, Zoidov K, Medkov A 2013 *Formation of a new evolutionary model of transport and communication interaction between Russia and China* (Moscow: CEMI RAN) 231
[29] Sharapov S, Novoselcev E, Holosha M 2018 *The Railway Transport Magazine* **11** 54–62
[30] Sharapov S, Novoselcev E, Holosha M 2018 Seaports **8 (169)** 20–23
[31] Sharapov S, Novoselcev E, Holosha M 2018 Seaports **9 (170)** 20–23
[32] Kurenkov P, Vakulenko S, Goncharenko S, Safronova A, Volodin A 2019 *Science Journal of Transportation* **9** 74-82
[33] Kurenkov P, Rudakova E, Pereslegina T, Matyash A 2019 *Transport Services Market (Problems of Increasing Efficiency)* **12** 96-103
[34] Arkhipov A, Grigoriev E, Sinitsyn M 2020 *E3S Web Conf.* **164** 11020

A method that improves the safety states of technical means of control and management

S Gromyshova[1] and V Gozbenko[1, 2]

[1] Irkutsk State Transport University, bld. 15, Chernyshevskogo str., Irkutsk, 664074, Russia
[2]Angarsk State Technical University, 60, Chaykovskogo str., Angarsk, 665835, Russia

E-Mail: ghromyshova7997mail.ru

Abstract. The main focal points of improving the traffic safety management system are as follows. First, one must coordinate the interaction among all functional branches, their structural divisions and subsidiaries involved in the transportation process, based on common approaches to managing processes related to traffic safety. Second, one should implement the approaches and requirements that are congruent with international standards; to maintain a positive level of traffic safety culture through awareness of the importance and social responsibility of the holding employees when performing work that affects traffic safety, with simultaneous integration of mandatory and strict rules of conduct in all production processes. Third, one is to improve the regulatory framework. Train traffic safety is ensured at all stages of the life cycle of rolling stock, traffic support systems and infrastructure elements. Nevertheless, failures and malfunctions are detected during the operation. This article presents one of the methods for increasing the train traffic safety in the automation and telemechanics system.

1. Introduction

Improving the railway transport safety is one of the most important goals of the reform, included in the structural adjustment program approved by the Government of the Russian Federation [1]. The strategic goal of the Russian Railways OAO is to achieve a traffic safety level corresponding to the status of a holding that takes a leading position among the world's railways in terms of the safety of the provided infrastructure services, security and reliability of all processes. To achieve this goal is to form a qualitatively new image of the holding by the middle of the next decade. In order to head over to a higher level of development in relations with the state, society, partners and employees of the holding in the field of traffic safety, the following goals have been determined: ensuring people's health and life safety; ensuring a given level of traffic safety that meets international and national requirements; ensuring cargo, rolling stock and infrastructure safety; minimization of the consequences of traffic accidents. Train traffic safety is ensured at all stages of the life cycle of rolling stock, traffic support systems and infrastructure elements. Nevertheless, failures occur and malfunctions can be detected during the operation [2].

Today, great attention is paid to ensuring a high level of traffic safety [1-3] and operational reliability. Transport accidents lead to large technical and material losses. The consequence of transport accidents is goods and equipment damage, destruction of tracks and contact networks,

Content from this work may be used under the terms of the Creative Commons Attribution 3.0 licence. Any further distribution of this work must maintain attribution to the author(s) and the title of the work, journal citation and DOI.
Published under licence by IOP Publishing Ltd

cessation of train traffic on the site, which leads to large economic and human losses. This risk gave impetus to the development and implementation of rolling stock control means, namely, the use of automated diagnostic systems [3].

The rolling stock control system performs such types of control as: control of the axle box unit temperature and wheel braking, drag control, control of the overall dimensions of cars, control of wheel defects along the rolling circle, control of the geometric parameters of the wheel, control of the wheel parameters, control of the parameters of the shock-traction mechanism, control of cars loading unevenness, control of axle box sliding from the axle neck.

The placement of such control devices on the tracks or approaches to the maintenance points of freight cars allows, even before the train arrives, to obtain reliable information about the nature of the arising defect, its location, and the process of its development. All this makes it possible to significantly improve the quality of maintenance and repair of rolling stock, and, consequently, the safety of train traffic.

The object of the research is "Multifunctional Technical Means Complex".

2. The multifunctional system of technical means (henceforth referred as MSTM)

It is designed to control the heating of the axle-box unit while the train is in motion with the formation of alarm indications with the levels "Alarm 0" (requires no train stop), "Alarm 1" (requires a train stop at the station) and "Alarm 2 "(requires a train stop at the station-to-station block). All devices are integrated into the automated control system of the rolling stock ACS RS, which allows monitoring the dynamics of heating of the axle-boxes and, when a predetermined threshold is exceeded, generating a signal "Alarm 0 (dynamics)". This is especially important at the sections of non-stop train operation, since Russian Railways OAO has established a procedure for presenting cars, which had alarm readings with the level "Alarm 0" along the route, for inspection at the maintenance depot [3,4].

As of January 01, 2019, 4802 control points are in operation. 5888 MSTM devices are located at these points, including 4611 MSTM-02 devices and 190 MSTM-03 devices. The average distance between the devices in the main and heavy traffic directions is 21.5 km [3].

In 2002 the multifunctional complex of technical means KTSM-02 was adopted. KTSM-02 is focused on automatic non-contact detection of: brake malfunctions in cars and locomotives, as well as overheating of axle boxes of wheelsets. Microprocessor-based diagnostic tools allow increasing the level of traffic safety through the introduction and accurate determination of the technical condition of the rolling stock In this case, the resource costs for maintenance of the rolling stock by employees are minimized, the quality of information equipment increases, and the labor supply of the operating staff increases [4].

The complex for monitoring KTSM-02 of the technical condition of the rolling stock while the train is in motion allows:

1 to minimize train delays, the need for spare parts associated with the operating costs of repair and maintenance;

2 to improve the technical condition of the rolling stock and the repair mechanism associated with failures of expensive scheduled preventive maintenance of rolling stock, from the periodic distraction of rolling stock from work on mileage and time;

3 to determine the technical condition of the rolling stock, as a result of which there is an increase in the reduction of resource consumption for maintenance of the rolling stock, the working conditions of workers are increased, the quality of service and operation is increasing.

The increase in the introduction of microprocessor-based diagnostics allows increasing the level of traffic safety.

Throughout the railway network, the largest number of failures (66% of the total) is accounted for by failures of the floor-mounted MSTM camera [5]. As the analysis of the operation of the floor-mounted equipment has shown, the weakest points of the floor-mounted MSTM camera are: the receiving capsule, the bolometer, and the sensors [5].

In the area of the floor-mounted cameras, the ballast layer is applied exclusively by hand. When heavy trains are in motion, pressure is exerted on the cameras, the fastening of the rail to the reinforced concrete sleepers weakens, the sleepers shift and clamp the floor cameras, which leads to wear of the shock absorbers, beating against the camera, i.e. mechanical stress occurs, which results in false readings of the hot boxes of the device. Tamping sleepers or installing an anti-theft system in these cases do not lead to the desired result. It is proposed to use metal plates 4 (Figure 2) on reinforced concrete sleepers 2 (Figure 2), used for the installation of insulated joints. One plate is solid; the second one is welded from two plates. To prevent the sleepers from moving, they are fastened in pairs on both sides with plates, by means of standard bolted joints that fasten rails 1 (Figure 2) to sleepers 2 (Figure 2). Installation in pairs of blocks 4 (Figure 1) (6-4 mm) on wooden sleepers 2 (Figure 1), fixed with self-tapping screws (70mm).

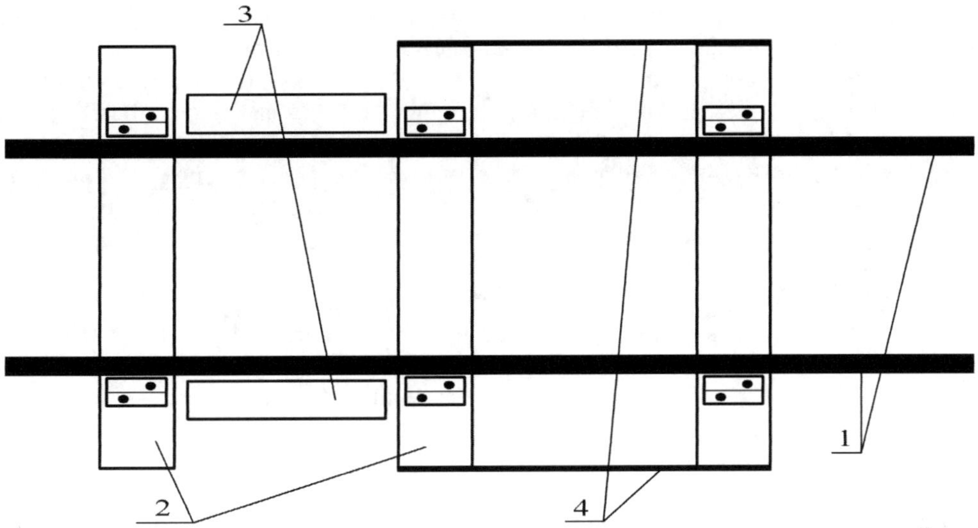

Figure 1. 1 - rails, 2 - wooden sleepers, 3 - floor-mounted cameras, 4 - wooden blocks.

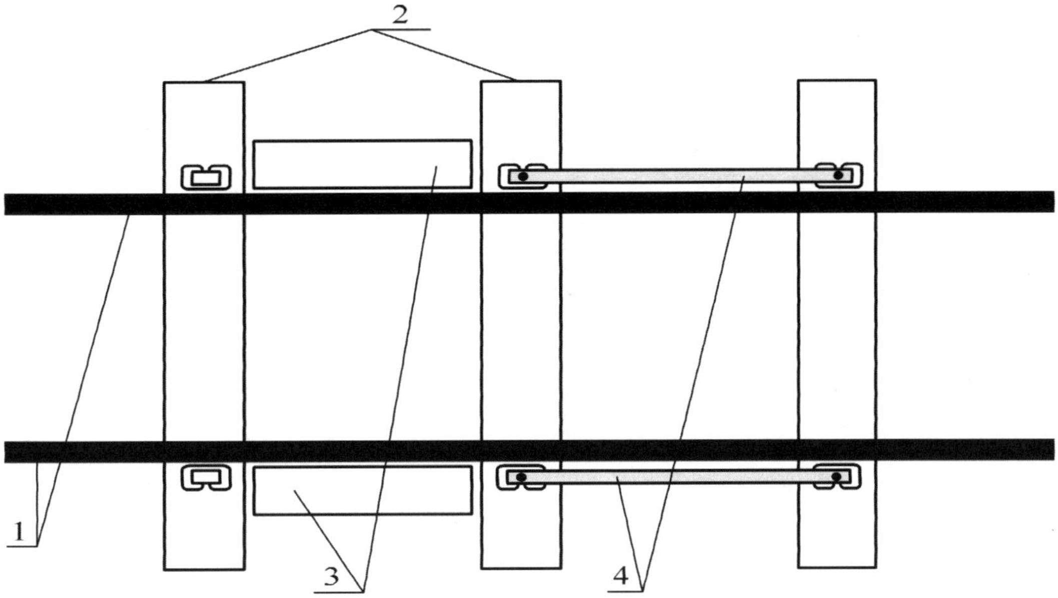

Figure 2. 1 - rails, 2 - reinforced concrete sleepers, 3 - floor-mounted cameras, 4 – metal plates.

3. Conclusion

Train traffic safety is the main condition for the operation of the railway, for the transport of passengers and goods. All organizational and technical measures in railway transport must meet the requirements for safe and uninterrupted train traffic. Traffic safety is ensured by the maintenance of all railway structures, tracks, rolling stock, equipment and mechanisms, signaling and communication devices in constant working order. Of no less importance in ensuring safety is the activities of railway personnel directly involved in the implementation of train traffic (machinists, station attendants, etc.). Not only accurate implementation, but, most importantly, the safety and reliability of the entire transportation process depends on their professional preparedness, experience, ability to quickly navigate and make the right decisions in difficult situations. Automation in railway transport, to some extent, will help to reduce the threshold of violations in the safety of train traffic and, automated enforcement of the technology of work in transport by collecting, processing and analyzing relevant information, both from individual rail transport enterprises and from the transportation management system in in general, it is the main purpose of the Multilevel Rail Transport Management System.

While improving the equipment, advanced technologies are being introduced, thereby reducing the number of unreliable operations on the control device. The commissioning of microprocessor-based diagnostic tools makes it possible to increase the level of traffic safety through accurate and modern diagnostics of the condition of the rolling stock. As a result, there is a reduction in resource costs for maintenance of the rolling stock. The quality of information provision and working conditions for workers are also improved.

With the expansion of the use of the multifunctional technical means complex, it is possible to reduce train delays, operating costs for maintenance and repair, and the need for spare parts. Monitoring the technical condition of the rolling stock will make it possible to organize a rolling stock repair and maintenance system based on the actual state of critical components. This will make it possible to dismiss an expensive scheduled preventive maintenance system with periodic distraction of the rolling stock from operation by mileage or time.

The reliability of the device (Figure 1, Figure 2) has been checked for a month. This design protects the cameras from mechanical stress, which is directly related to the safety of train traffic.

This construction is used at the Zilovskaya division of signaling and interlocking on the Trans-Baikal railway.

References

[1] *Russian Railways OAO: the official portal*, available at: http://www.rzd.ru

[2] Maryukhnenko V S and Pultyakov A V 2016 Features of monitoring the technical condition of rolling stock while the train is in motion *Automation in transport* **2(2)** 272-287

[3] 2019 *The concept for the development of diagnostic tools for rolling stock when the train is in motion.*

[4] The operating manual IN7.354000 2002 *Floor-mounted camera control unit "BUNK"* (Ekaterinburg: SPC INFOTEX Publ)

[5] Gromyshova S S 2019 Operation of devices of a multifunctional system of technical means in complex-structured transport systems as a factor in ensuring train traffic safety and operation reliability *Modern technologies. System Analysis. Modeling* **4(64)** 167–173

[6] Efanov D V 2016 Formation and prospects of systems of functional control and monitoring of devices of railway automation and telemechanics *Automation in transport* **1(1)** 124–148

[7] Shamanov V I 2016 Methods for optimization of maintenance of automation systems *Automation in transport* **2(4)** 481–496

Modelling the efficiency of infrastructure repair based on the risk process

Y M Krakovsky[1], N A Hoang[2] and Ya M Ivanyo[2]

[1]Irkutsk State Transport University, 15, Chernyshevskogo str., Irkutsk, 664074, Russia
[2]Irkutsk State Agrarian University named after A.A. Ezhevsky, Molodezhny settlement, Irkutsk, 664038, Russia

E-mail : 79149267772@yandex.ru

Abstract. To improve the efficiency of railway infrastructure repairs, it is proposed to introduce an insurance fund, which performs two functions: it accumulates payments with different frequency and cost for performing different types of repair work; it pays for these works as necessary. To mathematically describe the state of the insurance fund, it is suggested that the risk process of a special kind be used. This approach made it possible to introduce indicators of repairs in the form of resource-cost and financial risks, considering the possibility of non-performance of repairs due to lack of funds. To study these indicators, a modeling program has been created in the MATLAB environment, based on the event-driven approach. Experiments with the modelling program led to the conclusion that, in terms of a minimum_of risk values, preference should be given to the option when the shares and frequency of payments should depend on the type of repair work and initial data.

1. Introduction

The purpose of the maintenance and repair system is to manage the technical condition of the equipment within its resource, allowing one to: a) ensure that the equipment is ready for use as intended and operational during operation; b) minimize time, labor and equipment cost for repairs. The importance of introducing mathematical methods in improving the efficiency of repairs of various equipment, as well as in the creation of indicators for evaluating performance management systems, is confirmed by publications [1-3].

Maintenance and repair of railway infrastructure include operation and maintenance of the track, contact network and the infrastructure complex as a whole [3, 4]. Three types of repairs can be identified: 1) current, 2) emergency, 3) major. To make this work more effective, considering the developed system of diagnostics and monitoring [5], they need to be performed according to the actual condition. Repairs by actual condition take into account an important factor of freight transport - functioning under uncertainty and risk [6].

To improve the efficiency of repairs and their evaluation, it is planned to introduce an insurance fund, which performs two functions: 1) accumulation of payments with different frequency and cost; 2) paying for these works, as necessary. At the same time, taking into account the maintenance of the infrastructure according to the actual state in the conditions of uncertainty of freight transport, the time intervals between the repair works (day) and the cost of these works (million rubles) can be considered as random variables with known distribution functions up to the values of their parameters. This approach makes it possible to mathematically describe the state of the insurance fund when

Content from this work may be used under the terms of the Creative Commons Attribution 3.0 licence. Any further distribution of this work must maintain attribution to the author(s) and the title of the work, journal citation and DOI.
Published under licence by IOP Publishing Ltd

modeling the effectiveness of infrastructure repair works, using a special type of risk process, which is used in insurance mathematics [7], as well as when assessing the effectiveness of complex equipment repair works [8].

2. Mathematical description of performance indicators based on risk process

Considering that the work investigates three types of repair work, we define the risk process as follows:

$$R(t) = X_0 + Y_1(t) + Y_2(t) + Y_3(t) - YT(t) - YA(t) - YK(t), \tag{1}$$

where X_0 is the initial funds of the insurance fund; $Yj(t)$ is the total accumulation of payments by type of work, ($j = 1,2,3$); $YT(t)$ is the total cost for current work; $YA(t)$ is the total cost for emergency situations; $YK(t)$ is the total cost for major work.

When servicing the railway infrastructure during its operation, the amount of payments for various types of repairs for a year is initially formed (X , million rubles). Further, the annual amount of payments is distributed by type of work:

$$X^{(1)} = c_1 \cdot X, \; X^{(2)} = c_2 \cdot X, X^{(3)} = c_3 \cdot X; \; c_1 + c_2 + c_3 = 1. \tag{2}$$

Here c_1 is the coefficient that takes into account a part of the payments for current repair work; c_2 is the coefficient that takes into account a part of the payments for emergency repair work; c_3 is the coefficient that takes into account a part of the payments for major repair work; $X^{(j)}$ are annual payments to the insurance fund by type of work, ($j = 1,2,3$).

The cost of one payment to the insurance fund for the j-th type repair work is based on (2):

$$Y_j = h_j \cdot X^{(j)} / Tg = c_j \cdot h_j \cdot X / Tg, \; j = 1,2,3, \tag{3}$$

where Tg is the number of days in a year; h_j is the frequency of payments to the insurance fund (day) for the j-th type of work.

The total accumulations of payments to the insurance fund for j-th type repair works both based on (2) and (3) and assumptions of replenishment frequency are equal to:

$$Y_j(t) = Y_j \cdot N_j(t) = (c_j \cdot h_j \cdot X / Tg) \cdot N_j(t), \; j = 1,2,3, \tag{4}$$

where $N_j(t)$ is the number of payments to the insurance fund over time t for the j-th type of work.

For the random risk process (1), the time instant τ (day) is determined when the condition $R(t) < 0$ is fulfilled for the first time:

$$\tau = \min_t \{t : \mathrm{R(t)} < 0\}. \tag{5}$$

A moment of time (5) characterizes the efficiency of repair work in terms of the formation of payments by their types (3), and the probability of a special event is proposed to be considered as a resource-cost risk:

$$r_\tau = P(\tau < T_\tau), \tag{6}$$

where T_τ is set time (day).

The risk indicator (6) is called resource-cost because it evaluates the “Resource- cost” model for repair work related to the railway infrastructure operation process. In the absence of funds in the insurance fund, there is a risk of non-performance of these works.

In practice, a two-factor model is more widely used, where in addition to the probability of a negative event, the financial consequences of the event are considered. In this regard, the work introduces the concept of financial risk of the form:

$$R_{\phi} = r_{\tau} \cdot C_R, \text{ million rubles,} \tag{7}$$

where C_R is the losses from non-performing repairs, million rubles; r_{τ} is the value (6) that characterizes resource-cost risk. Values (6) and (7) are unknown; in simulation they are replaced by point and interval estimates:

$$\tilde{R}_{\tau} = k_{\tau} / n_0, \tag{8}$$

where k_{τ} is the number of process implementations (1) for which the condition (5) is fulfilled, n_0 is the total number of created implementations by simulation method;

$$\tau_1 = k_{\tau} / [k_{\tau} + (n_0 - k_{\tau} + 1) \cdot F_1(\nu_1, \nu_2)], \tag{9}$$

where $F_1(\nu_1, \nu_2)$ is the critical value for the F-distribution with ν_1 and ν_2 degrees of freedom and confidence probability $\gamma = 0,95$; $\nu_1 = 2 \cdot (n_0 - k_{\tau} + 1)$, $\nu_2 = 2 \cdot k_{\tau}$;

$$\tau_2 = \frac{(k_{\tau} + 1) \cdot F_2(\nu_3, \nu_4)}{[n_0 - k_{\tau} + (k_{\tau} + 1) \cdot F_2(\nu_3, \nu_4)]}, \tag{10}$$

where $F_2(\nu_3, \nu_4)$ is the critical value for the F-distribution with ν_3 and ν_4 degrees of freedom and confidence probability $\gamma = 0,95$; $\nu_3 = 2 \cdot (k_{\tau} + 1)$, $\nu_4 = 2 \cdot (n - k_{\tau})$:

$$\tilde{R}_{\phi} = \tilde{R}_{\tau} \cdot C_R; \tag{11}$$

$$R_1 = \tau_1 \cdot C_R, R_2 = \tau_2 \cdot C_R. \tag{12}$$

3. Selecting the original data and the task of a computational experiment

Based on the developed mathematical support described above, a modeling program for the study of rail infrastructure repair work has been created by the simulation method. The software is based on the MATLAB programming language, which has several advantages over other software environments designed to perform scientific and engineering calculations [9].

Table 1 shows the distribution laws and their numerical characteristics for the time intervals between the types of work and the costs of these works used in this study: mi, mz are the mathematical expectations, k_v is the coefficients of variation. The choice of distribution laws and their numerical characteristics was made with the help of experts and literary sources [10]. The software allows for changing these models.

Table 1. Distribution laws and their numerical characteristics

Works	Intervals of time, day		Costs, million rubles	
Current	Birnbaum-Saunders		Lognormal	
	mi=15.0	k_v=0.20	mz=3.5	k_v=0.20
Emergency	Weibull		Pareto	
	mi=45.0	k_v=0.30	mz=10.0	k_v=1.50
Major	Gamma		Normal	
	mi=60.0	k_v=0.25	mz=20.0	k_v=0.15

With the selected mathematical expectations (table 1), the average cost of repairs by type (current, emergency, major) per year is equal to:

$$P_j = Tg \cdot mz_j / mi_j, j = 1,2,3. \tag{13}$$

Considering (13) and the values of Table 1, average expenses are equal to:

$$P_1 = 84 \text{ million rubles}, \; P_2 = 80 \text{ million rubles}, \; P_3 = 120 \text{ million rubles}. \tag{14}$$

In total, these expenses are equal to the annual amount of payments X. Then the share of payments for the types of repairs (2) can be determined as:

$$c_j = P_j / X, j = 1,2,3; X = \sum_{j=1}^{3} P_j . \tag{15}$$

If we put in the formula (15) value (14), we will get the following shares of payments:

$$c_1 = 0,296; \; c_2 = 0,282; c_3 = 0,422. \tag{16}$$

Values (3) with a view (13) equal:

$$Y_j = h_j \cdot P_j / Tg, j = 1,2,3. \tag{17}$$

Values (4) per year are equal to:

$$Y_j(Tg) = Y_j \cdot Tg / h_j, j = 1,2,3. \tag{18}$$

The amount of payments for the year (18) and the amount of expenses for the year (13) for each type of repair should be, on average, equal to each other:

$$Y_j(Tg) = P_j, j = 1,2,3.$$

In risk theory, it has been proven that it is necessary to have an excess of income over expenses [6]. In our case, the condition must be met:

$$X_0 + \sum_{j=1}^{3} Y_j(Tg) > \sum_{j=1}^{3} P_j . \tag{19}$$

In the work [8], it is shown that the excess of the revenue side over the expenditure side is carried out at the expense of the initial annual funds of the insurance fund X_0.

In this paper, it is proposed to link the frequency of payments (h_j) with the mathematical expectations of the time intervals between types of work (mi_j, Table 1):

$$q_j = h_j / mi_j , \; j = 1,2,3. \tag{20}$$

To obtain practical results, using the created modeling program, three options were investigated. For each option, three cases are considered, when X_0 is equal to 5, 10 and 20 percent of annual payments of $X = 284.0$ million rubles: *a)* $X_0 = 14.2$ million rubles; *b)* $X_0 = 28.4$ million rubles; *c)* $X_0 = 56.8$ million rubles. This research includes three options:

- option A, when the share of payments is equal (16), the frequency of payments based on (20) is equal: $h_1 = 10$ days ($q_1 = 0.67$), $h_2 = 30$ days ($q_2 = 0.67$), $h_3 = 90$ days ($q_3 = 1{,}50$);
- option B, when the payment shares are the same: $c_1 = 0.333$; $c_2 = 0.333$; $c_3 = 0.334$; the frequency of payments based on (20) is equal: $h_1 = 10$ days ($q_1 = 0.67$), $h_2 = 60$ days ($q_2 = 0.67$), $h_3 = 90$ days ($q_3 = 1.50$);
- option C, when the share of payments is equal (16), the frequency of payments based on (20) is equal: $h_1 = 20$ days ($q_1 = 1.33$), $h_2 = 60$ days ($q_2 = 1.33$), $h_3 = 90$ days ($q_3 = 1{,}50$).

The considered options are modeled for four values of the quantity T_τ: 30 day, 90 day, 180 day, 360 day.

4. The results of a computational experiment

Table 2 shows the results of a computational experiment.

Analysis of the values in Table 2 allows us to conclude that, in terms of reducing resource-cost and financial risks, the share of payments should depend on the type repair work and determined by the formula (15) (for the selected initial data, they are equal (16)). The frequency of payments should depend on the mathematical expectations of the time intervals between the types of work, the coefficients of proportionality (20) should be close to each other. For the selected initial data, option C has the best frequency of payments. This option, when the shares of payments are equal (16), has the lowest risks in all cases.

The confidence of this conclusion is also justified by the fact that differences in point estimates are confirmed not by the intersection of confidence intervals of risks. This fact allows us to choose a statistical hypothesis that the risks are less for option C than for option B and option A.

Table 2. Results of point and interval risk estimates

V_1	V_2	T_τ	k_τ	$\tilde{R}_\tau$	τ_1	τ_2	$\tilde{R}_\phi$	R_1	R_2
A	a	30	45	0.002	0.002	0.003	0.563	0.432	0.721
		90	1303	0.065	0.062	0.068	16.288	15.575	17.023
		180	2973	0.149	0.145	0.153	37.163	36.131	38.213
		360	5442	0.272	0.267	0.277	68.025	66.731	69.331
	b	30	33	0.002	0.001	0.002	0.412	0.302	0.551
		90	653	0.033	0.031	0.035	8.162	7.652	8.698
		180	1668	0.083	0.080	0.087	20.850	20.051	21.671
		360	3583	0.179	0.175	0.184	44.788	43.675	45.917
	c	30	12	0.001	0.0	0.001	0.150	0.087	0.243
		90	223	0.011	0.010	0.012	2.788	2.489	3.113
		180	637	0.032	0.030	0.034	7.963	7.458	8.492
		360	1582	0.079	0.076	0.082	19.775	18.995	20.577
B	a	30	88	0.004	0.004	0.005	1.100	0.915	1.313
		90	2538	0.127	0.123	0.131	31.725	30.761	31.709
		180	4514	0.226	0.221	0.231	56.425	55.211	57.654
		360	6637	0.332	0.326	0.337	82.962	81.592	84.342
	b	30	85	0.004	0.004	0.005	1.063	0.881	1.272
		90	2083	0.104	0.101	0.108	26.038	25.154	26.943
		180	3638	0.182	0.177	0.186	45.475	44.356	46.611
		360	5212	0.261	0.255	0.266	65.150	63.874	66.439
	c	30	64	0.003	0.003	0.004	0.800	0.643	0.985
		90	1408	0.070	0.067	0.073	17.600	16.861	18.362
		180	2297	0.115	0.111	0.119	28.712	27.790	29.656
		360	3083	0.154	0.150	0.158	38.538	37.491	39.603
C	a	30	63	0.003	0.003	0.004	0.787	0.632	0.971
		90	1413	0.071	0.068	0.074	17.663	16.923	18.425
		180	3114	0.156	0.151	0.160	38.925	37.874	39.995
		360	5632	0.282	0.276	0.287	70.400	69.092	71.720
	b	30	38	0.002	0.001	0.002	0.475	0.356	0.622
		90	684	0.034	0.032	0.036	8.550	8.028	9.097
		180	1692	0.085	0.081	0.088	21.150	20.346	21.977
		360	3523	0.176	0.172	0.181	44.038	42.932	45.160
	c	30	16	0.001	0.0	0.001	0.200	0.125	0.304
		90	236	0.012	0.011	0.013	2.950	2.643	3.284
		180	671	0.034	0.031	0.036	8.388	7.870	8.930
		360	1633	0.082	0.078	0.085	20.413	19.621	21.226

Additionally, it can be argued that with an increase in the initial value, the resource-cost risk decreases, and with an increase in time T_τ, it increases. The latter is logically obvious, but the simulation analysis makes it possible to quantify these changes.

5. Conclusion

A study of the efficiency of railway infrastructure repairs has been carried out in the conditions of the insurance fund. The state of this fund is described by the risk process. A program based on an event approach has been created to model this process. The modeling program creates a sample of the moments of time when there are no funds for equipment repairs. Sampling values are processed according to the proposed algorithms for assessing resource-cost and financial risks. These indicators allow us to assess the efficiency of railway infrastructure repairs and to obtain practically important results. For example, computational experiments show that the criteria for minimum risk assessments should be preferred to the option when the share of payments should depend on the type of repair work, and the frequency of payments should depend on the mathematical expectations of the time intervals between the types of work.

6. References

[1] Lyfar V A, Safonova S A and Ivanov V G 2015 Development of a method for optimizing repair and restoration work taking into account risk indicators *Technological audit and production reserves* **2 (22)** 11–17

[2] Isaev D V 2020 Evaluating performance management systems development programs relying on the maturity model *Journal of Applied Informatics* **15(3)** 5–18

[3] Nachigin V A and Krakovsky Y M 2015 Formalization of selective technology of infrastructure maintenance and insurance fund *World of Transport* **13(1)** 94–99

[4] Bugaenko V M 2015 Monitoring and diagnostics of the infrastructure of high-speed mobile complexes *Path and track facilities* **4** 12–16

[5] Biryuzov V P 2016 On the main directions of development of systems for diagnostics and monitoring of track facilities until 2025 *Path and track facilities* **4** 4–8

[6] Krakowsky Y M, Kargapoltsev S K and Nachigin V A 2018 *Modeling of the transportation process by railway transport: analysis, forecasting, risks* (Saint Petersburg: LITEO)

[7] Korolev V Yu, Bening V E and Shorgin S Ya 2011 *Mathematical foundations of risk theory* (Moscow: FIZMATLIT)

[8] Krakovsky Y M and Hoang N A 2019 Assessment of resource-cost risk in the organization of repair work of complex robotic equipment *Bulletin of Voronezh State University. Series: systems analysis and information technology* **4** 29–35

[9] Dyakonov V P 2012 *MATLAB. Complete tutorial* (Moscow: DMK Press)

[10] Thomas M 2005 *Mathematics of risk insurance* (Moscow: CJSC "Olymp-Business")

Formation of algorithm of automatic parametric optimization of PI controller with variable parameters while using Internal Model Control

V V Kulikov[1], A P Kutsyi[1,2] and N N Kutsyi[1]

[1] Irkutsk National Research Technical University, Irkutsk, 664074, the Russian Federation
[2] Irkutsk State Transport University, Irkutsk, 664074, the Russian Federation

E-mail: tosha1993irk@yandex.ru

Abstract. Known by its advantages, a proportional-integral controller (henceforth – a PI controller) with variable or switchable parameters, related to a class of controllers with a variable structure ((henceforth – variable structure controllers, VSC) that do not use a sliding mode, is applied to control an object in the automatic system with a sufficient delay, reducing the performance quality of standard controllers (integral, proportionally integral, proportionally integral differential). The specifics of building this controller makes it more difficult to use analytical approaches to the parametric optimization of the system, which defines the use of algorithmic methods. The present work uses a gradient algorithm, in which the gradient constituents are calculated with the help of sensitivity functions with their definite advantages. To obtain initial values of VSC adjustable parameters, an adjustment method of a standard PI controller based on the internal model controller (IMC) is used. The gradient algorithm of an automatic parametric optimization (henceforth – an APO algorithm) has calculated optimal parameters of VSC starting from the closeness of the output coordinate of the object to the reference process with a minimum of the integral quadratic criterion. The correctness of the calculated vector of the controller adjustment by the formed APO algorithm is confirmed by necessary and sufficient extremum conditions of the chosen optimization criterion.

1. Introduction

Objects of transport systems with transport delay represent an important class [1-6]. With a ratio of $\tau_{ob}/T_{obm}>1$, τ_{ob} is the magnitude delay, $T_{obm}=max(T_{ob1}, T_{ob2},\ldots, T_{obn})$ is the constant of controlled object time, serial continuous controllers (integral, proportionally integral, proportionally integral differential) do not provide the required quality of transient processes [1,4,6]. A way to control these objects is to apply controllers with a delay link in their structure (e.g., a predictive proportionally integral controller) [6-18]. An alternative way of compensating the object delay is to apply the switching of parameters without a sliding mode (henceforth – VSC) in the PI controller [2]. To bypass the switching of VSC structures in a steady-state mode due to interference, it is necessary to introduce a dead zone [4,5]. The presence of switching elements, a filter and the dead zone in these controllers makes it more difficult to apply analytical approaches to solve the problems of parametric optimization, consisting in the calculation of the VSC adjustment vector $\mathbf{q}$ according to the extremum value of the accepted optimality criterion on the basis of the reference model.

Content from this work may be used under the terms of the Creative Commons Attribution 3.0 licence. Any further distribution of this work must maintain attribution to the author(s) and the title of the work, journal citation and DOI.
Published under licence by IOP Publishing Ltd

In the present work the parametric optimization algorithm, using the gradient procedure based on the sensitivity theory elements, is applied to VSC, and thus the parametric optimization problem for the considered automatic system is solved. To obtain the starting values of the VSC adjustment vector $\mathbf{q}$, it is supposed to use a method on the basis of building a controller with an internal model (IMC) [6,18-21].

2. Problem statement

The structural diagram of the considered automatic system is represented in Figure 1.

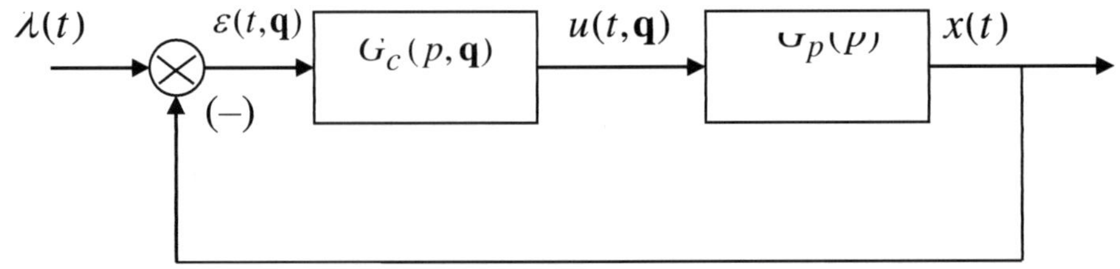

Figure 1. The structural diagram of the studied automatic system

Below we present a description of processes of the automatic system in Fig. 1 with the given PI controller with a dead zone [4,5]:

$$
\begin{gathered}
\varepsilon(t,\mathbf{q}) = \lambda(t) - x(t), \\
u(t,\mathbf{q}) = G_c(p,\mathbf{q}^i)\varepsilon(t,\mathbf{q}), (i = 1,2,3), \\
u(t,\mathbf{q}) = \begin{cases} u_1(t,\mathbf{q}) = \varepsilon(t,\mathbf{q})G_c(p,\mathbf{q}^1), ((\Psi(t,\varepsilon(t,\mathbf{q})) > 0) \vee (0 < t < t_{ob})) \wedge (|\varepsilon(t,\mathbf{q})| > \alpha\lambda(t)); \\ u_2(t,\mathbf{q}) = \varepsilon(t,\mathbf{q})G_c(p,\mathbf{q}^2), (\Psi(t,\varepsilon(t,\mathbf{q})) < 0) \wedge (|\varepsilon(t,\mathbf{q})| > \alpha\lambda(t)); \\ u_3(t,\mathbf{q}) = \varepsilon(t,\mathbf{q})G_c(p,\mathbf{q}^3), |\varepsilon(t,\mathbf{q})| < \alpha\lambda(t); \end{cases} \\
x(t) = G_p(p)u(t,\mathbf{q}),
\end{gathered} \tag{1}
$$

where $\varepsilon(t,\mathbf{q})$ is the switching system error; $\Psi(t,\varepsilon(t,\mathbf{q})) = \varepsilon(t,\mathbf{q})\dfrac{p\varepsilon(t,\mathbf{q})}{T_f p + 1}$ is the switching function; T_f is the filter parameter; $\mathbf{q} = (\mathbf{q}^1(q_1,q_2), \mathbf{q}^2(q_3,q_4), \mathbf{q}^3(q_5,q_6))$ is the vector of VSC parameters, with the given switching function $\Psi(t,\varepsilon(t,\mathbf{q}))$ $q_5 = 0, q_6 = 0$ are the non-optimizable parameters; α is the controller's dead zone, in the present work $\alpha = 5\%\lambda(t)$; $\lambda(t)$ is the setting action; $u(t,\mathbf{q})$ is the controlling action; $x(t)$ is the output coordinate of the automatic control system (ACS); $G_c(p,\mathbf{q}^1) = q_1 + \dfrac{q_2}{p}$, $G_c(p,\mathbf{q}^2) = q_3 + \dfrac{q_4}{p}$, $G_c(p,\mathbf{q}^3) = 0 + \dfrac{0}{p}$ is the operator of the classic PI controller with variable parameters; $G_p(p)$ is the controlled object operator; $p = d/dt$ is the differentiation operator; $\wedge$, $\vee$ are the logical operations.

Controlled object operation $G_p(p)$ is chosen in the form by which it is possible to describe processes of industrial facilities:

$$G_p(p) = \frac{k_{ob}}{(T_{ob1}p+1)(T_{ob2}p+1)} e^{-\tau_{ob} p} \quad , \tag{2}$$

where k_{ob} is the static amplification factor; T_{ob1}, T_{ob2} are the time constants; τ_{ob} is the delay time.

For further numerical experiments, the following basic values of parameters of the object (2) operator are taken, provided that the condition of the large delay of the object $\tau_{ob}/T_{ob2} > 1$ is satisfied:

$$T_{ob1} = 10, \; T_{ob2} = 40, \; k_{ob} = 1, \; \tau_{ob} = 50. \tag{3}$$

Requirements to the transient process (oscillability, response speed, maximal deviation, accuracy in the steady-state mode) can be implemented by the reference model presented by an oscillatory link [22]:

$$G_{et}(p) = \frac{k_{et}}{T_{et}^2 p^2 + 2T_{et}\zeta_{et} p + 1} e^{-\tau_{et} p}. \tag{4}$$

To provide further examples of the calculation experiment, the following parameters of the reference model (4) are taken: $\tau_{et} = \tau_{ob}$, $T_{et} = 40$, $k_{et} = 1$, $\zeta_{et} = 1{,}25$.

In the present work, the value of the parameter filter T_f is chosen to be equal to the value of $\frac{\tau_{ob}}{5 \div 6}$, which makes it possible to reduce phase retardation in the law of switching with a filter, compared to the law of switching without a filter, and not to "pass over" the first switching point t_0.

Let us represent an optimized integral criterion defining a degree of closeness of output coordinate $x(t)$ to the reference process $x_{et}(t)$:

$$I = \int_0^{\infty} (x_{et}(t) - x(t,\mathbf{q}))^2 dt \quad , \tag{5}$$

where $x_{et}(t) = G_{et}(p)\lambda(t)$.

3. Optimization algorithm

The basis of the formed APO algorithm is comprised by the gradient procedure, and it is necessary to calculate the gradient constituents of the chosen optimization criterion, which in the present work are created with the help of sensitivity functions. Let us provide the sensitivity equations for the system (1) with criterion (5) [22,23,24]:

$$\xi_j(t) = G_p(p)\frac{\partial u(t,\mathbf{q})}{\partial q_j}, (j = 1,2,...,4). \tag{6}$$

Let us accept the following notations in the present work to compactly represent expressions defining $\frac{\partial u(t,\mathbf{q})}{\partial q_j}$:

$$\begin{aligned}
\Psi^+ &= ((\Psi(t,\varepsilon(t,\mathbf{q})) > 0) \vee (0 < t < t_{ob})) \wedge (|\varepsilon(t,\mathbf{q})| > \alpha\lambda(t));\\
\Psi^- &= (\Psi(t,\varepsilon(t,\mathbf{q})) < 0) \wedge (|\varepsilon(t,\mathbf{q})| > \alpha\lambda(t));\\
\Psi^0 &= |\varepsilon(t,\mathbf{q})| < \alpha\lambda(t).
\end{aligned} \tag{7}$$

Due to the limited scope of the article, we present only an expression to calculate $\frac{\partial u(t,\mathbf{q})}{\partial q_1}$:

$$\frac{\partial u(t,\mathbf{q})}{\partial q_1}=\begin{cases}\frac{\partial u_1(t,\mathbf{q})}{\partial q_1}=-\xi_1(t)(q_1+\frac{q_2}{p})+\varepsilon(t,\mathbf{q}),\Psi^+;\\ \frac{\partial u_2(t,\mathbf{q})}{\partial q_1}=-\xi_1(t)(q_3+\frac{q_4}{p}),\Psi^-;\\ \frac{\partial u_3(t,\mathbf{q})}{\partial q_1}=-\xi_1(t)(0+\frac{0}{p}),\Psi^0.\end{cases} \quad (8)$$

According to the gradient algorithm in the process of optimization, the vector $\mathbf{q}$ of the adjustable parameters changes in accordance with the expression [22]:

$$\mathbf{q}[l]=\mathbf{q}[l-1]-\Gamma\nabla_{\mathbf{q}}I(x_{et}(t)-x(t,\mathbf{q}[l-1])),\quad (l=1,2,...), \quad (9)$$

where $\Gamma=\{\gamma_i\}$ is the vector of weighting factors; l is the algorithm step number; $\nabla_{\mathbf{q}}I(x_{et}(t)-x(t,\mathbf{q}))$ is the gradient vector (4). Let us represent an expression to define the gradient vector $\nabla_{\mathbf{q}}I(x_{et}(t)-x(t,\mathbf{q}))$, connected with the calculation of the vector of sensitivity function (7):

$$\frac{\partial I(x_{et}(t)-x(t,\mathbf{q}))}{\partial q_i}=-2\int_0^\infty(x_{et}(t)-x(t,\mathbf{q}))\xi_i(t)dt,(i=1,...4). \quad (10)$$

As is known while using the gradient procedure, it is necessary to indicate the initial values of the vector $\mathbf{q}$, to solve this problem in the present work, it is supposed to use the building of the standard PI controller based on the controller with an internal model (a PI-IMC controller). According to [6, 25], for the object (4), initial conditions of the vector $\mathbf{q}$ while using the PI-IMC controller will be the following:

$$q_1=q_3=\frac{T_{ob1}+T_{ob2}}{k_{ob}(2\eta+\tau_{ob})};q_2=q_4=\frac{1}{k_{ob}(2\eta+\tau_{ob})}, \quad (11)$$

where η is the parameter taking on values within the range from 0 to +∞.

Taking consideration of parameters of the object (3) and expressions (11) we define the boundaries of ranges of adjustable VSC parameters:

$$q_1,q_3\in(0;1];q_2,q_4\in(0;0.02]. \quad (12)$$

To make it possible to implement the problem to be solved using the computer, it is necessary to involve a condition of stopping the optimization process. In the present work, the stopping condition is formed according to the following principle [26]: in each *n* steps of algorithm (18), a minimum of the chosen optimization criterion (5) is checked, within the accuracy of two signs, compared to the previous value of the criterion. The algorithm (18) is performed while the condition is true:

$$I_{\min}(l\in[1;n])>I_{\min}(l\in[i\cdot n+1;(i+1)n])>...,(i=1,2,...), \quad (13)$$

where $I_{\min}$ is the minimal value of the criterion (5) for each *n* steps of the algorithm (10). In the course of testing the APO algorithm, the reference value of *n*=20 is chosen.

4. The optimization algorithm operability check

It is necessary to perform an operability test for the formed APO algorithm, which consists in the checking the credibility of the calculated values of the adjustable parameters of $\mathbf{q}^*$ from the point of finding a local minimum of the criterion (5).

While starting the APO algorithm from different initial values of the vector of adjustable parameters $\mathbf{q}_k^0 = (q_{1k}^0, q_{2k}^0, q_{3k}^0, q_{4k}^0)(k = 1,2,...)$, the corresponding final values at the optimum point $\mathbf{q}_k^* = (q_{1k}^*, q_{2k}^*, q_{3k}^*, q_{4k}^*)$, computed by the APO algorithm, should provide the fulfillment of the necessary extremum condition at these points:

$$\frac{\partial I(x_{et}(t) - x(t,\mathbf{q}^*))}{\partial \mathbf{q}} = (0 \pm \Delta), \tag{14}$$

where Δ is the calculation error.

Also, at the optimum point, from the position of the minimum of the criterion (4), a sufficient optimality condition should be performed:

$$\frac{\partial^2 I(x_{et}(t) - x(t,\mathbf{q}^*))}{\partial q_i \partial q_j} > 0, (i, j = 1,...,4). \tag{15}$$

The basis of the condition (14) is formed by a square Hessian matrix, consisting from second derivatives from the optimality criterion (5) by the adjustable parameters q_i:

$$\frac{\partial^2 I(x_{et}(t) - x(t,\mathbf{q}^*))}{\partial q_i \partial q_j} = \begin{bmatrix} \frac{\partial^2 I}{\partial q_1^2} & \frac{\partial^2 I}{\partial q_1 \partial q_2} & \frac{\partial^2 I}{\partial q_1 \partial q_3} & \frac{\partial^2 I}{\partial q_1 \partial q_4} \\ \frac{\partial^2 I}{\partial q_2 \partial q_1} & \frac{\partial^2 I}{\partial q_2^2} & \frac{\partial^2 I}{\partial q_2 \partial q_3} & \frac{\partial^2 I}{\partial q_2 \partial q_4} \\ \frac{\partial^2 I}{\partial q_3 \partial q_1} & \frac{\partial^2 I}{\partial q_3 \partial q_2} & \frac{\partial^2 I}{\partial q_3^2} & \frac{\partial^2 I}{\partial q_3 \partial q_4} \\ \frac{\partial^2 I}{\partial q_4 \partial q_1} & \frac{\partial^2 I}{\partial q_4 \partial q_2} & \frac{\partial^2 I}{\partial q_4 \partial q_3} & \frac{\partial^2 I}{\partial q_4^2} \end{bmatrix} \tag{16}$$

$$\frac{\partial I(\varepsilon(t,\mathbf{q}^*))}{\partial q_{ij}} = 2\int_0^\infty (\xi_i(t)\xi_j(t) - \xi_{ij}(t)(x_{et}(t) - x(t,\mathbf{q})))dt, (i, j = 1,...4). \tag{17}$$

where $\xi_{ij}(t), (i, j = 1,...,4)$ is the second-order sensitivity function.

Based on the things stated above in the present work, the indicator of performance of the formed APO algorithm is the fulfillment of criteria (14), (15) for $\mathbf{q}^*$.

5. Study results

Let us present the operability check for the APO algorithm by practical examples with the setting action $\lambda(t) = 1 \cdot 0.5(t)$. Figure 3 demonstrates the fact that the formed APO algorithm of a system with the controller (1) provides the calculation of $\mathbf{q}_k^* (k = 1,2,3)$ for different types of initial transient processes in the automatic system with different $\mathbf{q}_k^0 (k = 1,2,3)$. In this figure one can see that the obtained final processes sufficiently coincide with the reference model.

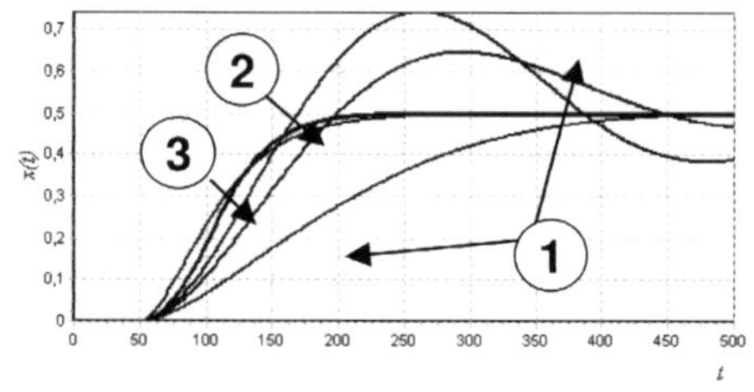

Figure 3. The graphs of transient processes at the initial (1) and final (2) points of the APO algorithm performance of the reference model (3)

Figure 3 shows that at the start of the APO algorithm from different initial values of the vector of adjustable parameters $\mathbf{q}_k^0 (k = 1,2,3)$, the corresponding final values at the optimum point $\mathbf{q}_k^* (k = 1,2,3)$, calculated by the APO algorithm, provide the fulfillment of the necessary condition of the extremum (14) of the criterion (5) at these points.

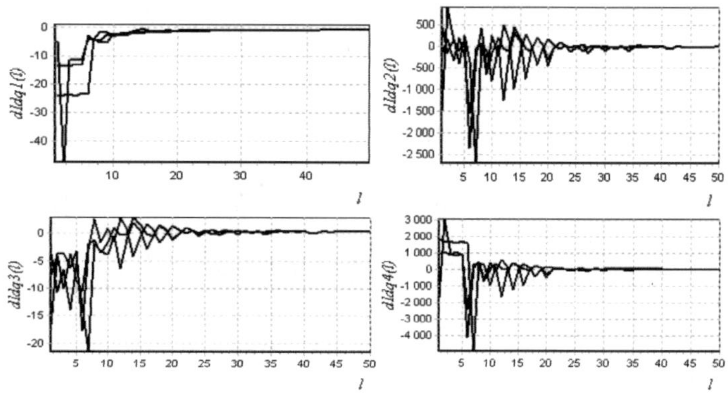

Figure 3. The values of the constituents of the gradient $dI/dq_1(l)$, $dI/dq_2(l)$, $dI/dq_3(l)$, $dI/dq_4(l)$ of the optimization criterion (5) in the course of work of the APO algorithm with the parameters of the object (3)

Table 1 reflects the results of the APO algorithm work at different $\mathbf{q}_k^0 (k = 1,2,3)$ and presents a Hessian matrix.

Table 1. Results of the APO algorithm performance

k	q_1^0	q_2^0	q_3^0	q_4^0	q_1^*	q_2^*	q_3^*	q_4^*	$I(\mathbf{q}^0)$	$I(\mathbf{q}^*)$	A Hessian matrix for the criterion (5)
1.	0,09	0,01	0,09	0,01	0,10588	0,01855	0,09553	0,00020	3,89	0,11	$11; 5{,}9\cdot10^2; 4{,}7; 1{,}3\cdot10^2;$ $5{,}9\cdot10^2; 1{,}5\cdot10^5; 1{,}2\cdot10^3; 1{,}1\cdot10^5;$ $4{,}7; 1{,}2\cdot10^3; 12; 7{,}7\cdot10^2;$ $1{,}3\cdot10^2; 1{,}1\cdot10^5; 7{,}7\cdot10^2; 9{,}4\cdot10^4;$
2.	0,095	0,013	0,095	0,013	0,11892	0,01840	0,10228	0,00032	8,27	0,12	$11; 5{,}8\cdot10^2; 4{,}5; 1{,}2\cdot10^2;$ $5{,}8\cdot10^2; 1{,}5\cdot10^5; 1{,}2\cdot10^3; 1{,}1\cdot10^5;$ $4{,}5; 1{,}2\cdot10^3; 12; 7{,}6\cdot10^2;$ $1{,}2\cdot10^2; 1{,}1\cdot10^5; 7{,}6\cdot10^2; 9{,}5\cdot10^4;$
3.	0,085	0,005	0,085	0,005	0,09585	0,01854	0,08900	0,00013	7,31	0,13	$11; 5{,}7\cdot10^2; 4{,}9; 73;$ $5{,}7\cdot10^2; 1{,}7\cdot10^5; 1{,}2\cdot10^3; 1{,}2\cdot10^5;$ $4{,}9; 1{,}2\cdot10^3; 11; 7{,}4\cdot10^2;$ $73; 1{,}2\cdot10^5; 7{,}4\cdot10^2; 1{,}1\cdot10^5;$

For the parameters of the object (3), according to Table 1, the Hessian matrix is positively defined, which is confirmed by the fact that the formed APO algorithm has found optimal parameters of the controller (1), based on the minimum of the criterion (5) while using the starting values of the vector $\mathbf{q}$, obtained with the help of the IMC method.

6. Conclusion

The present work, with the help of the formed gradient algorithm, solves a problem of parametric optimization of a PI controller with variable parameters for an object with large delay at the given requirements for the quality of the transient process with filtration error in the law of switching and using initial values of the vector $\mathbf{q}$, obtained with the help of the IMC method. The operability of the formed APO algorithm is proved by the fulfillment of the necessary and sufficient conditions of extremum of the criterion (5).

References

[1] Guretskiy Kh 1974 *Analysis and synthesis of control systems with delay* (Transl. From Polish. Moscow: Mashinostroenie)

[2] Kulakov G T et al 2017 *The theory of automatic control of power energy processes* (Minsk: Vysheishaia shkola)

[3] Denisenko V V, 2009 *Computer management of technological process, experiment, equipment* (Moscow: Goryachaya liniya - Telekom)

[4] Govorov A A 2002 *Methods and means of constructing controllers with expanded functional capabilities for continuous technological processes*, diss. of Dr. Sc. Engineering, Moskva

[5] Shigin E K 1966 Automatic regulation of the object with pure delay using the controller with switchable parameters II *Automation and telemechanics* **6** 72-81.

[6] Åström K J and Hägglund T 2001 The future of PID control *Control Engineering Practice* **9 (11)** 1163-1175

[7] Ramírez A, Mondié S and Garrido R 2013 Proportional integral retarded control of second order linear systems *IEEE 52nd Annual Conference on Decision and Control (CDC)* p 2239-2244.

[8] Ramírez A, Garrido R and Mondié S 2013 Integral Retarded Control Velocity Control of DC Servomotors *IFAC TDS Workshop* **46(3)** 558-563

[9] Silviu-Iulian Niculescu and Wim Michiel, 2004 Stabilizing a Chain of Integrators Using Multiple Delays *IEEE Transactions on Automatic Control* **49** 802-807.

[10] Villafuerte R, Mondié S and Garrido R 2013 Tuning of Proportional Retarded Controllers: Theory and experiments *IEEE Trans-actions on Control Systems Technology* **21(3)** 983 - 990.

[11] Ramírez A, Mondié S and Garrido R 2013 Integral retarded velocity control of dc servomotors *11th IFAC Workshop on Time-Delay Systems* **46 (3)** 558-563.
[12] Arousi F, Schmitz U, Bars R and Haber R 2008 PI controller based on first-order dead time model *Proceedings of the 17th World Congress* (Seoul, Korea) p. 5808-5813.
[13] Airikka P 2011 Extended predictive proportional-integral controller for typical industrial processes *18th IFAC World Cogress* (Milano, Italy) p. 7571-7576.
[14] Larsson P and Hagglund T 2012 Comparison between robust PID and predictive PI controllers with constrained control signal noise sensitivity *2nd IFAC Control Conference on Advances on PID Contro* (Brescia, Italy) p 175-180.
[15] Airikka P 2013 Stability analysis of a predictive PI controller *21st Mediterranean Conference on Control and Automation* (Saint-Raphaël, France) p 1380-1385.
[16] Airikka P 2014 Robust predictive PI controller tuning *19th World Congress* IFAC (Cape Town, South Africa) p 9301-9306.
[17] Prakash G and Alamelumangai V 2015, Design of predictive fractional order PI controller for the quadruple tank process *WSEAS Trans. Syst* **10** 258–265.
[18] Ujjwal Manikya Nath, Chanchal Dey, Rajani K and Mudi P 2020 Stabilized IMC-PI controller designing for IPDT processes based on gain and phage margin criteria *IFAC-PapersOnline* **53 (1)** 258–265.
[19] Suresh Kumar Chiluka, Seshagiri RaoAmbati, Muralimohan Seepana, Uday BhaskarBabu Gara 2020 A New VRFT Approach for IMC-PID Feedback-Feedforward Controller Design based on Robustness *IFAC-PapersOnline* **53 (1)** 147–152.
[20] Shuprajhaa T, Haseena B A and Srinivasan K 2020 Design and development of modified RTD-A controller for unstable bioreactor *IFAC-PapersOnline* **53(1)** 240–245.
[21] Pushkar Prakash Arya, Sohom Chakrabarty 2018 IMC based Fractional Order Controller Design for Specific Non-Minimum Phase Systems *IFAC-PapersOnline* **51(4)** 847–852,
[22] Gorodetskiy, V I, Zakharin F M, Rozenvasser E N, Yusupov R M 1971 *Methods of sensitivity theory in automatic control* (Moscow: E`nergoizdat)
[23] Kulikov V V, Kutsyi A P and Kutsyi N N 2020 The Gradient-Based Algorithm for Parametric Optimization of a Variable Structure PI Controller with Dead Band *Mekhatronika, Avtomatizatsiya, Upravlenie* **21(9)** 530-534.
[24] Kulikov, V V and Kutsiy N N 2018 Searchless algorithm of parametric optimization of a PI controller with semiconstant integration *The herald of Irkutsk State Technical University* **22(6)** 98-108
[25] Kulakov G T, Kulakov A T, Kravchenko V V, Kukhorenko A N et al. 2017 *The theory of automatic control of power energy processes* (Minsk: Vysheishaia shkola)
[26] Boyarinov A I and Kafarov V V 1969 *Optimization methods in chemical technology*. (Moscow: Khimiya)

IOP Conf. Series: Materials Science and Engineering 1151 (2021) 012032 doi:10.1088/1757-899X/1151/1/012032

Detailing the impact structure of the participants of the complex transport service

V V Zubkov

Deputy Head of the Production Infrastructure Department of JSC FGC, Ekaterinburg, 44d, Kuibyshev street, 620036, Russia

E-mail: zubkovvv1973@gmail.com

Abstract. During the reforms underway in the railway industry, the consolidation of companies and vehicles, the formation of new models of transportation services is achieved through the creation and implementation of integrated high-tech information systems. In this case, a holistic approach to the effective management of the production system should determine the extent of their responsibility. The article illustrates the main aspects of quality management of customer service as one of the key factors of transportation company efficiency in the segment of freight transportation. The procedure of quality estimation of the transportation service is presented in the Target model of the complex transport service by each of its parties in the different transportation categories. Specification of the impact structure of the parties of the complex transport service on its quality level through calculation of a responsibility ratio has been made.

1. Introduction

Most transport production processes in the railway industry consist of multiple transport services, in many cases not interconnected and not interacting, which in itself complicates a single technological process and reduces its efficiency. Currently, in the current conditions for the consumer of transport services there is a problem of determining the zones of influence of the participants of transport - providing functions on the quality of service and determining the degree of their influence on the quality criteria. To solve these problems, the author of the article proposed a target model of integrated transport services in the freight market segment. It uses the principles of consolidated, consistent and preferential application of various transport services in a single transport system and in different categories of transportation. The use of this model allows one to achieve optimal planning and quality management of transport services, while providing maximum aggregate profit. This article represents a logical development of the early editions. Thus, in [1] the model of complex transport service is proposed, in [2] the stages of formation of the target model of complex transport service in the field of freight transportation are presented. To improve the production process, in terms of improving the quality of transport services and reducing costs, it is necessary to develop new approaches. In this article the method of evaluation of quality of transport service in the target model of the complex transport service by detailing the structure of influence of subjects of the complex transport service on the level of its quality in different categories of transportations [2,3] is presented. At present, the main

Content from this work may be used under the terms of the Creative Commons Attribution 3.0 licence. Any further distribution of this work must maintain attribution to the author(s) and the title of the work, journal citation and DOI.
Published under licence by IOP Publishing Ltd

quality criteria in the market of transport services in the cargo transportation space are: the safety of the transported cargo and goods (K_{safe}), the safety of the transportation of finished products K_{fp}), the speed of delivery of goods and goods K_{sp}), constancy and mass transportation of products (K_{tr}), price availability and rationality of transport services K_{price}) [3,4]. These criteria actually give an idea of the consumers' attitude towards observance and effective improvement of transport services quality by all participants of the transport process. In order to ensure the level of each quality criterion, it is necessary to develop and implement a set of measures aimed at increasing the quality of transport services, which in turn require large financial investments.

2. Materials and methods

In the existing model of transport services in the area of cargo transportation, the level of mandatory quality and its significance according to the requirements of federal laws of the Russian Federation is established in the legal field of responsibility for failure to meet the criteria of quality of services [4,5].

Figure 1 shows the subjects of services in a descending order of the level and competence of responsibility for compliance with the requirements of the main quality criteria.

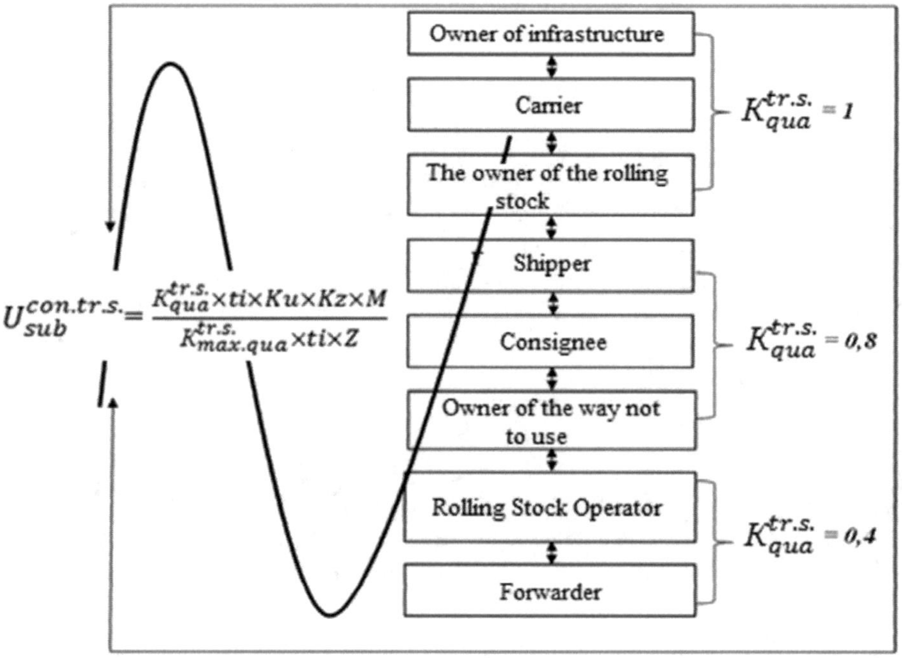

Figure 1. Spatial structure of the location of transport services subjects in a descending order by the level and competence of responsibility for compliance with the requirements of the main quality criteria

It can be seen that the carrier, the owner of infrastructure and the owner of the rolling stock [5] must meet five main quality criteria. Therefore, the quality factor ($K^{tr.s.}_{qua}$) is 1. The consignor, consignee and the owner of the track are responsible [6] for the 4 main quality criteria, resulting in a quality factor of 0.8. The rolling stock operator and freight forwarder [6, 7] actually fulfill two main quality criteria, respectively, the quality factor is 0.4. When organizing a single transportation process [7], each transport service subject is responsible for quality criteria in its area of influence. Coefficients of quality of transport services show full compliance with the conditions of transport service, but can not be part of a single quality compliance (as a result, do not give a total of 1 or

100%).

3. Results

Calculation of the quality factor provides an opportunity to assess the impact on compliance with the level of quality criteria by transport service providers, but does not provide a definition of their responsibility [7, 8]. For this purpose, it is necessary to detail the structure of the impact of the participants of the production transport process on the quality of services provided [8] by calculating the coefficient of maximum quality assurance by all subjects ($K^{tr.s.}_{max.qua}$). The coefficient of maximum quality assurance is defined as a sum of the coefficients of quality of the subjects of transport services participating in a single transport process [9]. The number of transport services subjects participating in a single process of goods - cargo movement may vary depending on the design of the transport services market:

$$K^{tr.s.}_{max.qua}=\sum K^{tr.s.}_{qua\ 1...i}=K^{tr.s.}_{qua1}+K^{tr.s.}_{qua\ i}\ . \tag{1}$$

To determine the amount of responsibility of the subjects of transport services, the responsibility coefficient is calculated by dividing the coefficient of quality of the subject of transport services ($K^{tr.s.}_{qua}$) by the coefficient of maximum quality assurance ($K^{tr.s.}_{max.qua}$):

$$K^{con.tr.s.}_{sub}=\frac{K^{tr.s.}_{qua}}{K^{tr.s.}_{max.qua}}, \tag{2}$$

To determine the quality level of transport and logistics services, the coefficient of the quality level of service is calculated by the following formula:

$$U^{con.tr.s.}_{sub}=\frac{K^{tr.s.}_{qua}\times ti\times Ku\times Kz\times M}{K^{tr.s.}_{max.qua}\times ti\times Z}, \tag{3}$$

where:

t_i - time for performing i - transport service;

K_u - correlative coefficient of the transport service provider's liability;

K_z - correlation coefficient of the Client's liability for transport services;

M - number of calendar days in the reporting period;

Z - rhythm of transport services provision is determined by the number of calendar days required to fulfill the transport services provision plan by the Provider.

The formula shows that the service quality level factor depends on the time of transport service performance, the correlation coefficients of responsibility of the Contractor and the Customer, the quality factor of the subject of transport services. It follows from this that the increase of responsibility of the subjects for compliance with the plan of transport services is the most important condition for improving the quality of service and reducing the cost of transport services. In the existing model, the coefficient of maximum quality assurance of transport services is 6.2. Provided that there is the impact of all participants of transport services allows providing functions on the quality of provided services in a single transportation process.

Figure 2 shows the dependence of the quality factor of the subjects of transport services on their responsibility factor.

In Figure 2, you can see that the dependency curve tends to decrease (curve 3) and increase (curve 1). Moreover, the higher the value of the responsibility factor is (curve 1, curve 2), the higher the value of the quality factor of the subject of transport services is, i.e., the more investments are made by the corresponding participants of the process to provide guaranteed transport services.

Thus, in the current model of transport services on the railway transport there are subjects, including those with a minimum share of responsibility for compliance with the main quality criteria, and therefore less effective in the development of transport services quality [10,11]. The possibility of effective and rational interaction with other modes of transport in a single transport system has been eliminated [11,12], which ultimately leads to higher prices for finished products and non-competitive consumers of services.

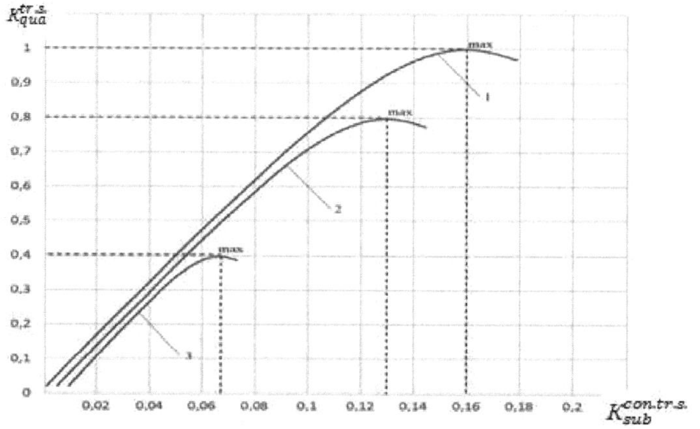

1 - a curve of dependence of the quality factor of the owner of the infrastructure, the carrier, the owner of the rolling stock on the factor of their responsibility.

2 - a curve of dependence of the quality factor of the consignor, consignee, owner of the way of non-use on their liability factor.

3 - a curve of dependence of the rolling stock operator's and freight forwarder's coefficient on the coefficient of their responsibility.

Figure 2. Dependence of the quality factor of transport service subjects on their responsibility factor

In conditions of declining growth rate of the country's economy, it is necessary to search for new, alternative ways to improve the quality of the transport process in the market of transport services in various categories of transport using the potential of a unified transport system [12,13].

The author of the article suggests a new target model of complex transport service, which presents new subjects of transport services and defines the areas of their impact by categories of transportation [13,14].

Figure 3 presents new subjects of transport services and their impact areas in categories of transportation in the proposed target model of the integrated transport service.

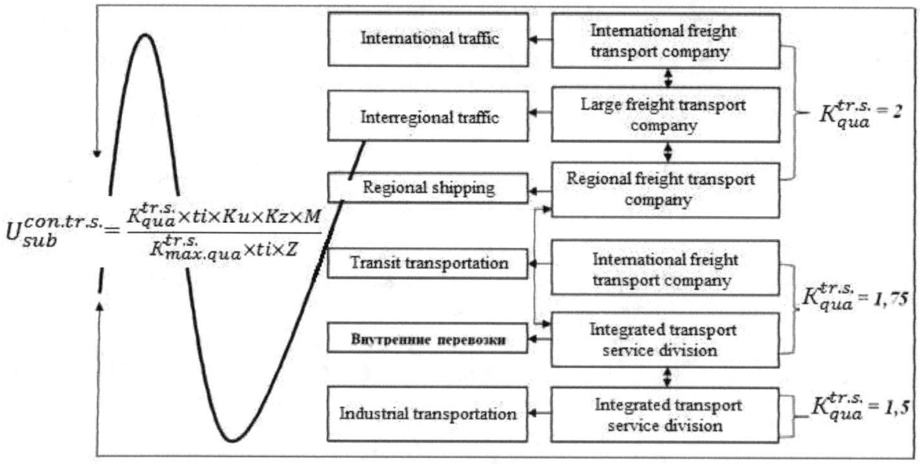

Figure 3. Spatial structure of new subjects of transport services and areas of their influence in the categories of transportation in the proposed target model of the complex transport service

Figure 3 shows that a regional freight transport company in the category of transportation (regional), a large-scale freight transport company in the category of transportation (interregional) and an international freight transport company in the category of transportation (international) ensure the fulfillment of all eight main criteria of quality of a complex transport service. Consequently, the quality factor ($K_{qua}^{tr.s.}$) is 2. Division of the complex transport service in the category of transportations (domestic), the international freight transport company in the category of transportations (transit) ensures the fulfillment of seven main quality criteria. The quality coefficient ($K_{qua}^{tr.s.}$) is 1.75. The division of the complex transport service in the category of transportations (industrial) ensures the fulfillment of six main criteria, respectively, the quality coefficient ($K_{qua}^{tr.s.}$) is equal to 1.5.

The coefficient of maximum quality assurance is defined as the sum of the coefficients of quality of the subjects of transport services participating in the complex transport service:

$$K_{max.qua}^{tr.s.}=\sum K_{qua\,1...i}^{tr.s.}=K_{qua1}^{tr.s.}+K_{qua\,i}^{tr.s.}. \tag{4}$$

The responsibility factor of the subjects in the model of the complex transport service is calculated as a partial division of the factor of quality of the complex transport service of the subject ($K_{qua}^{tr.s.}$) by the factor of maximum quality assurance of the complex transport service ($K_{max.qua}^{tr.s.}$):

$$K_{sub}^{con.tr.s.}=\frac{K_{qua}^{tr.s.}}{K_{max.qua}^{tr.s.}}. \tag{5}$$

The quality level coefficient of transport and logistics services is calculated by the formula:

$$U_{sub}^{con.tr.s.}=\frac{K_{qua}^{tr.s.}\times ti\times Ku\times Kz\times M}{K_{max.qua}^{tr.s.}\times ti\times Z}. \tag{6}$$

In the proposed model the coefficient of maximum quality assurance of the complex transport service is 11 under the conditions of the impact on the quality of services provided by all participants of the complex transport service and the use of two types of transport in the transport scheme, namely: road and rail. When used in the transport process [14,15] in more modes of transport, this coefficient increases.

As a result, the liability factor for an international freight transport company in the categories of transportation (international and transit) is calculated as follows:

$$K_{sub}^{con.tr.s.}=\frac{2+1.75}{11}=0.34.$$

For subdivision of the complex transport service in categories of transportation (internal and industrial) the coefficient of responsibility is calculated as follows:

$$K_{sub}^{con.tr.s.}=\frac{1.5+1.75}{11}=0.3.$$

For a regional freight transport company in the category of transportation (regional) and a large-scale freight transport company in the category of transportation (inter-regional) the calculation of the coefficient of responsibility looks like this:

$$K_{sub}^{con.tr.s.}=\frac{2}{11}=0.18.$$

Figure 4 shows the dependence of the quality coefficient of the subjects of transport services on the coefficient of their responsibility for the categories of transport in the target model of the complex transport service.

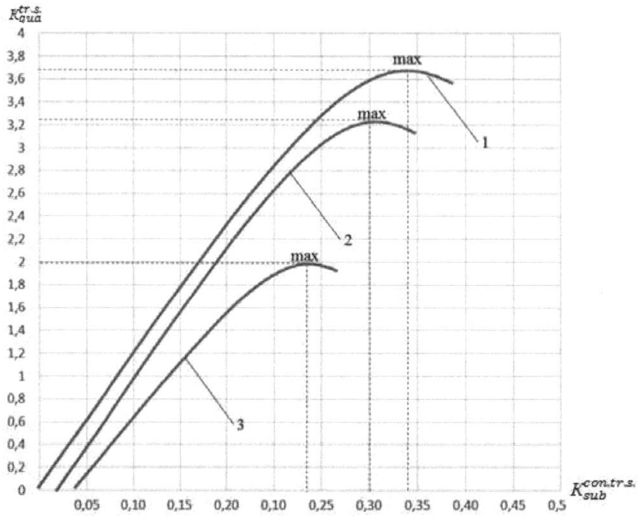

1 - Curve of dependence of the quality factor of an international freight transport company in the categories of transportation (international and transit) on the liability factor.
2 - Curve of dependence of the quality coefficient of the division of complex transport service in the categories of transportation (domestic and industrial) on the liability coefficient.
3 - Curves of dependence of the quality coefficient of a regional transport company in the category of transportation (regional) and a large-scale freight transport company in the category of transportation (interregional) on the liability coefficient.

Figure 4. Dependence of the quality factor of transport services subjects on the factor of their responsibility by transport categories in the target model of the complex transport service

Figure 4 shows that the curve of dependence between $K_{qua}^{tr.s.}$ and $K_{sub}^{con.tr.s.}$ has the desire to fill, and the higher the value of $K_{qua}^{tr.s.}$. , this filling occurs earlier (curve 1,2). In its turn, the coefficient of responsibility of transport service subjects depends on the coefficient of maximum quality assurance.

As a result, the parameters of conducted calculations show [15] that the processes of rendering transport services in the New Target Model have been formed effectively. The data on responsibility coefficients show that investment in ensuring the main criteria of quality by the subjects of transport services will be maximum. In practice, there may be cases in the process of rendering transport services in which a certain subject could not provide on the proper level those quality criteria that must be met. This leads to incomplete compliance with the quality index of complex transport services. It should be noted that this provision leads to failure to meet the requirements of transport and logistics services, and, therefore, compensation of losses to the consumer, the subject of the service because of which occurred failure to meet one of the quality indicators.

4. Conclusion

The presented in the article technique of transport service quality estimation in the Task model of the complex transport service allows one to trace and accordingly estimate influence of each subject on quality assurance of services with the help of maximum values of quality coefficients.

On the basis of the detailed structure of influence of participants of the complex transport service on improvement of the quality of service, coefficients of their responsibility for observance of the plan of transport services, calculated for each category of transportations, are defined.

By calculating the coefficients of the quality service, it is possible to assess the value of the contribution to the compliance with the requirements of transport and logistics services not only by the Executor, but also by the Customer of these services.

The presented Task Model of Integrated Transport Services excludes subjects of transport services with a minimum share of responsibility for compliance with the main quality criteria, and therefore less effective in the development of the quality of transport services. In this model we formed an effective and rational mutualization of different types of transport in a single transport system, which as a result reduces the rate of growth of value added of finished products and increases the competitiveness of consumers of integrated transport services.

The responsibility coefficients should be used rationally when determining the size of obstacles to the subjects of transport services, as well as when calculating the cost of transport services and income rate.

References

[1] Zhenhua C, Nobuhiko D and Jonathan L G 2017 Socioeconomic impacts of transportation public-private partnerships: A dynamic CGE assessment *Transport Policy* **58** 80-87

[2] Dementiev A 2018 Contracting out public transport services to vertical partnerships *Research in Transportation Economics* **69** 126-134

[3] Baklanov A, Cárdenas B, Tsz-cheung L, Leroyer S, Masson V, Luisa T M, Muller T, Ren C, Felix R V and James A V 2020 Integrated urban services: Experience from four cities on different continents *Urban Climate* **32** 100610

[4] Weber K, Heller-Schuh B, Godoe H and Roeste R 2014 ICT-enabled system innovations in public services: Experiences from intelligent transport systems *Telecommunications Policy* **38** 539-557

[5] Jittrapirom P, Marchau V, Heijden R and Meurs H 2014 Dynamic adaptive policymaking for implementing Mobility-as-a Service (MaaS) *Research in Transportation Business & Management* **27** 46-55

[6] Sun L, Wang S, Liu S, Yao L, Luo W and Shukl A 2018 A completive research on the feasibility and adaptation of shared transportation in mega-cities – A case study in Beijing *Applied Energy* **230** 1014-1033

[7] Kavanagh P 2016 A case for negotiated performance-based contracting rather than competitive tendering in government public transport (bus) service procurement *Research in Transportation Economics* **59** 313-322

[8] Gao X, Cao M, Yang T and Basiri A 2020 Transport development, intellectual property rights protection and innovation: The case of the Yangtze River Delta Region, China *Research in Transportation Business & Management* 100563.

[9] Lovelace R, Parkin J and Cohen T 2020 Open access transport models: A leverage point in sustainable transport planning *Transport Policy* **97** 47-54

[10] Kiba-Janiak M 2019 EU cities' potentials for formulation and implementation of sustainable urban freight transport strategic plans *Transportation Research Procedia* **39** 150-159

[11] Blyde J and Molina D 2015 Logistic infrastructure and the international location of fragmented production *Journal of International Economics* **95** 319-332

[12] Shahin M, Ali Babar M and Aufeef C M 2020 Architectural Design Space for Modelling and Simulation as a Service: A Review *Journal of Systems and Software* **170** 110752

[13] Rothengatter W 2019 Megaprojects in transportation networks *Transport Policy* **75** A1-A15

[14] Jokinen J-P, Sihvola T and Mladenovic M N 2019 Policy lessons from the flexible transport service pilot Kutsuplus in the Helsinki Capital Region *Transport Policy* **76** 123-133

[15] Nowakowska-Grunt J and Strzelczyk M 2019 The current situation and the directions of changes in road freight transport in the European Union *Transportation Research Procedia* **39** 350-359

A method for regulating modes of motion of a wheeled vehicle with installed thermal and reactive energy converters

V A Kim, I S Sazonov, O V Bilyk and E A Moiseev

Vehicle Maintenance Department, Belarusian-Russian University,43, pr. Mira, Mogilev, 212000, Republic of Belarus

E-mail: moisan@inbox.ru

Abstract. Understanding the interaction between an elastic tire and a bearing surface is important not only for studying motion of wheeled vehicles, their controllability and running smoothness, but also necessary for creating systems for car motion automatic control. At the same time, the interaction forces between wheels and the bearing surface attract special attention since they have a decisive effect on vehicle's controllability and stability. Indeed, all effects, including aerodynamic ones, are reflected in forces and moments arising in the wheel/bearing surface contact area, which, in turn, determine car motion parameters. Consequently, in creating a control system for vehicle motion, the primary task is to establish patterns of changes in force factors depending on characteristics of the bearing surface and parameters of the operator's control in order to develop a criterion for controlling signal formation. The analysis of experimental and theoretical studies of wheel rolling showed that all research results are reduced to establishing patterns of change in force factors (indirect parameters) in the wheel/bearing surface contact patch depending on the parameters that indirectly characterize the kinematic parameters. Meanwhile, establishing a regular pattern of changes in force factors in the wheel/bearing surface interface is the most important task in developing a criterion for the formation of control signals for active safety systems based on force analysis. The paper presents a method for increasing vehicle's traction and braking dynamics by partially converting thermal energy of the car into reactive tractive force (air propeller, gas turbine).

1. Introduction

The reactive force applied to a wheeled vehicle during slipping and sliding of the contact area of wheels and obtained by converting part of the thermal energy of the engine will allow maximum realization of the conditions of adhesion between vehicle tires and the bearing surface (road) and will increase the traction and braking dynamics, as well as the fuel efficiency of the machine [1].

Under braking, the reverse reactive force complements the braking force realized by the tire with the bearing surface.

2. Purpose of the research

The purpose of research was to develop a method for controlling the modes of motion of a wheeled vehicle by converting some part of the thermal energy of an internal combustion engine into the air reactive tractive force in order to increase traction and braking dynamics and ensure the stability of vehicle motion.

Content from this work may be used under the terms of the Creative Commons Attribution 3.0 licence. Any further distribution of this work must maintain attribution to the author(s) and the title of the work, journal citation and DOI.
Published under licence by IOP Publishing Ltd

IOP Conf. Series: Materials Science and Engineering 1151 (2021) 012033 doi:10.1088/1757-899X/1151/1/012033

3. Implementation of the purpose

The results of testing the process of VAZ - 2108 car emergency braking (Figure 1) show that the braking moments realized by the wheels at the contact patch between the tires and the bearing surface decrease after reaching the maximum (points A and B).

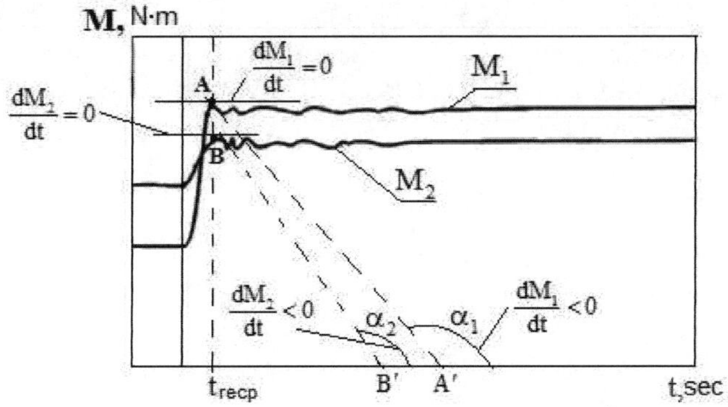

Figure 1. Test results of the process of VAZ-2108 car emergency braking.

The maximum of moments of forces is characterized by their first derivatives being equal to zero $\frac{dM_1}{dt}=0, \frac{dM_2}{dt}=0$. The decrease of the moments can be identified by the negative signs of their derivatives $\frac{dM_1}{dt}<0, \frac{dM_2}{dt}<0$.

Figure 2 shows the results of the change in the hook force of a wheeled tractor depending on the change in the slip of driving wheels [9]. It should be noted that the hook force of the tractor is proportional to the forces at the contact patch between the tractor driving wheels and the bearing surface.

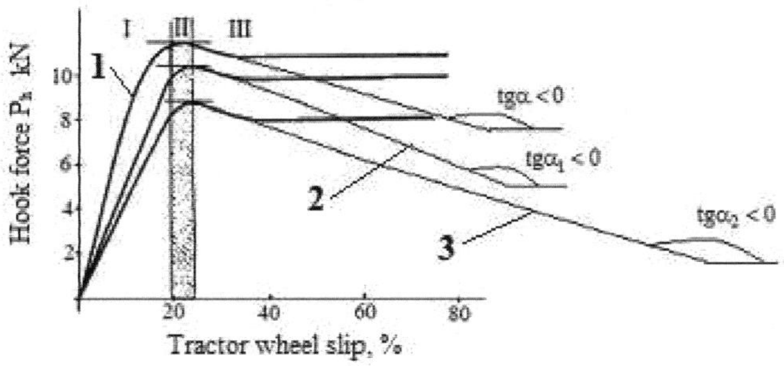

Figure 2. The results of changing the hook force of a wheeled tractor depending on the change in the slip of driving wheels: 1 - stubble field; 2 - light loam; 3 – sandstone.

Reactive tractive force can be obtained by using an air propeller driven by an electric motor powered by an electric generator of the ICE. The magnitude of the tractive force of the propeller can be calculated using the formula:

$$F_{prop} = \alpha \cdot \rho \cdot D_{prop}^4 \cdot \omega_{prop}^2 \quad (1)$$

where α is the coefficient of traction determined experimentally and it depends on the shape and number of blades, as well as on the pitch of the air propeller; ρ is the air density; D_{prop} is the propeller diameter; ω_{prop} is the angular speed of the propeller rotation [1].

The general strategy of the method for controlling the motion of the wheeled machine is schematically shown in Figure 3 and Figure 4.

The wheeled vehicle contains the skeleton of a wheeled vehicle 1, internal combustion engine 2 (ICE), transmission of the wheeled vehicle with the gearbox 3 (TRWV), brake actuator 4 (BA), vehicle's rear axle 5 (RA), rear axle shafts 6, driven wheels of the vehicle 7, driving wheels of the vehicle 8, wheel brakes 9 (WB), brake lines of the front and rear circuits 10, 11, storage battery 12 (SB), electric generator 13 (EG), source for generating a reactive force, for example, an air propeller 14, electric motor for a propeller drive 15 (EM), the unit for information processing and generating control signals by actuating mechanisms 16 (IP & GCSAM unit), sensors for measuring braking torques 17 (SBT), sensors for measuring tractive forces 18 (STF), voltage stabilizer 19 (VS),

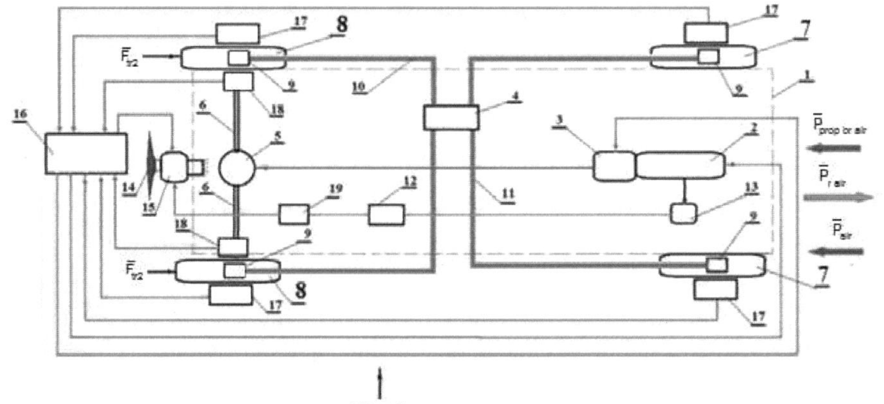

Figure 3. The diagram of the method for controlling the motion of a wheeled machine (top view).

In Figure 3, $\overline{F}_{tr2}$ is the tractive force vector created by the driving wheel; $\overline{P}_{air}$ is the force of air resistance to the vehicle motion; $\overline{P}_{prop\ air}$ is reactive tractive force; l, a are the geometric parameters of the machine; $\overline{P}_{r\ air}$ is the air resistance force, $\overline{P}_{prop\ br\ air}$ is the reactive braking force generated by the propeller during its reverse rotation (Figure 3).

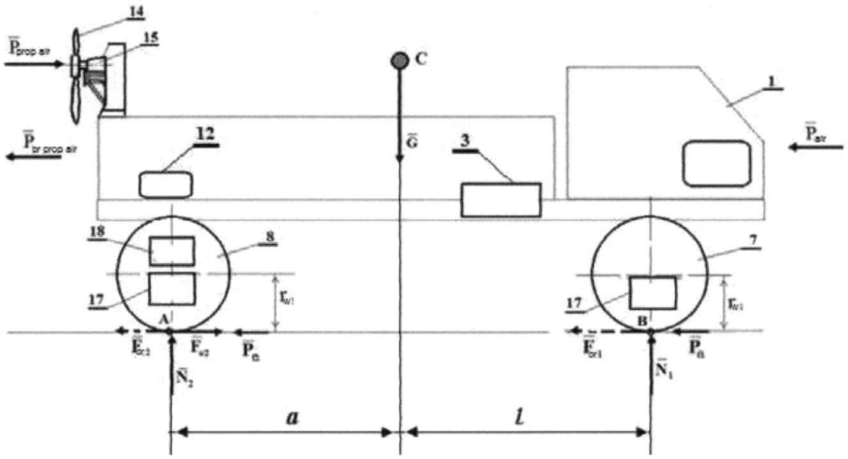

Figure 4. The diagram of the method for controlling wheeled car motion (side view).

Figure 4 presents the diagram showing the arrangement of elements which make it possible to implement the proposed method of controlling the motion of the wheeled vehicle and the forces acting on it: $\overline{P}_{r\ air}$ is the force of air resistance to the machine motion; $\overline{P}_{prop\ br\ air}$ is the propeller's reactive force which slows the vehicle down; $\overline{F}_{tr_2}$ is the tangential tractive force at the contact area between the driving wheels and the road; $\overline{P}_{f_2}$, $\overline{P}_{f_1}$ are the forces of resistance to wheels' rolling; $\overline{N}_1$, $\overline{N}_2$ are normal reactions of the bearing surface of the wheels of car axles; $\overline{F}_{br1}$, $\overline{F}_{br2}$ are braking forces at the contact patches of the wheels; l, a are the geometric parameters of the car.

When a wheeled car accelerates, the sensors for measuring tractive forces (STF) 18 continuously generate electrical signals proportional to the torques M_r realized by the driving wheels 8 with the bearing surface A. The electrical signals proportional to M_r are fed to IP & GCSAM unit 16. The IP unit differentiates signals M_r and identifies negative signs of time derivatives of M_r. When identifying negative signs of derivatives of torques $\frac{M_r}{dt} < 0$, the GCSAM unit generates a signal to turn on the EM 15. As a result of the rotation of the propeller screw 14, the reactive tractive force P_{prop} is created, which "pushes" the vehicle.

When identifying $\frac{M_r}{dt} = 0$, which means that there is no increase in the acceleration in the vehicle's motion, i.e. $\frac{d\ddot{x}}{dt} = 0$, the control operation ends. Besides, with the equality $\frac{M_r}{dt} = 0$, the IP unit 16 fixes the current value of the measured torque M_{rbr}, which characterizes the maximum torques.

Tracking the sliding of the contact patches of car driving wheels 8 will allow reducing fuel consumption of the ICE 2 by limiting the wheel slip at large torques transmitted by the ICE 2 through the transmission (TRWV) 3 to the driving wheels 8, which reduce the coefficient of adhesion at the driving wheel-bearing surface interface.

When the vehicle slows down, the sensors SBT 17 which measure braking torques, actually realized by the wheels (Figure 3, 4), continuously generate electrical signals proportional to the actually realized braking forces F_{br1} and F_{br2} (proportional to the torques $M_{br1} = F_{br1} \cdot r_{w1}$, $M_{br2} = F_{br2} \cdot r_{w2}$) realized at the braking wheels/bearing surface interface. Electrical signals proportional to the braking forces are fed to the IP and GCSAM unit 16 and brake actuators BA of the wheel brakes WB 9. The IP unit 16 also performs differentiation of signals F_{br1} and F_{br2} and identification of their negative signs of time derivatives. When the negative sign of the derivative $\frac{dF_{br1}}{dt} < 0$, $\frac{dF_{br2}}{dt} < 0$ is identified, the IP unit generates a signal to turn on the EM 15 of the propeller drive to create a reactive reverse braking force $\overline{P}_{prop\ br\ air}$. The force $\overline{P}_{prop\ br\ air}$ is independent of the conditions of wheel adhesion. When the force $\overline{P}_{prop\ br\ air}$ arises, the total braking force of the wheeled machine increases. Besides, with $\frac{dF_{br1}}{dt} < 0$, $\frac{dF_{br2}}{dt} < 0$, the current magnitude of the braking forces $F_{br1} = \frac{M_{br1}}{r_{w1}}$, $F_{br2} = \frac{M_{br1}}{r_{w2}}$ is stored in the controller, which is residently installed in the IP & GCSAM unit for subsequent assessment of the adhesion conditions of braking wheels. Based on the current magnitudes of the braking torques F_{br1}, F_{br2}, the GCSAM unit generates a control signal for the BA

4, which brings the pressure in the brake drive of the wheel brakes WB 9 in accordance with adhesion conditions of the vehicle's braking wheels in such a way that $F_{br1} = N_1 \cdot \varphi_{adh1}$; $F_{br2} = N_2 \cdot \varphi_{adh2}$.

There were repeated attempts to install an air propeller on cars in different periods of the development of road transport [6, 7, 8, 9]. The peculiarity of using the propeller in the design of automobiles is associated with energy recuperation, setting the vehicle into motion, its acting as an additional mover aimed at increasing fuel efficiency and speed of car travel.

In the science laboratory of the Vehicle Maintenance Department of the Belarusian-Russian University, the system incorporating the VV-1200-6 screw propeller and the FF 37 D90S4 cylindrical geared motor is manufactured and installed on the Daewoo Matiz II car. The technical characteristics of the geared motor with an installed electric motor are as follows: power N=1.1 kW, rotational speed of the output shaft n = 98 min^{-1}, the force developed on the output shaft F = 3,680 N, the torque on the output shaft of the geared motor M=108 Nm, 12 V consumed voltage). The installed electric motor uses a standard generator and a car battery, since in terms of power consumption it is comparable with an electric starter.

Figure 5. The vehicle equipped with an air propeller: 1– vehicle; 2 - air propeller.

The technical characteristics of the VV-1200-6 screw propeller are presented in Table 1 [10]

Table 1. The technical characteristics of the VV-1200-6 screw propeller.

Parameter	Magnitude
Diameter, mm	1,200
Hub diameter, mm	244
Number of blades	6
Permissible power, hp	150
Allowable number of revolutions, min^{-1}	2,400
Mass, kg	14.0±0.3
Guaranteed service life, h	1,000
Temperature range, degrees	-40/+40
Direction of rotation	left, right
Material	aluminium alloy

The method developed for controlling the modes of motion of the vehicle equipped with thermal and reactive energy converters can be used when accelerating a car at the moment of slipping, thereby creating an additional force spent on overcoming the resistance forces acting on the car during acceleration, which shortens engine run time in the acceleration mode, thus reducing the amount of fuel consumed for the transition to the specified driving mode [7].

In the process of braking with the help of a screw propeller driven by the electric motor, tractive force is directed opposite to the movement of the car. Hence, additional air resistance arises, which contributes to reducing the braking distance.

The differential equations which describe the braking process of the automobile either excluding the air reverse reactive force of braking or taking it into account, take the form:

$$\begin{cases} \ddot{x}_{1br} = -\dfrac{1}{M}\left(F_{br1} + F_{br2} + G\cdot(f_1 + f_2) + F_w\right); \\ \ddot{x}_{2br} = -\dfrac{1}{M}\left(F_{br1} + F_{br2} + G\cdot(f_1 + f_2) + F_w + P_{prop\ br}\right), \end{cases} \tag{2}$$

where $\ddot{x}_{1br}$ is the acceleration of deceleration of the wheeled vehicle without taking into account reactive force of the air propeller;

$\ddot{x}_{2br}$ is the acceleration of deceleration of the wheeled vehicle taking into account the reactive force of the air propeller;

M is the car mass (mass of the Daewoo Matiz II car during testing was 890 kg);

G is the car weight (weight of the Daewoo Matiz II car during testing was G=8730 N);

$\overline{F}_{br1}$, $\overline{F}_{br2}$ are the braking forces at the contact patches of the wheels;

P_{propbr} is the braking force arising from reverse rotation of the propeller;

F_w is the air resistance to car motion;

g is the acceleration of gravity;

f_1, f_2 are the coefficients of resistance to the rolling of car wheels based on the characteristics of the road and tires;

$\ddot{x} = -\dfrac{d^2x}{dt^2}$ is the acceleration of deceleration of the wheeled vehicle.

The maximum braking force at contact patches of the car wheels with the road $F_{br\,max}$ at full use of the coefficient of adhesion between the wheel and the bearing surface is equal to:

$$F_{br\ max} = G\cdot\varphi_{adh} \tag{3}$$

where φ_{adh} is the coefficient of adhesion between the car wheels and the bearing surface, and it depends on the characteristics of the road surface (in what follows, assuming that the coefficients of adhesion for the wheels-road interface are the same $\varphi_{adh1} = \varphi_{adh2}$).

Taking into account (2), the system of equations can be rewritten as follows:

$$\begin{cases} \ddot{x}_{1br} = -\left(g\cdot(\varphi_{adh} + f_1 + f_2) + \dfrac{1}{M}\cdot F_w\right); \\ \ddot{x}_{2br} = -\left(g\cdot(\varphi_{adh} + f_1 + f_2) + \dfrac{1}{M}\cdot(F_w + P_{prop\ br})\right). \end{cases} \tag{4}$$

From the system (4), let us find the value of the increase in deceleration with the propeller operating in the reverse mode:

$$\Delta\ddot{x} = \frac{1}{M}\cdot P_{br\ air}. \tag{5}$$

By integrating the system of equations (4), the laws of speed change during braking of the wheeled car can be obtained:

$$\begin{cases} \dot{x}_{1br} = -\left(g\cdot(\varphi_{adh} + f_1 + f_2) + \dfrac{1}{M}\cdot F_w \right)\cdot t + V_0; \\ \dot{x}_{2br} = -\left(g\cdot(\varphi_{adh} + f_1 + f_2) + \dfrac{1}{M}\cdot (F_w + P_{prop\ br}) \right)\cdot t + V_0, \end{cases} \tag{6}$$

where V_0 is the speed before braking;

t is the point of time.

The braking time is determined from the condition that at the end of braking, the speed of the wheeled vehicle is zero taking into account the force $P_{prop\ br}$ and excluding the reverse force of the propeller.

Consequently, when the air reactive reverse force generated by the propeller is applied, the time of car braking decreases by the value:

$$\Delta t_{br} = \frac{M\cdot V_0}{P_{prop\ br}}, \text{ s.} \tag{7}$$

$$\begin{cases} t_{1br} = \dfrac{V_0}{\left(g\cdot(\varphi_{adh} + f_1 + f_2) + \dfrac{1}{M}\cdot F_w \right)}; \\ t_{2br} = \dfrac{V_0}{\left(g\cdot(\varphi_{adh} + f_1 + f_2) + \dfrac{1}{M}\cdot (F_w + P_{prop\ br}) \right)}. \end{cases} \tag{8}$$

To obtain the braking distance of the machine, let us integrate the system of equations (6):

$$\begin{cases} x_{1br} = -\dfrac{1}{2}\cdot\left(g\cdot(\varphi_{adh} + f_1 + f_2) + \dfrac{1}{M}\cdot F_w \right)\cdot t^2 + V_0\cdot t; \\ x_{2br} = -\dfrac{1}{2}\cdot\left(g\cdot(\varphi_{adh} + f_1 + f_2) + \dfrac{1}{M}\cdot (F_w + P_{prop\ br}) \right)\cdot t^2 + V_0\cdot t. \end{cases} \tag{9}$$

Whence, when the reversing air force of the propeller is applied, the braking distance of the machine will be reduced by the value:

$$\Delta x_{br} = \frac{P_{prop\ br}}{2\cdot M}\cdot t^2, \text{ m.} \tag{10}$$

Therefore, the braking performance of the wheeled vehicle is increased by using the reversing force of the air propeller.

The traction force of the VV-1200-6 screw propeller is calculated by the formula (1):

$$P_{propbr} = 0.17\cdot 1.225\cdot 1.2^4\cdot 10.26^2 = 45.46\ N.$$

When substituting the magnitude of the air reverse force of the propeller into the expression (10), we can determine the magnitude by which the braking distance of the machine will be reduced by using the screw propeller. The Daewoo Matiz II car was bench-tested using a roller brake tester at the testing ground of the *Vehicle Maintenance* Department. These tests determined the braking forces developed at the contact patches of the wheels (for wheels of the front axle F_{br1}=1.85 kN, for wheels of the rear axle F_{br2}=1.44 kN). Also, the *EFFECT-02* device made it possible to obtain the following—

the acceleration of deceleration, the magnitude of braking distance, the time of braking (the initial speed of travel V_0=25 km/h, the braking time t=3.2 s), then the use of the VV-1200-6 screw propeller reduces the braking distance by:

$$\Delta x_{br} = \frac{45.46}{2 \cdot 890} \cdot 3.2^2 = 0.26\ m.$$

With the measured braking distance being x_{1t} = 14.0 m, a 1.9% reduction of the braking distance occurred.

4. Conclusions

The proposed method for controlling the process of wheeled vehicle acceleration makes it possible to increase:

- Traction dynamics of the wheeled vehicle due to the additional propeller-generated traction force, which is independent of the adhesion conditions of the driving wheels.
- Fuel efficiency in the traction mode of wheeled vehicle motion due to the reduction of acceleration time provided by an additional reactive force generated by the air propeller, which is independent of the conditions of adhesion between the driving wheels of the vehicle and the bearing surface.
- Stability of the traction mode of wheeled vehicle motion by limiting the sliding of contact patches of the driving wheels in the longitudinal direction, which ensures that the adhesion coefficients of the driving wheels are preserved in the lateral direction.

References

[1] Electronic Stability Control System ESP 1998 *J. Autom. Ind. Abroad* ***11*** 16-17
[2] Yablonsky A A 1977 *Theoretical Mechanics Course Part II Dynamics* (Moscow: Vysshaya Shkola)
[3] Kim V A et al. 2003 *Methodology for Creating Active Safety Systems for Vehicles Based on Force Analysis* (Mogilev: Belarusian-Russian University)
[4] Kim V A 1998 Control Method for Vehicle's Anti-Lock Brake System Bulletin WO 98/01332, MKI6 B *60* T 8/52
[5] Guskov V V 1982 *Tractor Theory* (Moscow: Mashinostroyeniye)
[6] Propeller Driven Cars 2020 Available at: https://www.drive2.ru/b/1427221/
[7] Gordon Murray finally launches his wildly innovative T.50 masterpiece Available at: https://newatlas.com/automotive/gordon-murray-t-50-launch/
[8] Nenashko E F 1992 *Automobile* Patent RU 2 048 358 C1 MKP B62D
[9] Medvedev V P 2000 *Automobile* Patent RU 2 189 922 C2 MKP B62D
[10] VV-1200-6 screw propeller Available at: http://vint-nn.ru/vozdushnye-vinty-i-ventilyatory/vinto-ventilyator-vv-1200-6.html
[11] Onishchenko G B 2015 *Electric Drive Theory* (Moscow: INFRA-M)

A technique of synthesis of unmanned aircraft control algorithms of a flat-symmetric scheme on the basis of the method of inverse problems of dynamics

V N Sizykh[1], A V Daneev[1], M V Bakanov[1] and R A Daneev[2]

[1] Irkutsk State Transport University, 15 Chernyshevsky str., Irkutsk, 664074, Russia
[2] East-Siberian Institute of Russian MIA, 110 Lermontova str., Irkutsk, 664074, Russia

E-mail: daneev@mail.ru

Abstract. The article discusses the development of non-traditional adaptive control systems for such a type of vehicle as an unmanned aerial vehicle based on the concepts of inverse problems of dynamics, adjacent to the class of absolutely stable dynamic systems. The paper proposes to use the idea of dividing spatial motion into isolated longitudinal and lateral motion (horizontal decomposition) to solve the problem of developing mathematical and information support for the trajectory control loop of unmanned aerial vehicles (AC), to synthesize a control algorithm by the method of inverse dynamics problems (OZD) and obtain analytical solution of the reference model. The relatively small tail surfaces of the unmanned aircraft lead to an insignificant mutual influence of the parameters in the longitudinal and lateral channels during the spatial movement of the aircraft, which makes it possible to accept the longitudinal and lateral channels of the aircraft movement as autonomous. The algorithm presented in the work can be considered as the mathematical support of the trajectory control loop of an unmanned aircraft. Based on the formulas given in the work, as well as mathematical dependencies that determine the optimization of the trajectory contour, a functional diagram can be built that determines the architectural appearance of a promising automatic control system for unmanned aircrafts.

1. The choice of a mathematical model of the unmanned aircraft

Relatively small tail surface areas of the unmanned aircraft result in a slight mutual influence of parameters in the longitudinal and lateral channels during the spatial movement of the aircraft. This allows the longitudinal and lateral channels of the aircraft to be condsidered as autonomous [1, 3].

Content from this work may be used under the terms of the Creative Commons Attribution 3.0 licence. Any further distribution of this work must maintain attribution to the author(s) and the title of the work, journal citation and DOI.
Published under licence by IOP Publishing Ltd

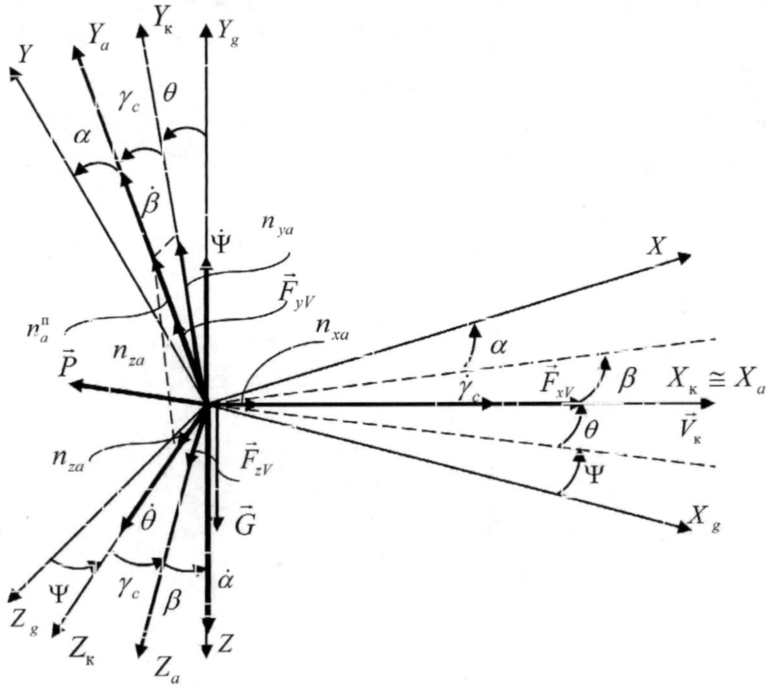

Figure 1. The coordinate systems used in the basic mathematical model of the movement of the unmanned aircraft in space

We will specify two main factors justifying the division of the spatial movement of the aircraft into isolated longitudinal and lateral movements [1, 2, 3].

1. With low angular speeds of the aircraft it can be assumed that aerodynamic coefficients related to longitudinal and lateral movements are autonomous:

$$c_x,\ c_y,\ m_z = f_{\text{long}}(\alpha, \omega_z, \delta_{\text{alt}}, \delta_{\text{bg}}),$$

$$c_z,\ m_x,\ m_y = f_{\text{lat}}(\beta, \omega_x, \omega_y, \delta_{\text{ail}}, \delta_{\text{dir}}), \quad (1)$$

where α, β are the angle of attack and sliding angle respectively, $\omega_i\ (i = x, y, z)$ are the angular velocities to body-axis coordinate system $OXYZ$ [2], $\delta_i\ (i = \text{alt}, \text{bg}, \text{el}, \text{dir})$ are the deviations of steering surfaces to longitudinal and lateral movements. The inertial couplings of movements are dismissively small.

2. The inconsequential influence of kinematic couplings of movements substantiated by the relative turns of different coordinate systems, particularly couplings caused by the wind (Figure 1).

The following coordinate systems are used in the basic mathematical model (BMM) [3].

The normal coordinate system (NCS) is $OX_gY_gZ_g$. The OY_g axis is directed vertically upwards and passes through the center of gravity of the aircraft. The OX_g axis coincides with the projection of the aircraft symmetry axis on the horizontal plane. The OZ_g axis complements the coordinate system to the right triple of vectors. Coordinates and required aircraft overloads are determined in the normal coordinate system.

The flight path coordinate system is $OX_{к}Y_{к}Z_{к}$. Its beginning is located in the center of gravity of the aircraft. The $OX_{к}$ axis is directed along the vector of its velocity, the $OY_{к}$ axis is located in a vertical plane and is directed upwards, the $OZ_{к}$ axis complements the coordinate system to the right one.

The wind-axis coordinate system (WACS) is $OX_aY_aZ_a$. Its beginning is also located in the center of gravity of the aircraft. The OX_a axis coincides with the air speed vector of the aircraft. The OY_a axis coincides with the full vector of the specified transverse overload of n_a, the OZ_a axis forms the right coordinate system. The wind-axis coordinate system determines the aerodynamic forces acting on the aircraft during the flight.

The body-axis coordinate system is $OXYZ$. Its beginning is located in the center of gravity of the aircraft. The OX axis coincides with the longitudinal axis of the aircraft's body, is directed from the tail surfaces of the aircraft to its fore-body and is in the plane OX_aY_a of the wind-axis coordinate system. The OY axis lies in the plane OX_aY_a of the wind-axis coordinate system, it is perpendicular to the OX axis and directed upwards. The OZ axis complements the coordinate system to the right one. The body-axis coordinate system is necessary to determine the current position of the aircraft engine thrust vector and aerodynamic moments on the respective axes.

In the flight path coordinate system, the preset overloads are formed and equations of isolated movements, corresponding to projections of forces acting on the aircraft during the flight, are written down in the simplest technique (Figure 1):

$$\begin{aligned} &m\frac{dV_{к}}{dt} = F_{xV} - G\sin\theta + P\cos\alpha, \\ &mV_{к}\frac{d\theta}{dt} = F_{yV}\cos\gamma_{c} - G\cos\theta + P\sin\alpha, \\ &mV_{к}\frac{d\Psi}{dt}\cos\theta = -F_{zV}\sin\gamma_{c}, \end{aligned} \tag{2}$$

where F_{xV}, F_{yV}, F_{zV} are the aerodynamic force of the frontal drag, lifting and lateral forces respectively, P is the thrust of the aircraft aero engine, $G = mg$ is the force of gravity, m is the weight of the aircraft, g is the acceleration of gravity, $V_{к}$ is the velocity of the aircraft concerning the Earth's surface, θ, Ψ are the tilt angle of a trajectory and traveling corner respectively, γ_{c} is the air-path bank angle (the angle of the bank at zero sideslip).

We will write down equations of trajectory movement (2) in the form of Cauchy's equations, through overload in the wind-axis coordinate system:

$$\begin{aligned} &\dot{V}_{к} = \frac{dV_{к}}{dt} = (n_{xa} - \sin\theta)g + \frac{P}{m}\cos\alpha, \\ &\dot{\theta} = \frac{d\theta}{dt} = \frac{g(n_{ya}\cos\gamma_{c} - \cos\theta)}{V_{к}} + \frac{P}{mV_{к}}\sin\alpha, \\ &\dot{\Psi} = \frac{d\Psi}{dt} = -\frac{n_{za}g\sin\gamma_{c}}{V_{к}\cos\theta}, \end{aligned} \tag{3}$$

where $n_{xa} = \frac{F_{xV}}{G}$, $n_{ya} = \frac{F_{yV}}{G}$, $n_{za} = \frac{F_{zV}}{G}$ are the components of the overload vector in the wind-axis coordinate system.

Dynamic equations of the rotational motion are described by equations (4):

$$\begin{bmatrix} \dot{\omega}_x \\ \dot{\omega}_y \\ \dot{\omega}_z \end{bmatrix} = \begin{bmatrix} \frac{(J_y - J_z)}{J_x}\omega_y\omega_z \\ \frac{(J_z - J_x)}{J_y}\omega_x\omega_z \\ \frac{(J_x - J_y)}{J_z}\omega_x\omega_y \end{bmatrix} + qS\begin{bmatrix} \frac{lm_x}{J_x} \\ \frac{lm_y}{J_y} \\ \frac{b_a m_z}{J_z} \end{bmatrix}, \quad (4)$$

where $J_i\ (i = x, y, z)$ are the inertia moments in body-axis coordinate system, $q = \frac{\rho V^2}{2}$ is the high-speed pressure, ρ is the air density, V is the air velocity, S, l, b_a are the area, length and average aerodynamic chord of a wing of the plane.

A peculiar feature of the construction of the aircraft's basic mathematical model is the fact that the connection between the speed and flight path coordinate systems is determined only by the air-path bank angle γ_c, longitudinal axes of OX_a and $OX_к$ coincide and are directed along the vector of the Earth's speed of $\vec{V}_к$. The vector $\vec{V}_к$ in the plane of the aircraft's symmetry coincides with the air speed vector $\vec{V}$.

In the basic mathematical model, instead of kinematic equations in the classical form it is preferable to use kinematic equations in the directional cosines (Poisson's equation) [3]:

$$\dot{\varepsilon}^{\mathrm{T}} = \Omega\varepsilon^{\mathrm{T}} = \begin{bmatrix} 0 & \omega_z & -\omega_y \\ -\omega_z & 0 & \omega_x \\ \omega_y & -\omega_x & 0 \end{bmatrix} \begin{bmatrix} \varepsilon_{xx} & \varepsilon_{yx} & \varepsilon_{zx} \\ \varepsilon_{xy} & \varepsilon_{yy} & \varepsilon_{zy} \\ \varepsilon_{xz} & \varepsilon_{yz} & \varepsilon_{zz} \end{bmatrix}, \quad (5)$$

where $\varepsilon_{ij}\ (i = x, y, z;\ j = x, y, z\ )$ are the elements of a matrix of the directing cosines. The simulated angular movements have no limitations, and it is possible not to calculate Euler's angles directly.

Kinematic translational equations, linking changes in the coordinates X_g, Y_g, Z_g of the center of the aircraft masses with velocities, are described by equations:

$$\begin{bmatrix} \dot{X}_g \\ \dot{H}_g \\ \dot{Z}_g \end{bmatrix} = \varepsilon \begin{bmatrix} V_{кx} \\ V_{кy} \\ V_{кz} \end{bmatrix} = \begin{bmatrix} \varepsilon_{xx} & \varepsilon_{xy} & \varepsilon_{xz} \\ \varepsilon_{yx} & \varepsilon_{yy} & \varepsilon_{yz} \\ \varepsilon_{zx} & \varepsilon_{zy} & \varepsilon_{zz} \end{bmatrix} \begin{bmatrix} V_{кx} \\ V_{кy} \\ V_{кz} \end{bmatrix}, \quad (6)$$

where $V_{кi}\ (i = x, y, z\ )$ are the projections of the Earth's speed $\vec{V}_к$ to axes of the body-axis coordinate system.

Calculations (3)-(6) are supplemented by the expression of the definition of the air-path bank angle through projections of longitudinal overloads of the aircraft on the axis of the body-axis coordinate system:

$$\cos\gamma_c = \frac{n_{ya}}{\sqrt{n_{xa}^2 + n_{ya}^2}}. \quad (7)$$

The aircraft's basic mathematical model equation system (3)–(7) closes through constraint equations for angular velocities without taking into account the impact of the kinematic couplings of the wind components. This allows one to calculate the current angles of attack, sliding and air-path bank angle without the use of Euler's angles, which greatly simplifies the mathematical model of the aircraft at the command control level:

$$\begin{aligned}
&\dot{\alpha} = \omega_z - \dot{\Psi}(\sin\beta\cos\theta + \sin\gamma_c\cos\beta\cos\theta) - \dot{\theta}\cos\gamma_c\cos\beta,\\
&\dot{\beta} = \omega_x(\sin\alpha - tg\gamma_c\cos\alpha\cos\beta) + \omega_y(\cos\alpha + tg\gamma_c\sin\alpha\sin\beta) + \omega_z tg\gamma_c\cos\beta -\\
&-\dot{\theta}\cos\gamma_c\cos\beta,\\
&\dot{\gamma}_c = \omega_x\cos\alpha\cos\beta - \omega_y\sin\alpha\cos\beta + \omega_z\sin\beta - \dot{\Psi}\sin\theta.
\end{aligned} \quad (8)$$

Thus, the aircraft basic mathematical model is fully described by equations (1), (3)-(8), on the basis of which the algorithms of the digital part of the aircraft control system can be developed.

2. Synthesis of the unmanned aircraft control law by the method of inverse problems of dynamics

On the basis of equations (3) of the basic mathematical model, let us perform a synthesis of aircraft overload (aerodynamic forces) control laws providing the optimal trajectory movement of the aircraft. The preset overloads are formed by the method of inverse problems of dynamics using the reference models of the aircraft's trajectory movement and are the setting actions at the pilot-centered (executive) level of control. The right parts of the reference models take into account the preset movement of the aircraft.

2.1. Formulation of a control synthesis problem

We believe that t_0, t_f are moments in time that corresponds to the beginning and final of the control process.

For controlled trajectory movement (3) which at the current moment of time $t \in [t_0, t_f]$ is characterized by the state:

$$y = (V_к, \theta, \Psi) = (y_1, y_2, y_3). \quad (9)$$

It is necessary to find such overloads n_{xa}, n_{ya}, n_{za} in which the system (3) moves from the state $y_s(t_0)$ to the new state of $y_s = y_s^0(t)$. At the same time, it is necessary that the coordinates of the controlled movement with the specified accuracy follow the coordinates:

$$y^* = (V_к^*, \theta^*, \Psi^*) = (y_1^*, y_2^*, y_3^*), \quad (10)$$

defined by the reference model [4]:

$$\ddot{y}_s^* + \chi_{s1}\dot{y}_s^* + \chi_{s0}y_s^* = \chi_{s0}y_s^0, \quad (11)$$

where $s = 1...3$ $\chi_{s1}, \chi_{s0} > 0$ are the adjusting coefficients, $y_s^0 = y_s^0(t)$ are the coordinates of the reference trajectory of the aircraft movement.

The degree of approximation of the controlled process $y_s^*(t) \to y_s^0$ is estimated by the functional:

$$G(n_a) = \frac{1}{2}\sum_{s=1}^{3}[\dot{y}_s^* - \dot{y}_s(t, n_a)]^2, \tag{12}$$

characterizing the energy of translational acceleration in the vicinity of the phase trajectories of the reference model. Here $n_a = (n_{xa}, n_{ya}, n_{za})$ is the vector of current overloads in the wind-axis coordinate system.

2.2. *Formation of preset overloads*

Let n_a^* be an overload vector which implements the absolute minimum of functional (12), that is, $G(n_a^*) = 0$ at any $t > 0$.

An effective control algorithm is synthesized from the condition so that the value of $G(n_a)$ at each moment of time $t > 0$ belongs to a small neighborhood of extremum – minimum. The laws of control of n_{sa} by each degree of freedom will be obtained by applying a scheme of gradient hill-climbing technique:

$$\frac{dn_{sa}(y)}{dt} = -\sum_{j=1}^{3}\tilde{r}_{sj}\frac{\partial G(n_a)}{\partial n_{ja}}, \tag{13}$$

where $\tilde{r}_{sj} = const$ $j = 1, 2, 3$ $1 \to x, 2 \to y, 3 \to z$.

Using formulas (3), (12), we will calculate gradients in the equation (13):

$$\frac{\partial G(n_a)}{\partial n_{xa}} = g\ , \quad \frac{\partial G(n_a)}{\partial n_{ya}} = \frac{g\cos\gamma_c}{V_к}; \quad \frac{\partial G(n_a)}{\partial n_{za}} = \frac{-g\sin\gamma_c}{V_к\cos\theta}. \tag{14}$$

From the formula (13), taking into account the expressions (14), laws of controlling preset overloads in the wind-axis coordinate system are formed:

$$\dot{n}_{xa}(V_к, \theta, \Psi) = -g\tilde{r}_{11}(\dot{V}_к^* - \dot{V}_к) - \frac{g\cos\gamma_c}{V_к}\tilde{r}_{12}(\dot{\theta}^* - \dot{\theta}) + \frac{g\sin\gamma_c}{V_к\cos\theta}\tilde{r}_{13}(\dot{\Psi}^* - \dot{\Psi}), \tag{15}$$

$$\dot{n}_{ya}(V_к, \theta, \Psi) = -g\tilde{r}_{21}(\dot{V}_к^* - \dot{V}_к) - \frac{g\cos\gamma_c}{V_к}\tilde{r}_{22}(\dot{\theta}^* - \dot{\theta}) + \frac{g\sin\gamma_c}{V_к\cos\theta}\tilde{r}_{23}(\dot{\Psi}^* - \dot{\Psi}), \tag{16}$$

$$\dot{n}_{za}(V_к, \theta, \Psi) = -g\tilde{r}_{31}(\dot{V}_к^* - \dot{V}_к) - \frac{g\cos\gamma_c}{V_к}\tilde{r}_{32}(\dot{\theta}^* - \dot{\theta}) + \frac{g\sin\gamma_c}{V_к\cos\theta}\tilde{r}_{33}(\dot{\Psi}^* - \dot{\Psi}). \tag{17}$$

Since the amount of longitudinal overload, due to the assumptions made for the aircraft's mathematical model, does not depend on the amount of transverse overload $n_a^{tr} = \sqrt{n_{ya}^2 + n_{za}^2}$, it is possible to accept that the gain factors $\tilde{r}_{12}$, $\tilde{r}_{13}$ being equal to zero. Then the preset longitudinal overload n_{xa} will be determined from (15) by the expression:

$$\dot{n}_{xa}(V_\text{к}) = -g\tilde{r}_{11}(\dot{V}_\text{к}^* - \dot{V}_\text{к}) \tag{18}$$

or after integration

$$n_{xa}(V_\text{к}) = -g\tilde{r}_{11}(V_\text{к}^* - V_\text{к}).$$

In the equations that characterize the transverse overload n_a^{tr}, the gain factors $\tilde{r}_{21}$, $\tilde{r}_{31}$ can also be assumed to be equal to zero. Then formulas (16), (17) will be rewritten as:

$$\dot{n}_{ya}(V_\text{к},\theta,\Psi) = -\frac{g\cos\gamma_\text{c}}{V_\text{к}}\tilde{r}_{22}(\dot{\theta}^* - \dot{\theta}) + \frac{g\sin\gamma_\text{c}}{V_\text{к}\cos\theta}\tilde{r}_{23}(\dot{\Psi}^* - \dot{\Psi}), \tag{19}$$

$$\dot{n}_{za}(V_\text{к},\theta,\Psi) = -\frac{g\cos\gamma_\text{c}}{V_\text{к}}\tilde{r}_{32}(\dot{\theta}^* - \dot{\theta}) + \frac{g\sin\gamma_\text{c}}{V_\text{к}\cos\theta}\tilde{r}_{33}(\dot{\Psi}^* - \dot{\Psi}). \tag{20}$$

After the integration of equations (18) – (20), we define overloads in the body-axis coordinate system, which are preset for the contour of the trajectory control and are of the form of:

$$n^{giv} = D_{wa}^{ba} n_a^{giv}, \tag{21}$$

where $n^{giv} = (n_x^{giv}, n_y^{giv}, n_z^{giv})$ are the vectors of angular velocities and overloads in the body-axis coordinate system, $n_a^{giv} = (n_{xa}^{giv}, n_{ya}^{giv}, n_{za}^{giv})$ are the overload vectors in the wind-axis coordinate system, calculated according to equations (18)-(20):

$$D_{wa}^{ba} = \begin{bmatrix} \cos\alpha\cos\beta & \sin\alpha & -\cos\alpha\sin\beta \\ -\sin\alpha\cos\beta & \cos\alpha & \sin\alpha\sin\beta \\ \sin\beta & 0 & \cos\beta \end{bmatrix}$$

is the matrix of the transition from the wind-axis coordinate system to the body-axis coordinate system.

3. Obtaining an analytical solution to the reference model

To implement the control laws of the preset overloads (18 – 20) one needs to choose the parameters of the reference model (11) and obtain its analytical solution.

The reference model of the aircraft's trajectory movement (11) in detail is of the form:

$$\begin{cases} \ddot{V}_\text{к}^* + \chi_{11}\dot{V}_\text{к}^* + \chi_{10}V_\text{к}^* = \chi_{10}V_\text{к}^0(t), \\ \ddot{\theta}^* + \chi_{21}\dot{\theta}^* + \chi_{20}\theta^* = \chi_{20}\theta^0(t), \\ \ddot{\Psi}^* + \chi_{31}\dot{\Psi}^* + \chi_{30}\Psi^* = \chi_{30}\Psi^0(t), \end{cases} \tag{22}$$

where $V_\text{к}^0, \theta^0, \Psi^0$ are the aircraft's basic movement coordinates.

The method of obtaining a reference model analytical solution is as follows:

1. Through the definition of the coordinates of the reference trajectory, the right parts of reference model are formed. To do this, the transition from the normal system of coordinates to the trajectory one for the basic trajectory movement is carried out:

$$\begin{bmatrix} \dot{x}^0 \\ \dot{y}^0 \\ \dot{z}^0 \end{bmatrix} = V_{\text{к}}^0 \begin{bmatrix} \cos\theta^0 \cos\Psi^0 \\ \sin\theta^0 \\ -\cos\theta^0 \sin\Psi^0 \end{bmatrix}. \tag{23}$$

By squaring the right and left parts of the formulas (23) and summing them up among themselves, we obtain expressions to describe the basic movement in the trajectory system of coordinates:

$$V_{\text{к}}^0 = \sqrt{(\dot{x}^0)^2 + (\dot{y}^0)^2 + (\dot{z}^0)^2}\,,\ \theta^0 = \arcsin\frac{\dot{y}^0}{V_{\text{к}}^0}\,,$$
$$\Psi^0 = \arccos\frac{\dot{x}^0}{V_{\text{к}}^0 \cos\theta^0}\,, \tag{24}$$

where linear speeds of $\dot{x}^0, \dot{y}^0, \dot{z}^0$ take into account the preset (basic) movement of the aircraft.

2. The parameters of the reference model are calculated [5]:

$$\chi_{11} = 2\xi\frac{1}{T_{V_к}},\quad \chi_{10} = \frac{1}{T_{V_к}^{\ 2}},\quad T_{V_к} = \omega_{Vestab}^{\ -1},\quad \xi = \frac{\sqrt{2}}{2} = 0{,}707,$$
$$\chi_{21} = 2\xi\frac{1}{T_\theta},\quad \chi_{20} = \frac{1}{T_\theta^{\ 2}}.\quad T_\theta = \omega_{\theta estab}^{\ -1}, \tag{25}$$
$$\chi_{31} = 2\xi\frac{1}{T_\Psi},\quad \chi_{30} = \frac{1}{T_\Psi^{\ 2}},\quad T_\Psi = \omega_{\Psi estab}^{\ -1},$$

where the factor of relative damping of the ideal oscillation link is taken as the decrement of damping ξ, constants of time and natural oscillation frequencies of variables $V_к, \theta, \Psi$ are predetermined by mathematical modeling of the aircraft's spatial movement (2)-(8) by the reaction to single jumps of the rudders $\delta_{\text{alt}}, \delta_{\text{ail}}, \delta_{\text{dir}}$.

3. The calculated parameters determine the roots of the characteristic polynomial of the reference model. To do this, three similar systems are written out of two algebraic equations:

$$\chi_{11} = \lambda_{1V} + \lambda_{2V},\quad \chi_{10} = \lambda_{1V}\ \lambda_{2V},$$
$$\chi_{21} = \lambda_{1\theta} + \lambda_{2\theta},\quad \chi_{20} = \lambda_{1\theta}\ \lambda_{2\theta}, \tag{26}$$
$$\chi_{31} = \lambda_{1\Psi} + \lambda_{2\Psi},\quad \chi_{30} = \lambda_{1\Psi}\ \lambda_{2\Psi},$$

the solution of which has two roots. For reasons of sustainability of such a solution, its negative roots are chosen:

$$\lambda_{2V}^{(1,2)} = \frac{\chi_{11} \pm \sqrt{\frac{\chi_{11}^2}{2} - 2\chi_{10}}}{2}; \quad \lambda_{1V}^{(1,2)} = \chi_{11} - \lambda_{2V}^{(1,2)},$$
$$\lambda_{2\theta}^{(1,2)} = \frac{\chi_{21} \pm \sqrt{\frac{\chi_{21}^2}{2} - 2\chi_{20}}}{2}; \quad \lambda_{1\theta}^{(1,2)} = \chi_{21} - \lambda_{2\theta}^{(1,2)}, \tag{27}$$
$$\lambda_{2\Psi}^{(1,2)} = \frac{\chi_{31} \pm \sqrt{\frac{\chi_{31}^2}{2} - 2\chi_{30}}}{2}; \quad \lambda_{1\Psi}^{(1,2)} = \chi_{31} - \lambda_{2\Psi}^{(1,2)}.$$

4. The analytical solutions of the reference model are written out:

$$V_{к}^*(t) = V_{к}^0(t) - c_{1V}e^{\lambda_{1V}t} - c_{2V}e^{\lambda_{2V}t}, \quad \dot{V}_{к}^*(t) = \dot{V}_{к}^0(t) - c_{1V}\lambda_{1V}e^{\lambda_{1V}t} - c_{2V}\lambda_{2V}e^{\lambda_{2V}t},$$
$$\theta^*(t) = \theta^0(t) - c_{1\theta}e^{\lambda_{1\theta}t} - c_{2\theta}e^{\lambda_{2\theta}t}, \quad \dot{\theta}^*(t) = \dot{\theta}^0(t) - c_{1\theta}\lambda_{1\theta}e^{\lambda_{1\theta}t} - c_{2\theta}\lambda_{2\theta}e^{\lambda_{2\theta}t}, \tag{28}$$
$$\Psi^*(t) = \Psi^0(t) - c_{1\Psi}e^{\lambda_{1\Psi}t} - c_{2\Psi}e^{\lambda_{2\Psi}t}, \quad \dot{\Psi}^*(t) = \dot{\Psi}^0(t) - c_{1\Psi}\lambda_{1\Psi}e^{\lambda_{1\Psi}t} - c_{2\Psi}\lambda_{2\Psi}e^{\lambda_{2\Psi}t}.$$

Consequently, the movement of the aircraft with the laws of overload control (18) – (20) over time is carried out strictly along the designated basic trajectory.

Formulas (18) – (20), (24) – (28) are implemented in the control unit of the basic mathematical model of the aircraft's longitudinal and lateral movement.

Thus, the given algorithm can be considered as a mathematical support for the trajectory circuit of the control of the unmanned aircraft. On the basis of the above formulas, as well as mathematical dependencies that determine the optimization of the trajectory circuit, one can build a functional scheme defining the architectural design of the prospective system of automatic control of the unmanned aircraft.

References

[1] Dil V F, Sizykh V N and Daneev A V 2017 Optimization of the control processes of the spatial motion of an aircraft based on the equations of nonlinear dynamics *Izvestia of Samara Scientific Center of the Russian Academy of Sciences* **19 (1)** 195

[2] Dil V F, Sizykh V N, Daneev A V and Tarmaev A A 2017 Methodology for constructing a nonlinear control system on the basis of the optimization of the trajectory contour of an aircraft *Izvestia of Samara Scientific Center of the Russian Academy of Sciences* **19(6)** 130

[3] Krasovskiy A A, Bukov V N and Shendrick V S 1977 *Universal algorithms for optimal control of continuous processes* (Moscow: Nauka Publ.)

[4] Krutko P D 2004 *Inverse problems of dynamics in the theory of automatic control* (Moscow: Mashinostroenie Publ.)

Simulation of conditions for the occurrence and elimination of asynchronous mode in the power supply system

A P Afanasyev

Sholom-Aleichem Priamursky State University, 70a, Shirokaya St., Birobidjan, 679015, Russia

E-mail: mr.preceptor@yandex.ru

Abstract. This article discusses the phenomenon of a two-frequency asynchronous mode as the most typical and common in power systems. The analysis of methods of registration and elimination of the asynchronous mode and a description of their principles of operation of these methods are carried out. The conditions for the occurrence of an asynchronous mode are considered and analyzed, the main signs of the transition of an electrical network to an asynchronous mode are highlighted. On the basis of the RUSTab software package, an experiment was modeled that simulates the asynchronous mode and synchronous oscillations in the network section included in the IES of the East. The critical values of the parameters of the elements of the experimental network are estimated. The composition and parameters of the control action have been determined, which makes it possible to exclude the asynchronous mode without changing the network topology. The results obtained show the efficiency of using asynchronous mode methods based on the constructed model.

1. Introduction

Power supply systems are among the most complex man-made objects. In modern industrial life, power supply systems play a leading role for normal life support. The complex heterogeneous structure of power supply systems [1] requires looking for new approaches to ensuring energy security and reliability.

Asynchronous running is the most severe mode after a short circuit, as it is accompanied by significant fluctuations in current and voltage, which can lead to equipment damage and loss of load stability. Asynchronous mode (AM) is usually considered on a two-machine system [3], on which it is possible to simulate a two-frequency asynchronous operation (Figure 1).

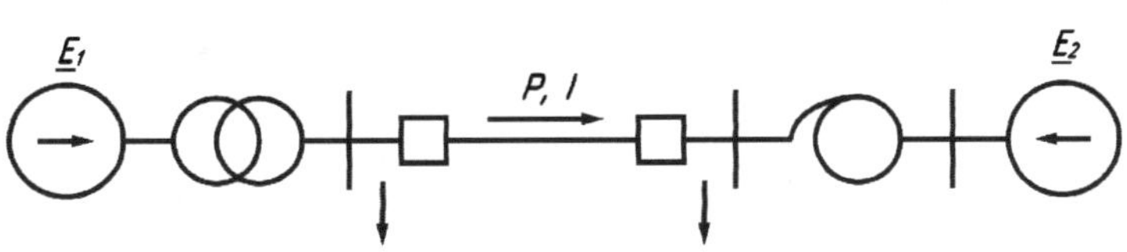

Figure. 1. Two-machine power supply system with power take-off in the intermediate unit

In the event of a violation of stability, the transfer of active power is stopped, and since the power of the turbines remains the same, the speed of rotation of the turbines and generators of one power

Content from this work may be used under the terms of the Creative Commons Attribution 3.0 licence. Any further distribution of this work must maintain attribution to the author(s) and the title of the work, journal citation and DOI.
Published under licence by IOP Publishing Ltd

plant (with a lower load) increases (generators are accelerated), and at another power plant (where there is a shortage of generating power) the speed of the turbines decreases.

The first characteristic feature of AM is a periodic change in the angle δ between asynchronous EMF from zero to 360° [3].

Figure 2 shows a vector diagram of the EMF and power transmission voltages during the asynchronous run at a fixed time (in order to simplify the consideration of the EMF vector of the system and the generator, they are considered equal in modulus).

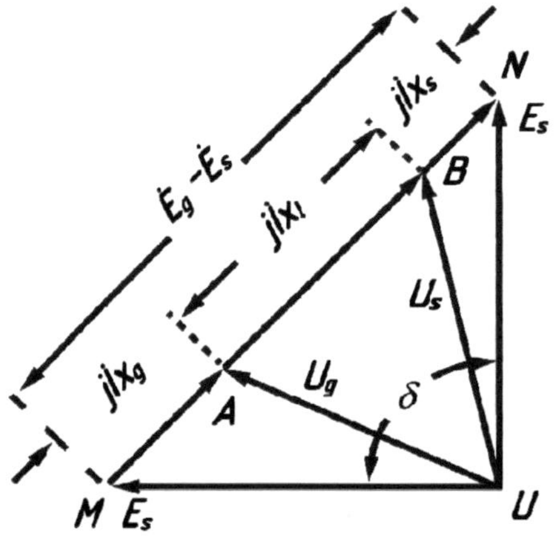

Figure. 2. Vector diagram of EMF and voltages

Each power transmission has a characteristic point at which the voltage at an angle of δ = 180° takes a minimum value or drops to zero. This point is called the electrical swing center (ESC).

Figure 3 illustrates the dependence of the voltage in the ESC and at the generator on δ.

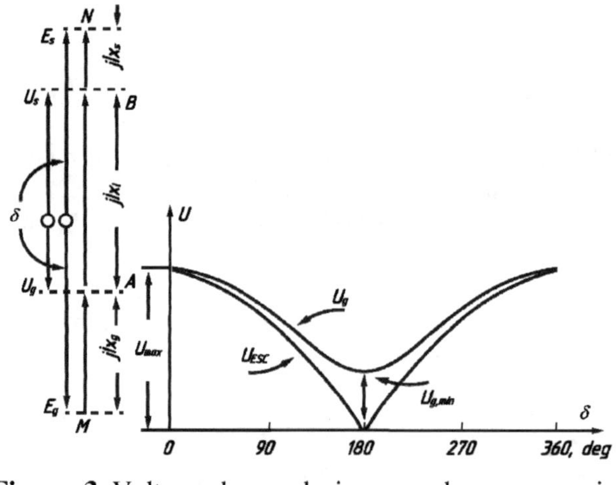

Figure. 3. Voltage change during asynchronous running

As the angle grows, the current through the electric power line increases, and at a value of the angle δ = 180°, the electric power transmission line turns on for a voltage twice rated. The magnitude of the current can cause both electrical and mechanical damage to the generators.

The moment of passage of the EMF vectors through the angle δ = 180° and the voltage distribution along the bond is shown in Figure 4.

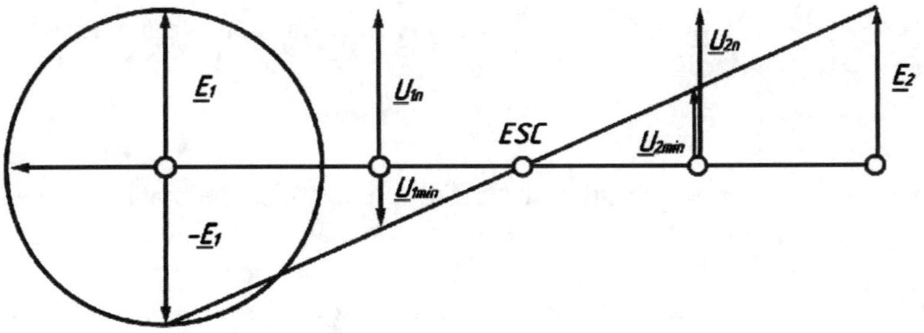

Figure. 4. Voltage change along the power transmission in asynchronous mode at an angle of δ = 180°

Figure 4 shows a planar projection of the voltage distribution at the moment of passing the angle δ of the 180°.

Let us note that at the points of power transmission, the voltages at AM have different values; to the left of the ESC, the voltage vector makes a full revolution with a slip frequency, and to the right it oscillates in magnitude and angle, and the swing increases as it approaches the ESC.

Let us consider a section at the ESC point for an arbitrary angle δ. The relative position of the EMF vectors and their difference, as well as the voltage at this point, is shown in Figure 5.

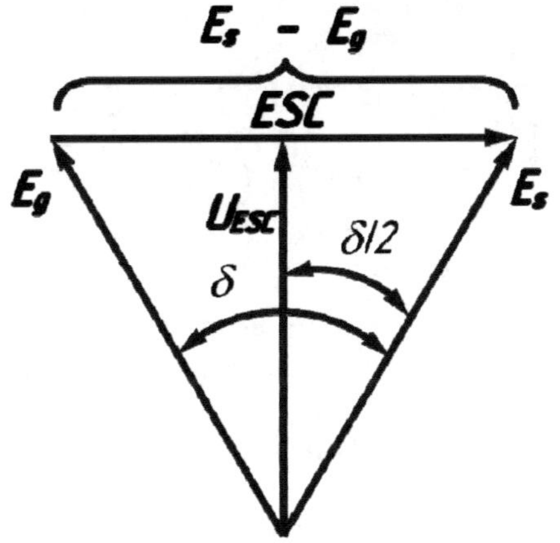

Figure. 5. Mutual arrangement of EMF and voltage vectors in ESC

It can be seen from the above diagram that when the EMF of the system and the generator is equal, the dependence of the voltage in the electric swing center on the angle δ can be represented as:

$$U_{EC} = E \cdot \cos\left(\frac{\delta}{2}\right);$$

the modulus of the difference vector is

$$E_c - E_g = 2 \cdot E \cdot \sin\left(\frac{\delta}{2}\right);$$

and, accordingly, the asynchronous mode current

$$I_{AM} = \frac{2 \cdot E}{X_\Sigma} \cdot \sin\left(\frac{\delta}{2}\right),$$

where X_Σ is the total resistance of the system under consideration.

The value of the active power P transmitted by the power transmission is equal to:

$$P_{AM} = \frac{E_G \cdot E_S}{X_\Sigma} \cdot \sin(\delta).$$

From the above ratios it can be seen that the active power, like the current, depends on the angle between the EMF vectors of the system and the generator, but in contrast to the current and voltage, which depend on the half angle, and the active power depends on the total angle δ.

Periodic change in active power occurs with double slip frequency, i.e. in one cycle of the asynchronous mode, the sign of the active power changes twice.

Physically, this means that the generators of power plants operate in the generator mode during the first half of the swing, and in the motor mode during the second half.

The average power for the swing period, transmitted through the power transmission, is zero [4].

The relationships for current and active power are illustrated in Figure 6.

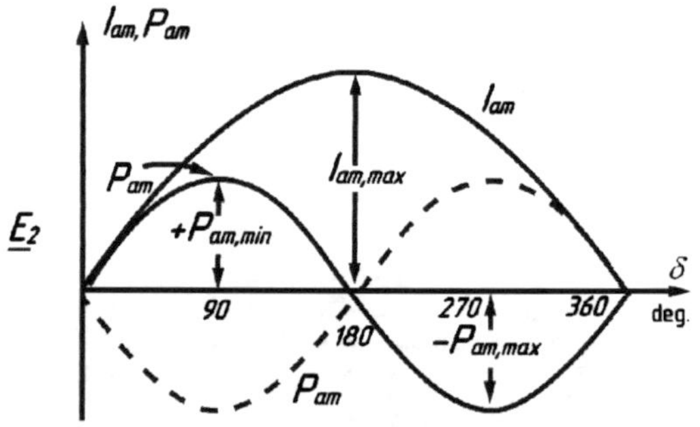

Figure. 6. Relationships for current and active power in asynchronous mode

Thus, summarizing the above, we can note four signs of the asynchronous mode [5], according to which the automatic elimination of the asynchronous mode (AEAM) detects the transition of the system to asynchronism:

1. periodic change of angles between power plants;
2. periodic change in the voltage swing along the system transit;
3. periodic variation of the current swing in transit;
4. periodic change in active power transmitted between power plants.

2. Purpose of the Study

Development of criteria allowing using parameters mode measured directly at the substation with installed control devices determines in real time moments of opening and closing cycles of switches power lines to control asynchronous running on power system interconnection lines [6].

In the course of research on the topic under consideration, the following tasks were set:

- clarification of the structure of indicators of quality and efficiency of AEAM;
- identification of resources and ways to improve the efficiency of AEAM;
- determination of generalized parameters of two-frequency asynchronous mode;
- improvements in how to detect asynchronous mode.

3. Modelling transition process

All existing AEAM devices are capable of detecting precisely the two-frequency asynchronous mode, discussed above, since it is based on the theory developed on a two-machine system.

The method for calculating the transient process used in the RUSTab software package allows simulating and analyzing the asynchronous mode with high accuracy, provided that the computational model is formed correctly [7]. This makes it possible for the AR to jointly simulate the operation of devices for automatic elimination of the asynchronous mode – AEAM.

The computational model contains assumptions that do not allow simulating the exact response of real devices for detecting asynchronous mode.

The simulation is performed in root mean square values, thus, the harmonic composition of the analyzed quantities is not taken into account.

The three-phase system appears to be symmetrical, which excludes the analysis of unbalanced mode.

The computational model is based on a two-machine system.

First, we consider simulating a transient that starts with a synchronous swing and goes into asynchronous mode.

This experiment allows you to get an idea of the principles of operation and tuning of the AEAM model.

The study area contains Nizhne-Bureyskaya hydroelectric power station (NBHES) located in Amur Region and Primorskaya GRES (PRIM.GRES), located in the near Luchegorsk city. The intermediate node is substation TS 220 Birobidzhan.

The power of the generator in the NBHES unit is approximately three times greater than that of the unit. Prim.GRES. The load of Prim.GRES exceeds the generating capacity at this node.

Three power lines are involved in the scheme. Two parallel lines have the same parameters, the third line has about twice the resistance of the first two.

The AEAM device is located at Substation 220 Birobidzhan. Figure 7 shows steady state operation, excluding the operation of compensating devices providing a fixed voltage.

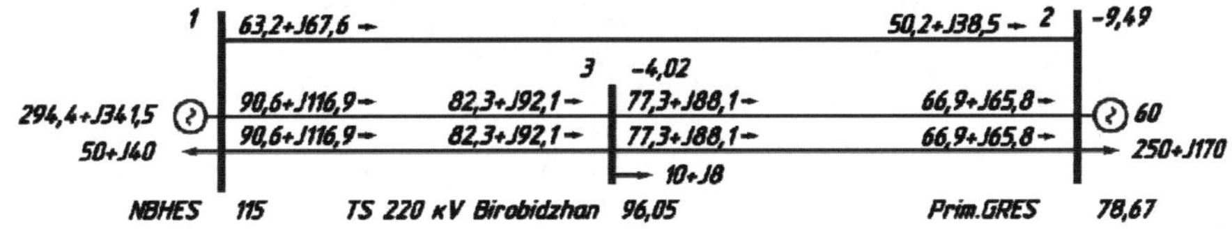

Figure. 7. Design scheme for the steady state

As a perturbation, the disconnection of two strong links with the balancing of the circuit for a weak link is modeled. Initially, to assess the parameters of the transient process, the ALAR devices are not involved and the transient process itself will develop only under the control of the speed regulators on the generators.

The graph of transient processes in frequency in the generating units is shown in Figure 8.

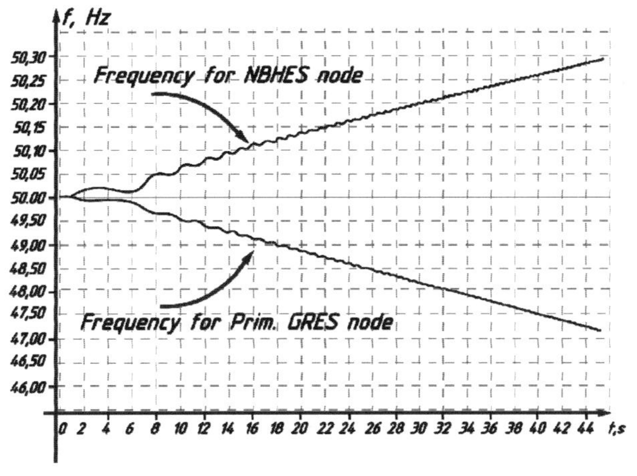

Figure. 8. Transient processes in frequency in generating units

It can be seen from the graph that in the event of an emergency situation similar to the one simulated in the simulation experiment, the internal capabilities of the generators to maintain the frequency are insufficient.

After the sixth second, an asynchronous run occurs, the difference between the frequencies of the generators increases, the system transitions to an unstable state.

To return the system to a stable state, it is necessary to have a control action from the side of the emergency automatics equipment.

The active and reactive power graph, see Figure 9, recorded at the intermediate node TS 220 Birobidzhan, gives an idea of the periodic change in power during synchronous swing and asynchronous run.

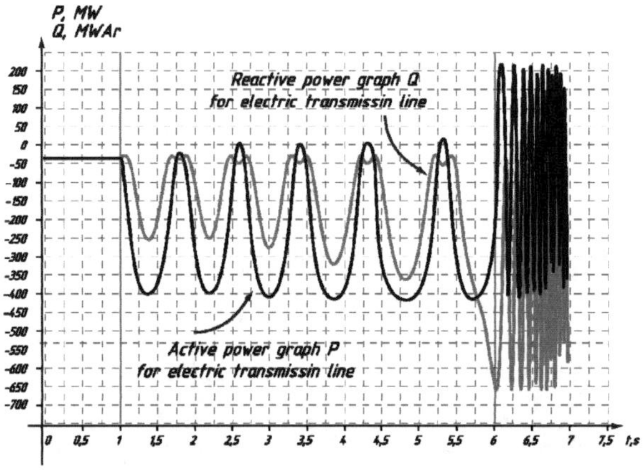

Figure. 9. Active and Reactive power graph

4. Simulation of AEAM-TS work

To detect asynchronous modes, AEAM-Ts have an angular detecting tool that predicts the occurrence of AM and triggers the device before the first asynchronous cranking [2]. This detecting device is used in cases where there is a real danger of a multi-frequency AM and when, during the first asynchronous rotation, deep voltage drops at substations located in the immediate vicinity of the electrical swing center are unacceptable

As a control action on the part of AEAM-Ts, shutdown of one generator unit per 100 MW at the NBHES node and disconnection of the load at the intermediate node TS 220 Birobidzhan are set.

Figure 10 illustrates the frequency graphs when simulating the operation of the AEAM-Ts.

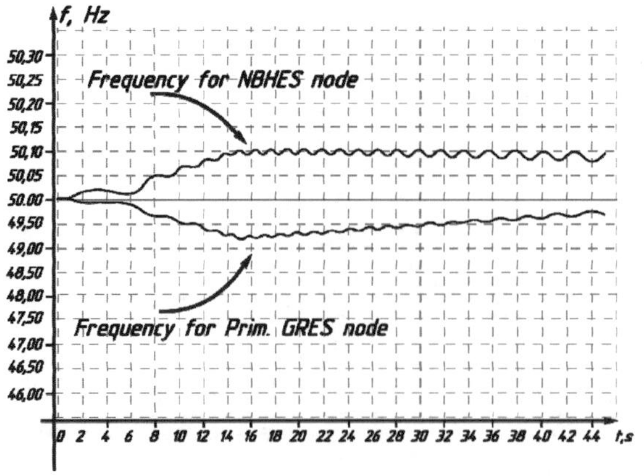

Figure. 10. Frequency graph after AEAM-Ts action

It can be seen from the given graphs that the control actions launched the processes of the system exit from the asynchronous mode and further stabilization under the influence of the frequency and power control systems of generators.

5. Conclusion

Asynchronous modes of operation of power grids are rare. The consequences of asynchronous modes are associated with a high risk of damage to equipment and entail serious economic costs.

To effectively counter situations with the emergence of an asynchronous mode, it is necessary to comprehensively study the conditions for its occurrence, which is possible only with the help of simulation experiments [8, 9 and 10].

During the simulation experiment, the following results were obtained:

- An effective method has been developed to identify the moment of occurrence of a two-frequency asynchronous mode in a section of the power system of the Far East.
- The control actions necessary to eliminate the asynchronous mode without changing the network topology have been determined.
- It has been shown experimentally that a violation of connectivity and stability in the power system can lead to an avalanche-hazardous increase in voltage and frequency in asynchronous mode.
- Criteria for the effectiveness and optimality of the use of emergency equipment were obtained.

From the analysis of the experimental results, it follows that in order to reduce the costs associated with the operation of the power system in the AM, switching equipment with a higher speed is needed, since the transition of the system to a steady state proceeds rather slowly, and, accordingly, the duration of overvoltage in the system can lead to significant economic losses.

In general, there is a need to perform a wide range of simulation studies to analyze possible scenarios for emergencies and select measures to prevent them and reduce economic costs.

References

[1] Afanasyev A P 2019 *Application Weighted Voronoi Diagrams to Determine the Areas of Responsibility of Transformer Substation, with Accounting Heterogeneity of Load Density, 2019 International Multi-Conference on Industrial Engineering and Modern Technologies (FarEastCon)*, Vladivostok, Russia, 2019 1-4, doi: 10.1109/FarEastCon.2019.8934008.

[2] Ivanov K M and Pastukhov V S 2009 *Sposob effectivnogo viyavleniya momenta vozniknoveniya asinkhronnogo rezhima v energosisteme. Nauchnyye problem transporta Sibiry i Dalnego Vostoka* **2** 382-384

[3] Jiaming L 2018 *Measurement and Analysis of Overvoltages in Power Systems* (Singapore: John Wiley & Sons Singapore Pte Ltd)

[4] Khrushchev Y V, Kuleshova Y O and Shandarova Y B 2017 *Elektromekhanicheskiye perekhodnyye protsessy v elektroenergeticheskikh sistemakh* (Tomsk: Izd-vo TPU)

[5] McDonald J D 2017 *Electric Power Substation Engineering* (Boca Raton: CRC Press Taylor & Francis Group)

[6] Narovlyanskiy V G, Vaganov A B and Ivanov I A 2010 *Rezultaty ispytaniy ustroystva likvidatsii asinkhronnogo rezhima ALAR-M na elektrodinamicheskoy modeli energosistemy. Releynaya zashchita i avtomatizatsiya* **1** 16-20

[7] Salam A A and Malik O P 2018 *Electric Distribution Systems* (New Jersey: John Wiley & Sons)

[8] Scutariu M and MacDonald M 2009 *Industrial power system protection against transmission system blackouts 44th International Universities Power Engineering Conference (UPEC)*, (Glasgow) 1-5.

[9] Sokolov A P and Gazizova O V 2018 *Improving the Accuracy Mathematical Modeling of Transients Emergency Mode Industrial Facilities Distributed Generation International Youth*

Scientific and Technical Conference Relay Protection and Automation (RPA) 1-15 doi: 10.1109/RPA.2018.8537190.
[10] Zhang C, Li Y, Yu Z and Tian F 2016 *Feature selection of power system transient stability assessment based on random forest and recursive feature elimination IEEE PES Asia-Pacific Power and Energy Engineering Conference (APPEEC)*, Xian,1264-1268, doi: 10.1109/APPEEC.2016.7779696

IOP Conf. Series: Materials Science and Engineering 1151 (2021) 012036 doi:10.1088/1757-899X/1151/1/012036

Evaluation of the compatibility of the power traction supply system in the field of implementing the train traffic interval regulation with a use of a "virtual coupling" technology

V A Olentsevich[1], N P Astashkov[1], A R Akhmetshin[2], K V Suslov[3], M G Shtayger[4], A I Karlina[3]

[1] Irkutsk State Transport University, 15 Chernyshevsky Street, Irkutsk, 664074, Russia
[2] Kazan State Power Engineering University, Kazan, Russia
[3] Irkutsk National Research Technical University, 83 Lermontov Street, Irkutsk, 664074, Russia
[4] MC Mechel Steel, 1 Krasnoarmeysky Street, 125167, Moscow, Russia

E-mail: olencevich_va@irgups.ru, karlinat@mail.ru

Abstract. This article will propose the practical use of the interval control technology that will reduce the interval of passing trains and will increase the capacity of railway lines. In general, the effectiveness of implementing the train traffic interval regulation with a use of a "virtual coupling" technology is presented in the works of many authors; however, the issues of the use of the existing infrastructure of the traction power supply system are not considered in the works. Comparing the "virtual coupling" technology with connected trains driving consists of the capacity reserve and train schedules evaluation. The use of modern software tools for mathematical modeling will allow you to solve not only the tasks set, but also to evaluate the compatibility of the currently used traction power supply system. According to the results of authors meant below, it can be seen that the temperature in the contact system and the suction line, the load limiting factor and the energy losses of the traction network trains movement in the "virtual coupling" are lower than those during the movement of the united train. When using trains with a virtual coupling, the voltage in the contact network is significantly higher than when using connected trains. Based on the above, it can be concluded that the virtual coupling trains use on the section under consideration is more appropriate than the use of the connected trains.

1. Introduction

In accordance with the scientific basis of the railway transport development strategy accepted for the period up to 2030, the tasks of industry main costs minimization are brought to the fore and are aimed at the technological equipment operation and repair [1]. Development of the transportation processes optimization policy is possible when we take into account such moments as the use of modern technical means and technological processes, which are compatible with operational conditions of electrical equipment and with power supply systems.

Today the use of new technologies in the transportation process is a necessity due to the sharp increase in cargo traffic towards the Far East. Over the past 15 years, the volume of freight traffic in this direction has increased more than 2.5 times, while the railway transport infrastructure in operation has not changed for the most part. Since the new railway lines construction is not economically justified, it

Content from this work may be used under the terms of the Creative Commons Attribution 3.0 licence. Any further distribution of this work must maintain attribution to the author(s) and the title of the work, journal citation and DOI.
Published under licence by IOP Publishing Ltd

is necessary to find alternative ways to meet the growing demands of local businesses and industrial enterprises, one of which is the "virtual coupling" technology, which can potentially increase the maximum of the possible cargo delivery volume more than five times [2, 3, 4].

Nowadays one of the main directions in the field of the transportation process improvement is the development of interval control technology, the use of which will allow one to determine the distance between trains not by free block sections, but by focusing on the current location of the rolling stock. The practical use of this technology will reduce the interval of passing trains and will increase the capacity of railway lines and other transport infrastructure. The technology is primarily aimed at solving the problem of the railway lines sections carrying capacity increase without upgrading the existing infrastructure, which is especially important in modern conditions with the growth of traffic volumes on the entire railway network [5, 6].

2. Traction calculation of freight trains on the Achinsk – Mariinsk section

Reliable power supply provision for train traction and non-traction power consumers, maintenance of the specified capacity and carrying capacity of railway sections without main elements overload, ensuring the standard quality of electricity [7-10], taking into account the use of the considered (in the article) system, will allow us to justify the expediency of its practical application.

When calculating acceptable inter-train intervals, we used a conditional parallel traffic schedule with an average distribution of the largest and average trains mass on each track along the entire section. Taking into account that only certain inter-station zones and tracks are the busiest at different time periods, calculations based on the specified schedules provide an acceptable margin of the capacity utilization degree to compensate for intra-day fluctuations in the traffic size [11, 12, 13].

On the basis of the mode maps, the article considers the calculated section of the Mariinsk – Achinsk railway, which track profile is presented in the "KORTES" program (Figure 1).

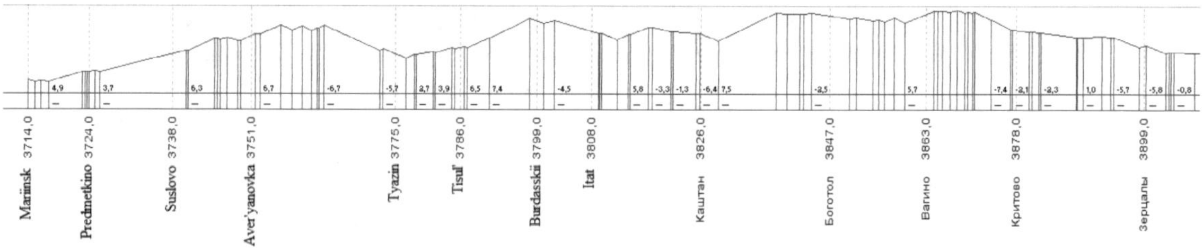

Figure 1. Longitudinal profile of the Mariinsk – Achinsk section.

Taking into account the speed limits of the presented section, which are 80 km/h in both even and odd directions, determines the current consumption of the train. The presented calculations are focused on a series of electric locomotives 2(3)ES5K, due to their use in the considered area.

The forecast values of freight train weight standards and the freight train traffic value for 2025 are shown in table 1.

Table 1. Freight train traffic sizes up to 2025

Section name	Weight standards of freight trains, tons	Size of freight trains, trains/day	
		there	back
Mariinsk – Achinsk	7100	7	1
	6000-6300	14	10
	4000-4200	19	19
	3000	10	10
	3000	1	1
	1500-1700	-	10
	Total	51	51

The traction calculations results with a use of the LVI-table in the program complex "KORTES", were acquired total current variation graphs from the train coordinates for the considered types of trains in the odd and even directions, see figure 2. These relationships indicate a possible omission of the all freight trains categories within the existing power supply system parameters and class insulation of the electric locomotive traction equipment [14, 15, 16, 17].

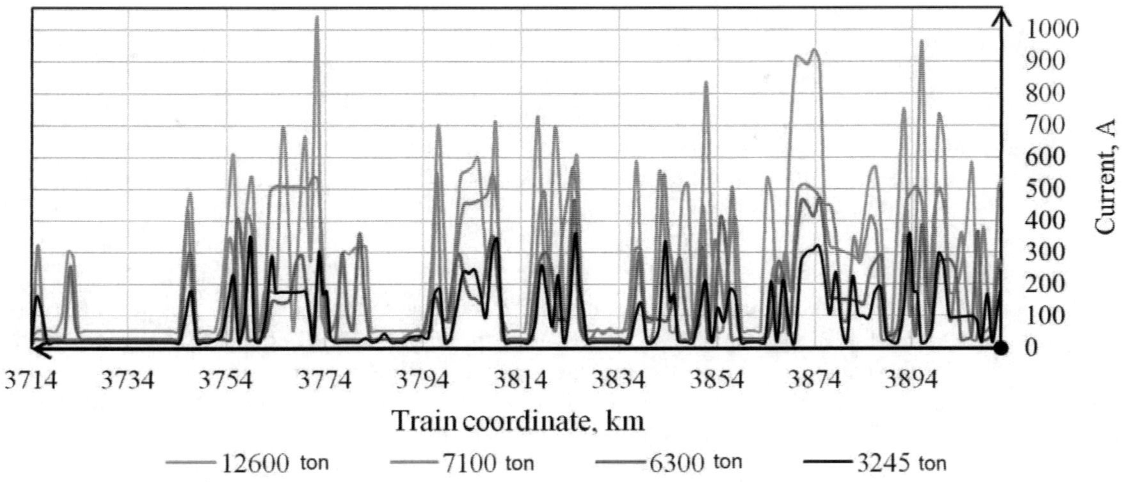

(a)

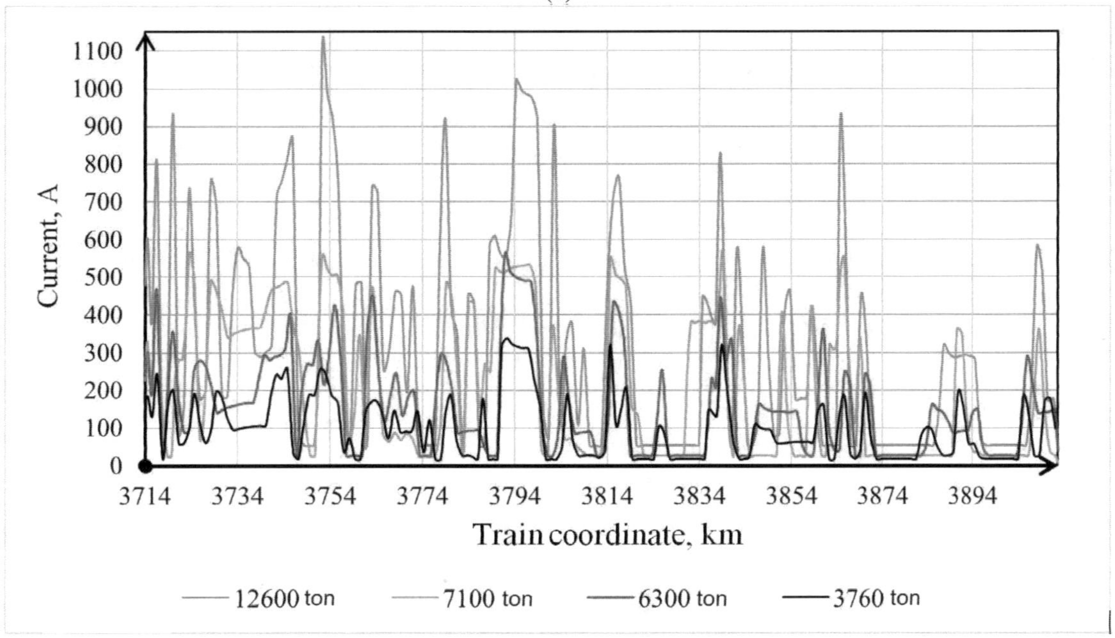

(b)

Figure 2. The total current dependences graphs on the coordinates of odd (a) and even (b) trains of different masses on the Achinsk – Mariinsk section.

According to the conducted scientific research, traction calculations were performed for the presented section in order to determine the value of the traction load in the power supply system [18-21]. The calculations consider high and medium weight freight trains in even and odd directions and simulate the movement of a connected train with a weight norm of 12.600 tons. On the basis of the data in table 1, the average freight trains weight is calculated for the specified weight standards without taking into account the number of trains with increased weight. The calculated value for the railway line sections is:

– odd direction of the Mariinsk – Achinsk section:

$$m_{mid} = \frac{4200 \cdot 19 + 3000 \cdot 10 + 3000 \cdot 1 + 1700 \cdot 10}{19 + 10 + 1 + 10} = 3245 \text{ tons}; \quad (1)$$

– even direction of the Mariinsk – Achinsk section:

$$m_{mid} = \frac{4200 \cdot 19 + 3000 \cdot 10 + 3000 \cdot 1}{19 + 10 + 1} = 3760 \text{ tons} \quad (2)$$

The parameters shown in table 3 indicate that the load limiting factor for the two types of trains does not exceed the permissible value, and the transformer oil temperature does not exceed the permissible value.

3. Conclusion

On the basis of the obtained results, it is proved that it is reasonable to move trains in the "virtual coupling" mode, since the energy consumption, losses in the traction network, the limiting load factor, the temperature in the contact network and the suction line are lower than when the connected train technology is used. When trains are used in the «virtual coupling» mode the voltage in the contact network is also significantly higher than when using connected trains.

The "virtual coupling" technology assumes that radio modems installed on the electric locomotives maintain communication between the master and slave trains via a secure digital channel and transmit energy-optimal driving modes for locomotives that also ensure the minimum safe interval between trains. As a result, the section capacity increases: the inter-train interval with the new technology will be 5 minutes, and the distance between trains will be reduced to the required stopping distance.

References

[1] *Development strategy of the Russian Railways Holding for the period up to 2030* 2013 (Moscow: JSC Russian Railways)

[2] *Official website of Russian Railways:* http://rzd.ru

[3] Taranets I 2019 Digital technologies will reduce the time of registration of transportation by five times *Newspaper "Gudok"*

[4] Klimova E V, Pilipushka L E, Ryabov V S 2019 The technology of "virtual coupling" of trains as a tool for increasing the carrying and carrying capacity of the line *Transport infrastructure of the Siberian region* Vol. 1 (Irkutsk: IrGUPS) pp 60-64

[5] Rosentals E M, Buinova N V 2019 *Interval regulation: innovations and development prospects: a thematic selection* (Krasnoyarsk: KrTsNTIB) p 232

[6] Cherpanov A V, Kutsy A P, Esaulenko A S 2020 Application of virtual coupling technology for high-mass trains *Young Science of Siberia: electron. scientific. zhurn* **2**

[7] Belogolov Yu I, Olentsevich V A, Astashkov N P 2018 Improving the operational management of transport processes in railway transport *Proceedings of 6th International Symposium on Innovation and Sustainability of Modern Railway ISMR 2018* (Irkutsk: ISTU) pp 602-609

[8] Olentsevich V A, Belogolov Yu I, Kramynina G N 2019 Set of organizational, technical and reconstructive measures aimed at improvement of section performance indicators based on the study of systemic relations and regularities of functioning of railway transport system *IOP Conf. Series: Materials Science and Engineering* **832** 012038

[9] Olentsevich V A, Upyr R Yu, Gladkih A M 2020 Computational procedure for preparing the technical conditions for stowage and securing cargo in rail cars and containers *Journal of Physics: Conference Series* **1615** 012029

[10] Olentsevich V A, Gozbenko V E, Kargapoltsev S K, Kramynina G N 2019 Complex of

organizational, technical and reconstructive measures aimed at improving the performance of the section based on the study of system connections and regularities of the functioning of the railway transport system *Modern technologies. System analysis. Modeling* Vol. 62 **3 (63)** pp 171-179

[11] Tumanin A S 2019 *Technological process of the traffic control center of the Krasnoyarsk traffic control directorate - a branch of Russian Railways* p 94

[12] Arkhangelsky E V, Vorobiev N A, Drozdov N A, Miroshnichenko R I, Segal L G 1997 *Calculation of the throughput of railways* (Moscow: Transport)

[13] Gromyshova S S, Astashkov N P, Olentsevich V A, Lobanov O V 2019 Assessment of the level of security of complex-structured transport systems in order to increase the level of their competitiveness in the market of transport services *Modern technologies. System analysis. Modeling* Vol. 62 **2** pp 250-259

[14] Astrakhantsev L A, Astrakhantseva N M, Astashkov N P 2012 Development of resource-saving electrified technological processes *Bulletin of KrasGAU* **8 (71)** 166-169

[15] Astrakhantsev L A, Astashkov N P, Ryabchenok N L, Alekseeva T L, Astrakhantseva N M 2012 Increasing the electromagnetic compatibility of rolling stock *Safety of regions - the basis of sustainable development* **1-2** 92-94

[16] Ryabchenok N, Alekseeva T, Astrakhancev L, Astashkov N, Tikhomirov V 2020 Energy-saving driving of heavy trains *Advances in Intelligent Systems and Computing* **982** 491-508

[17] Upyr' R Yu, Goncharova N Yu, Eremenko M N 2020 Heavy traffic - a reserve for increasing the carrying capacity *Railway transport* **7** 10-13

[18] Nazarychev S A, Akhmetshin A R and Gaponenko S O 2019 Full compensation of reactive power in electric networks 0.4-10kV *Journal of Physics: Conference Series, XI Scientific Technical Conference on Low Temperature Plasma during the Deposition of Functional Coatings.* **1588** 012036 DOI: 10.1088/1742-6596/1588/1/012036

[19] Suslov K V , Solonina N N, Smirnov A S 2014 Distributed power quality monitoring, *Proceedings of International Conference on Harmonics and Quality of Power, ICHQP* pp 517-520 DOI: 10.1109/ICHQP.2014.6842882

[20] Akhmetshin A, Marin G and Mendeleev D 2020 Modeling of asynchronous motor operation modes for the correct selection of voltage regulation devices *2020 High Speed Turbomachines and Electrical Drives Conference (HSTED 2020) E3S Web of Conferences* **178** 01015 DOI: 10.1051/e3sconf/202017801015

[21] Suslov K, Piskunova V, Gerasimov D, Ukolova E, Akhmetshin A, Lombardi P, Komarnicki P 2019 Development of the methodological basis of the simulation modelling of the multi-energy systems *In Proc. International Scientific and Technical Conference Smart Energy Systems 2019* (SES-2019) *E3S Web of Conferences* **124** 01049 DOI: 10.1051/e3sconf/201912401049

IOP Conf. Series: Materials Science and Engineering 1151 (2021) 012037 doi:10.1088/1757-899X/1151/1/012037

Checking the operation parameters of the traction power supply system with a long-term train schedule on the rated section of the railway line

K V Suslov[1], A R Akhmetshin[2], N P Astashkov[3], V A Olentsevich[3],M G Shtayger[4] and A I Karlina[1]

[1] Irkutsk National Research Technical University, 83, Lermontova str., Irkutsk, 664074, Russia
[2] Kazan State Power Engineering University, 51, Krasnoselskaya str., Kazan, 420066, Russia
[3] Irkutsk State Transport University, 15, Chernyshevskogo str., Irkutsk, 664074, Russia
[4] MC Mechel Steel, 1 Krasnoarmeysky str., 125167, Moscow, Russia

E-mail: souslov@ex.istu.edu, karlinat@mail.ru

Abstract. The handling of trains of increased weight and length, consisting of rolling stock with an axle load of 25 tons, and up to 27 – 29 tons in the future, is a difficult technical and economic task facing Russian Railways OAO today. A heavy train cannot be slowed down often, much less stopped, since this leads to a significant loss of energy. In the article, the authors considered the issue of strengthening the traction power supply system on the rated section of the railway line of the Eastern polygon in connection with an increase in the amount of cargo turnover by 2025. Based on the results of the traction calculation, the predicted values of the indicators were obtained. They are the total travel time of the train along the section of the railway line, the free running time of the train, the consumption of active and total electricity, the specific consumption of active and total electricity for each inter-substation zone of the railway line section. The operation parameters of the traction power supply system with a long-term train schedule on the rated section of the railway line are determined. The research results obtained by the authors made it possible to determine the limiting sections and objects of transport infrastructure for the handling of heavy-haul and long freight trains on the rated direction.

1. Introduction

In connection with the implementation of development programs in the energy complex, metallurgical and forestry industries on the territory of the Eastern polygon, a positive trend in the volume of cargo loading and expansion of the geography of railway transportation in general is expected in the long term. Thus, an increase in the volume of coal transportation for export after the implementation of the 1st and 2nd stages of development of the Eastern polygon in 2025 will increase the income of Russian Railways OAO (RZD OAO) and rolling stock operators by more than 180 billion rubles [1, 3, 5, 7].

The development of increasing volumes of freight traffic on the Eastern polygon is impeded by the shortfall in the carrying and throughput capacity of railway lines (which is more than 70% of the entire length of the Eastern polygon), as well as the presence of sections with insufficient track development of railway stations [2, 5, 11]. Intensive growth in traffic volumes also causes rapid deterioration of the

Content from this work may be used under the terms of the Creative Commons Attribution 3.0 licence. Any further distribution of this work must maintain attribution to the author(s) and the title of the work, journal citation and DOI.
Published under licence by IOP Publishing Ltd

main equipment of the infrastructure of this direction, which is already operated in difficult natural and climatic conditions: power supply and railway tracks. An increase in traffic volumes without the construction of additional main tracks leads to a decrease in the electricity quality and to significant losses both in the traction system and in external power supply. In this connection, these objects require regular strengthening, updating and technical development.

The handling of trains of increased weight and length, consisting of rolling stock with an axle load of 25 tons, and up to 27 – 29 tons in the future, is a difficult technical and economic task facing Russian Railways OAO today. In order to eliminate the identified problems, a set of measures is being implemented to develop the railway infrastructure of the polygon, included in the investment project "Modernization of the railway infrastructure of the Baikal-Amur and Trans-Siberian railways with the development of throughput and carrying capacity" [2, 4].

Increasing the weight norms is one of the priority directions, allowing one to increase the throughput and carrying capacity of the Eastern polygon, to increase the efficiency of work in these conditions. By 2025, it is planned to complete work on lengthening the reception and departure sidings of railway stations to the standard of 1050 meters along the routes of moving bulk cargo to ports and border crossings. With that, a decision was made to develop regulatory requirements for the arrangement and maintenance of railway infrastructure facilities on the sections of the running of freight trains weighing over 6000 tons, including the development of a method for monitoring their condition [4-6].

In the 1980s the Americans switched to an axle load standard of 27 – 29 tons for eight years. A special track was built on the experimental circle, in which new rails and fasteners were laid. The ones that were more reliable were chosen. The construction gang worked in the same way with the new rolling stock. They strengthened the existing power supply system. As a result, they mastered the technology of motion with high axle loads, and with it, they already headed over to the mainlines. The result turned out to be efficient. Then they spent the same amount of time switching to axle loads over 30 tons. We should have the same approach [7, 19].

2. Checking the existing operating parameters of the traction power supply system of the Gidrostroitel – Korshunikha section of the Eastern polygon

In the article, the authors considered the issue of strengthening the traction power supply system (TPSS) [3, 5, 12] and at the Gidrostroitel-Korshunikha section of the Eastern polygon due to the increase in the amount of the cargo turnover by 2025 (see Table 1).

The total length of the rated section is about 200 km. The traction calculation was carried out using the KORTES software package. The parameters of a given power supply section (separate points, speed limits, longitudinal profile, categories and types of trains), mass of trains and types of electric rolling stock used in the calculation were used as the initial data. For further calculation of the optimal load conditions of the traction power supply system at a given throughput capacity, it is necessary to make a traction calculation for the base train weights of 3000 tons and 6300 tons [8-10, 13, 16, 18].

Based on the results of the traction calculation, the predicted values of the following indicators were obtained:

– total running time of the train along the section of the railway line;
– free running time of the train;
– consumption of active and total electrical energy;
– specific consumption of active and total electrical energy for each inter-substation zone of the railway line section.

The results of the traction calculation for the up and down directions of movement of trains weighing 3000 and 6300 tons are presented in Table 2, and the specific consumption of active and total energy is shown in Table 3.

Table 1. Long-term volumes of traffic of a section of the railway line for 2025

Section of the railway line	Weight trains, tons	Traffic volumes, train pairs per day: freight		Traffic volumes, train pairs per day: passenger		Volumes of train traffic during intensive hour: of freight trains		Volumes of train traffic during intensive hour: of passenger trains in one direction		Train interval
	7100		35				4			
	6000	8	4			1	1			
	4000	4	5							
	3000 con.	1	1			5	1			
Korshunikha-Gidrostroitel	3000 ass.	5	5	8	5			0	0	10
	1500 - 1700 emp.	34	2							
	Total	52	52			6	6			

Table 2. Results of traction calculation by the directions of movement on the rated section of the railway line

Designation of the station-to-station block	Length, km	Full running time, min: 3000	Full running time, min: 6300	Free running time, min: 3000	Free running time, min: 6300	Active electricity consumption, kW*h: 3000	Active electricity consumption, kW*h: 6300	Full electricity consumption, kW*A*h: 3000	Full electricity consumption, kW*A*h: 6300
for down direction									
Gidrostroitel – Zyaba	25.1	20.6	20.6	7.0	7.9	482.1	545.1	436.0	648.8
Zyaba-Kezhemskaya	39.0	33.3	33.5	11.7	12.3	1290.2	1560.9	1535.0	1857.9
Kezhemskaya-Vidim	58.1	52.5	56.5	34.6	39.8	4012.0	5769.9	4773.7	6869.5
Vidim-Sokhatyi	19.0	16.2	17.1	6.5	7.7	739.0	1132.6	879.2	1348.5
Sokhatyi – Chyornaya	21.0	18.0	19.3	16.5	19.3	1505.5	2127.1	1791.6	2530.3
Chyornaya-Sredneilimskaya	33.0	36.3	37.6	31.8	32.4	3557.6	4818.5	4233.6	5738.6
Sredneilimskaya-Korshunikha	17.5	18.1	17.1	3.3	2.6	382.5	338.2	455.4	402.8
for up direction									
Gidrostroitel – Zyaba	25.1	22.9	24.6	20.7	22.0	1948.1	2921.6	2318.3	3477.8
Zyaba-Kezhemskaya	39.0	35.3	38.3	26.5	29.9	3037.1	4214.2	3613.9	5017.5
Kezhemskaya-Vidim	58.1	50.0	52.3	19.5	23.2	2253.5	3222.4	2681.0	3836.1
Vidim-Sokhatyi	19.0	18.5	20.2	15.8	17.7	1858.5	2632.8	2211.2	3135.2
Sokhatyi – Chyornaya	21.0	19.5	19.1	2.1	1.7	111.7	120.8	133.1	143.8
Chyornaya-Sredneilimskaya	33.0	31.0	30.8	1.9	0.8	233.3	117.8	277.5	140.2
Sredneilimskaya-Korshunikha	17.5	16.5	15.8	13.4	15.8	1464.7	1670.6	1742.6	1987.3

IOP Conf. Series: Materials Science and Engineering 1151 (2021) 012037 doi:10.1088/1757-899X/1151/1/012037

To check the operation parameters of the traction power supply system of the Gidrostroitel – Korshunikha section, the authors performed electrical calculations of the operating modes and characteristics of the throughput capacity of railway sections electrified using the 27.5 kV alternating current traction power supply system. The characteristics of the throughput capacity are determined depending on the problem to be solved under the design modes. In this case the calculation was performed in mode B "Determination of the smallest permissible intervals during intensive periods of section operation" and the load mode with a parallel train traffic schedule on the section. The calculation is implemented with two operating transformers and switched on devices for compensation of reactive power of the series capacitor bank at the Kezhemskaya traction substation [14, 15, 17-19]. Figure 1 shows an existing version of the train traffic schedule on the section with the current power supply system.

The results of calculating the minimum allowable train interval for the existing power supply scheme for the existing amount of traffic are presented in Table 4.

Table 3. Energy consumption depending on the series of locomotives.

Locomotive Series / Directions of traffic	2ES5K	2ES5K	2ES5K	2ES5K
	Up		Down	
Train weight, tons	6300	3000	6300	3000
Consumption of active energy, W*h/t*km	10.2	11.0	11.3	12.0
Consumption of total energy, V*A*h/t*km	12.2	13.1	13.5	14.3

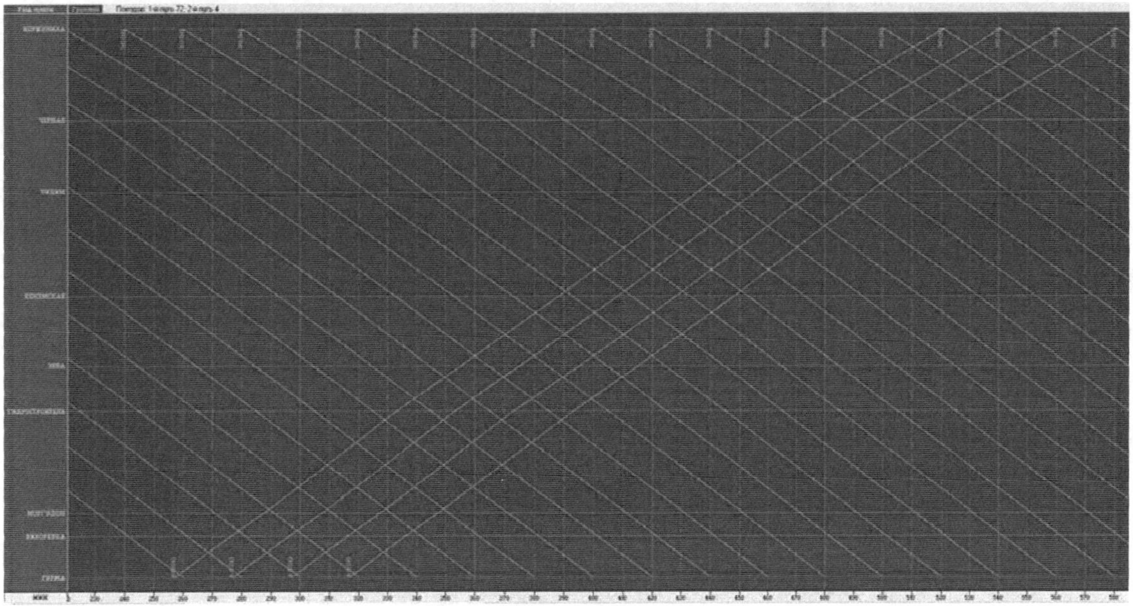

Figure 1. The long-term version of the train traffic schedule on the rated section.

Table 4. Allowable train intervals under the existing version of the train traffic schedule on the rated section

Designation of the inter-substation zone	Interval value, min, limited			
	By power of the step-down transformers	By overhead contact system voltage	By heated wires of the overhead contact system	The resulting value
Gidrostroitel – Zyaba	7	6	6	7
Zyaba - Kezhemskaya	8	10	6	10
Kezhemskaya - Vidim	8	16	6	16
Vidim – Chyornaya	7	11	6	11
Chyornaya - Korshunikha	6	13	6	13

3. Checking the operation parameters of the traction power supply system with a long-term train traffic schedule on the rated section of the railway line

Figure 2 shows a long-term version of the train schedule on the section with the current power supply system. The results of calculating the minimum allowable train interval for the existing power supply scheme for the long-term amount of traffic are presented in Table 5.

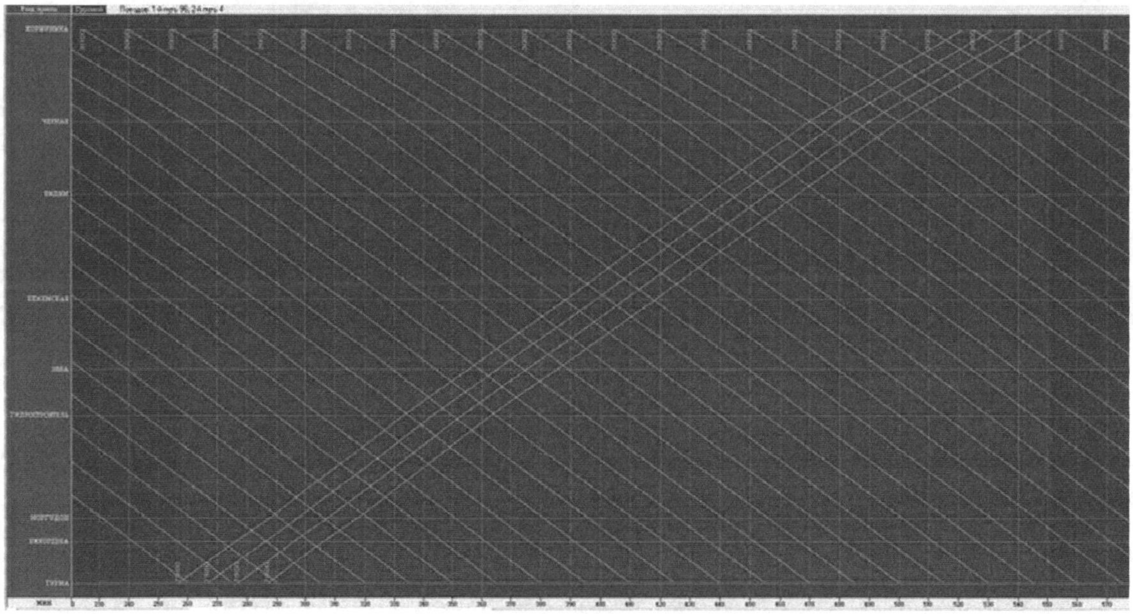

Figure 2. The long-term version of the train traffic schedule on the rated section.

Table 5. Allowable train intervals for the long-term version of the train traffic schedule for the rated section

Designation of the inter-substation zone	Interval value, min, limited			
	By power of the step-down transformers	By overhead contact system voltage	By heated wires of the overhead contact system	The resulting value
Gidrostroitel – Zyaba	6	6	6	6
Zyaba - Kezhemskaya	12	12	6	12
Kezhemskaya - Vidim	12	24	6	24
Vidim – Chyornaya	6	16	6	16
Chyornaya - Korshunikha	6	13	6	13

The calculation of the minimum voltage on the current collector of the locomotive with the available power supply scheme for the existing and long-term amount of train traffic on the section, as well as the calculation of the loads of transformers and heating of the catenary wires [14, 17, 18]. Based on the obtained values, a comparative analysis of the minimum voltages across the current collectors of locomotives for up (1) and down (2) railway tracks, respectively, with the existing and long-term train schedule was carried out. U1 is the voltage across the current collector under the long-term train schedule. U2 is the voltage across the current collector under the existing train schedule (Figures 3 and 4).

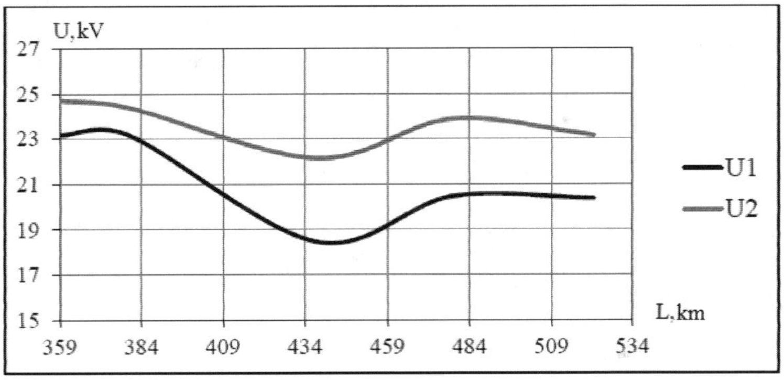

Figure 3. Minimum voltage across current collector 1, main track.

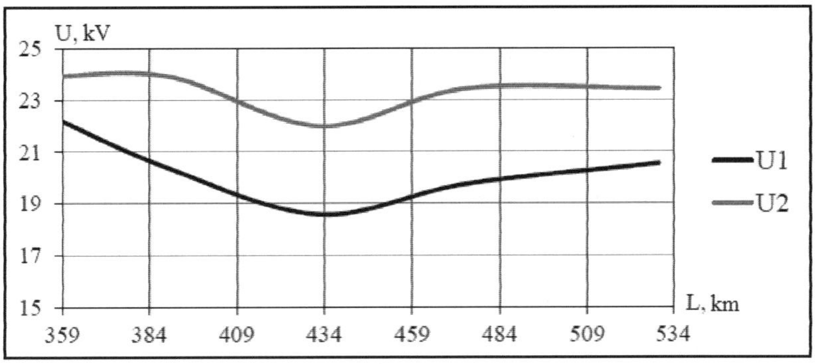

Figure 4. Minimum voltage across current collector 2, main track.

Figure 5 shows a comparative analysis of oil heating in a traction transformer under the existing and long-term train schedule. T1 is the oil temperature of the traction transformer under the long-term train schedule. T2 is the oil temperature of the traction transformer under the existing train schedule.

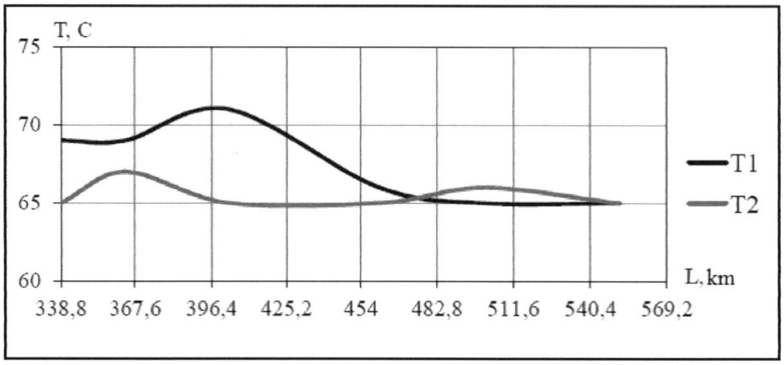

Figure 5. Oil temperature of traction transformer.

4. Conclusion

The research results obtained by the authors made it possible to determine the limiting sections and objects of transport infrastructure for the handling of heavy-haul and long freight trains on the rated direction. Thus, after the calculations under the existing scheme with a predetermined train interval of 10 minutes, on the section of the railway line Gidrostroitel – Korshunikha for the amount of traffic 3000 tons on the down track and 7.100 tons on the up track, there are four limiting station-to-station blocks of Zyaba – Kezhemskaya – Vidim, Vidim – Chyornaya, Chyornaya – Korshunikha. The train interval on this section was 12, 24, 16, 13 minutes, respectively.

Considering the obtained calculation results, we can conclude that to ensure the passage of trains of a given amount of traffic with a given inter-train interval of 8 minutes, measures are required to increase the voltage level at the limiting station-to-station blocks [12, 14, 18].

References

[1] Pekhterev F S 2015 Prospects for the development of the railway infrastructure of the Eastern polygon of the Russian Railways network *Economy of railways* **2** 60

[2] Pekhterev F.S. *Catching behind the cargo base: an interview*, available at: https://www.rzd-partner.ru/zh 10.10.2017

[3] *Railway transport in Russia: challenges until 2025* , available at: www.ipem.ru

[4] The Order of the Government of the Russian Federation No. 2116-r 2014 *Passport of the investment project "Modernization of the railway infrastructure of the Baikal-Amur and Trans-Siberian railways with the development of carrying and throughput capacities".*

[5] The order of the Government of the Russian Federation No. 877-r 2008 *The strategy for the development of railway transport in the Russian Federation until 2030*

[6] Gozbenko V E, Belogolov Yu I and Olentsevich V A 2018 Analysis of the level of reliability and stability of organizational and technical systems of the transportation process of railway transport *Modern technologies. System Analysis. Modeling* **1 (57)** 147-156

[7] *The official website of the Russian Railways,* available at: http://rzd.ru

[8] 1985 *Traction calculation rules for train operation* (Moscow: Transport Publ.)

[9] The Order of Russian Railways OAONo. 1799r 2016 *On approval of the Instruction for organizing the circulation of freight trains of increased weight and length on public railways of Russian Railways*

[10] Osipov S I and Osipov S S 2000 *The basics of train traction. A textbook for students of technical schools and colleges of the railway transport* (Moscow: UMK Ministry of Railways of Russia Publ.)

[11] Olentsevich V A, Belogolov Yu I and Grigoryeva N N 2020 Analysis of reliability and

sustainability of organizational and technical systems of railway transportation process *IOP Conference Series: Materials Science and Engineering* 012061
[12] Olentsevich V A, Gozbenko V E, Kargapoltsev S K and Kramynina G N 2019 Complex of organizational and technical and reconstructive measures aimed at improving the performance of the section based on the study of system links and patterns of the railway transport system *Modern technologies. System Analysis. Modeling* **3 (63)** 171-179
[13] Rosenfeld V E, Isaev V P, Sidorov N N and Ozerov M I 1995 *Electric traction theory*. (Moscow: Transport Publ.)
[14] 2007 *Rules for the arrangement of the traction power supply system of the railways of the Russian Federation TsE-462.* (Moscow: Transport Publ.)
[15] Alekseeva T L, Ryabchyonok N L, Astrakhantsev L A, Tikhomirov V A, Astashkov N P, Martusov A L and Alekseev M E 2020 Parallel operation of an inverter with an electrical ac network *IOP Conference Series: Materials Science and Engineering* 012003
[16] Ivankova L N, Ivankov A N and Fufacheva M V 2012 The stage-by-stage mastery of transportation on double-track lines when running long-haul freight trains *Modern technologies. System Analysis. Modeling* **2 (34)** 162–165
[17] Ryabchenok N, Alekseeva T, Astrakhancev L, Astashkov N and Tikhomirov V 2020 Energy-saving driving of heavy-haul trains. *Advances in Intelligent Systems and Computing* **982** 491-508
[18] Order of the Ministry of Transport of the Russian Federation No. 286 2010 *Rules for the technical operation of the railways of the Russian Federation,* available at: http://docs.cntd.ru/document/902256286.
[19] Muginstein L A 2018 Fundamentals of experimental methodology for determining critical norms for the mass of freight trains *Railway transport* **12** 22-30

Calculation of operating modes of the traction power network system and the capacity of the 27.5 kV railway line section systems for batch train schedules

K V Suslov[1], **Yu N Zatsarinnaya**[2], **A G Logacheva**[2], **V A Olentsevich**[3], **A I Karlina**[1], **M G Shtayger**[4]

[1] Irkutsk National Research Technical University, 83 Lermontov Street, Irkutsk, 664074, Russia
[2] Kazan State Power Engineering University, Kazan, Russia
[3] Irkutsk State Transport University, 15 Chernyshevsky Street, Irkutsk, 664074, Russia
[4] MC Mechel Steel, 1 Krasnoarmeysky Street, 125167, Moscow, Russia

E-mail: olencevich_va@irgups.ru

Abstract. Taking into account the relatively low profitability of railway transport, need of high capital intensity and the same infrastructure payback period, obviously the implementation of such large-scale projects is a difficult task. Therefore, the railway industry relies heavily on new modern technological processes of train traffic organization on the basis of the existing infrastructure or its partial modernization. The freight trains connection technology via virtual coupling will allow you to dynamically change the number and composition of trains that are part of a combined group, eliminate the inherent problems of the connected trains. But most important is that it creates a capacity reserve due to compaction of the train schedule, if there are no restrictions in the traction power network. By the example of the train schedule organization of the Mariinsk – Achinsk railway line section, the authors of this article analyzed the operating modes and capacity of the 27.5 kV system for a batch train schedule. The results show that connected trains require a longer minimum inter-train interval to restore the traction network. With the same inter-train interval for two trains' types, the main operating mode characteristics of trains with a virtual coupling are better than those for connected trains. Using freight trains in a virtual coupling allows you to reduce the inter-train interval about 1.5-2 times.

1. Introduction

The purpose of the freight trains connection technology via virtual coupling is not to imply physical coupling of this trains category using a special auto-coupling device, but to perform a process in which trains are located at a constantly equal remote distance from each other, which reduces the inter-train interval and significantly increases the railway lines capacity. This technology will allow you to dynamically change the number and composition of trains included in the combined group [1-3, 14].

The analog system successfully operates on all new and reconstructed trans-European railway corridors of the European Union – the European Rail Traffic Management System (ERTMS). It is a set of common standards developed in the framework of international cooperation for railway automation, telemechanics, communications and dispatching control [5, 6. 18].

Content from this work may be used under the terms of the Creative Commons Attribution 3.0 licence. Any further distribution of this work must maintain attribution to the author(s) and the title of the work, journal citation and DOI.
Published under licence by IOP Publishing Ltd

2. Automated intelligent driving system of trains of increased mass and length

A distinctive system advantage is the possibility of automated asynchronous and synchronous control of power plants of various types of electric locomotives in traction and electric braking mode, coordinated brake systems operation with combined brake lines, as well as the implementation of the automated synchronous or time-delayed control function of locomotives from the train head [3, 4, 7]. However, this technology has a number of disadvantages (figure 1).

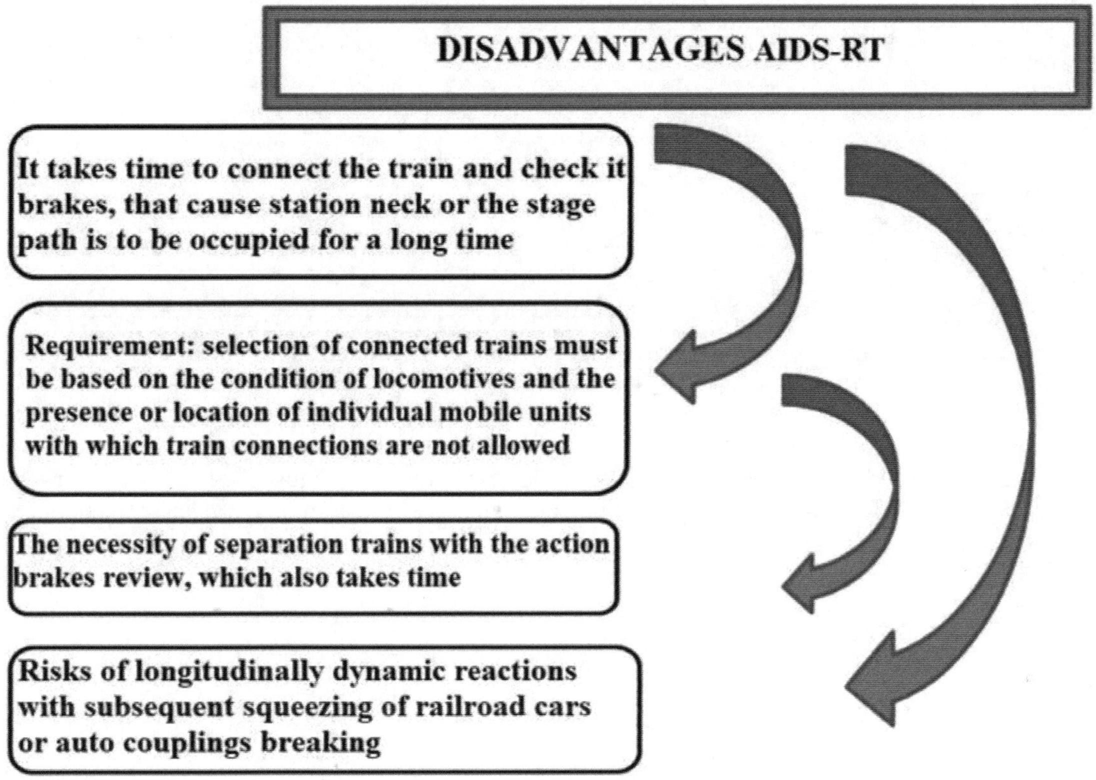

Figure 1. The shortcomings of intelligent systems for automated driving trains ISIT-RT.

By the example of the train schedule organization of the Mariinsk – Achinsk railway line section, the authors of this article analyzed the operating modes and capacity of the 27.5 kV system for a batch train schedule [8-10, 14-17].

3. Calculation of capacity and operating modes of the 27.5 kV system for a batch schedule

A batch schedule is a schedule where trains pass sequentially at a specified interval. For calculations we selected heavy freight trains package, consisting of 10 trains skipped in the even and odd travel directions. Calculations were made of a train schedules for 10 connected trains weighing 12600 tons (figures 2, 5) and for 10 trains in the virtual coupling mode weighing 6300-6300 tons (figures 3, 4). [11-13, 16-17].

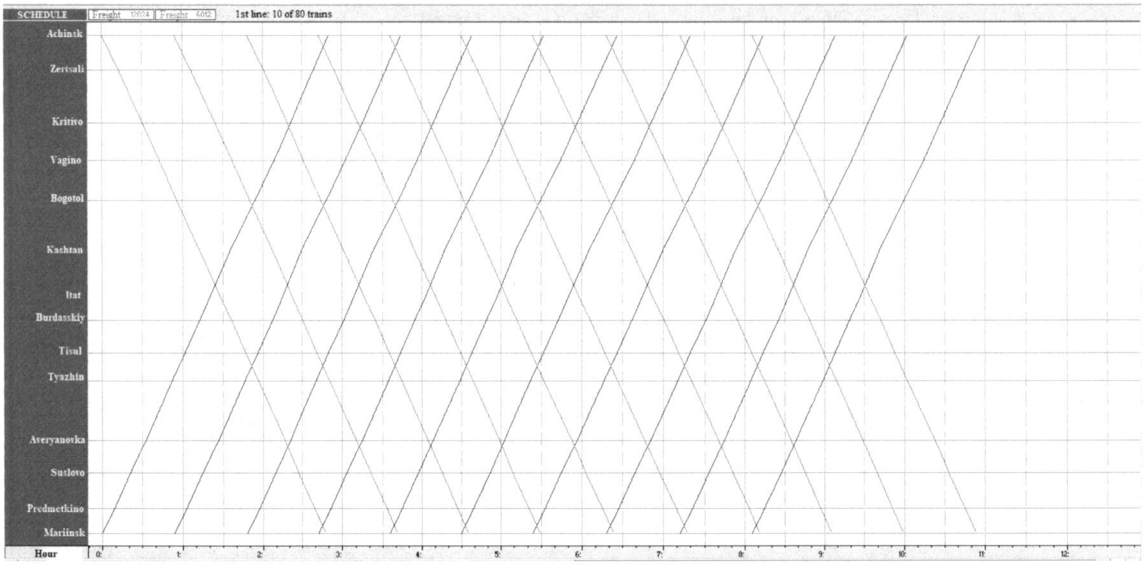

Figure 2. Batch schedule of connected trains of the first and second railway lines on the Mariinsk – Achinsk section.

According to the calculations, the minimum inter-train interval for a batch train schedule consisting of connected trains on the Mariinsk – Achinsk section is 54 minutes (figure 2). Let's model the traffic schedule for trains with a virtual coupling while preserving the inter-train interval set for the combined trains' batch schedule (figure 3).

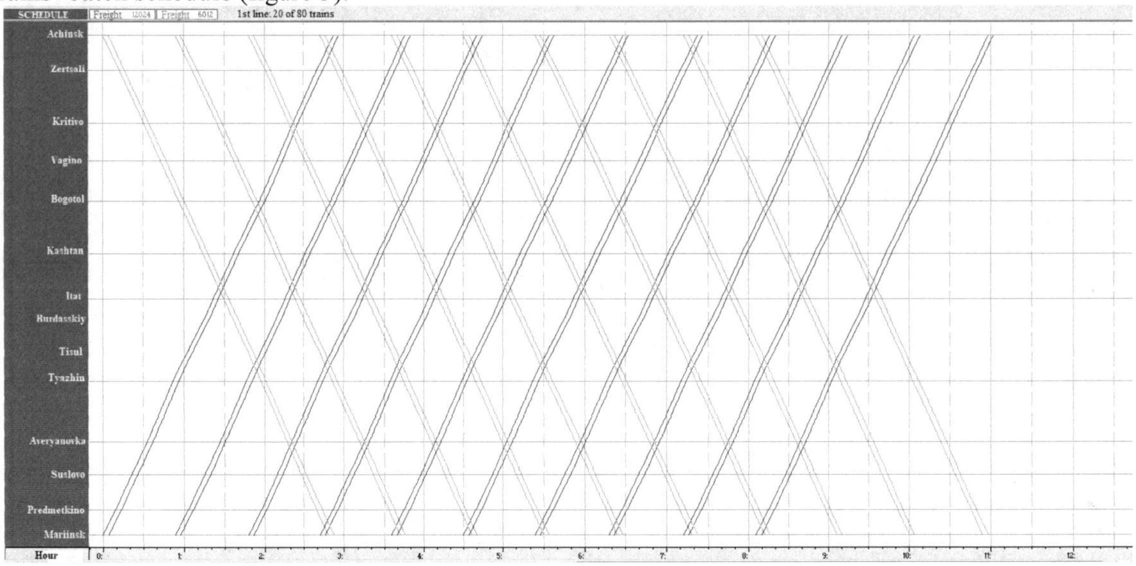

Figure 3. Batch schedule of trains in the virtual coupling mode of the first and second railway lines on the Mariinsk – Achinsk section with a 50-minute inter-train interval.

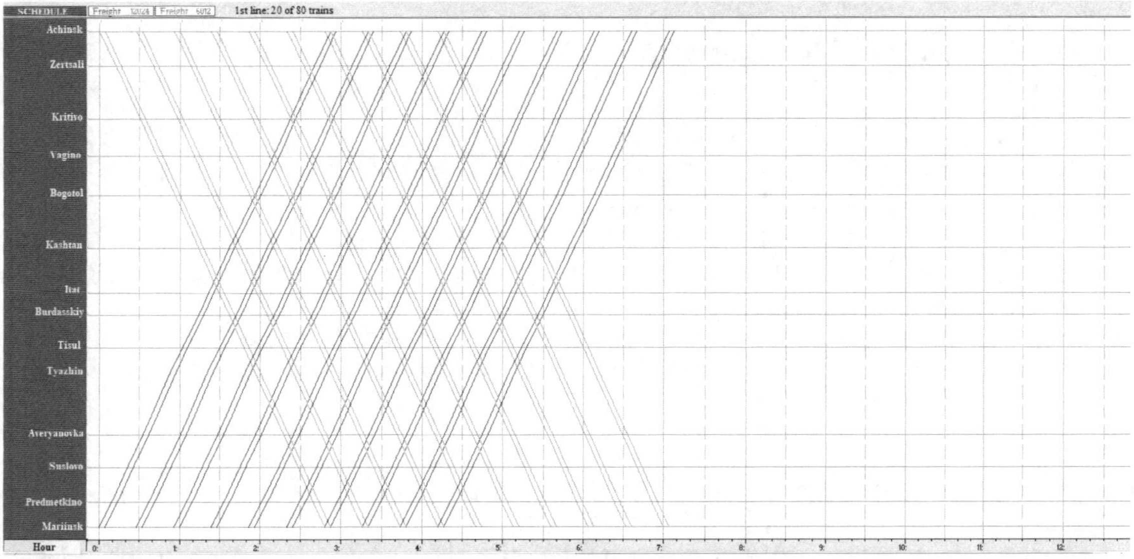

Figure 4. Batch schedule of trains in the virtual coupling mode of the first and second railway lines on the Mariinsk – Achinsk section with a 24-minute inter-train interval.

According to the calculations, the minimum inter-train interval is 24 minutes (figure 4).

According to the calculations, the minimum inter-train interval for a batch schedule consisting of connected trains on the Achinsk – Krasnoyarsk-Vostochny section is 42 minutes (figure 5).

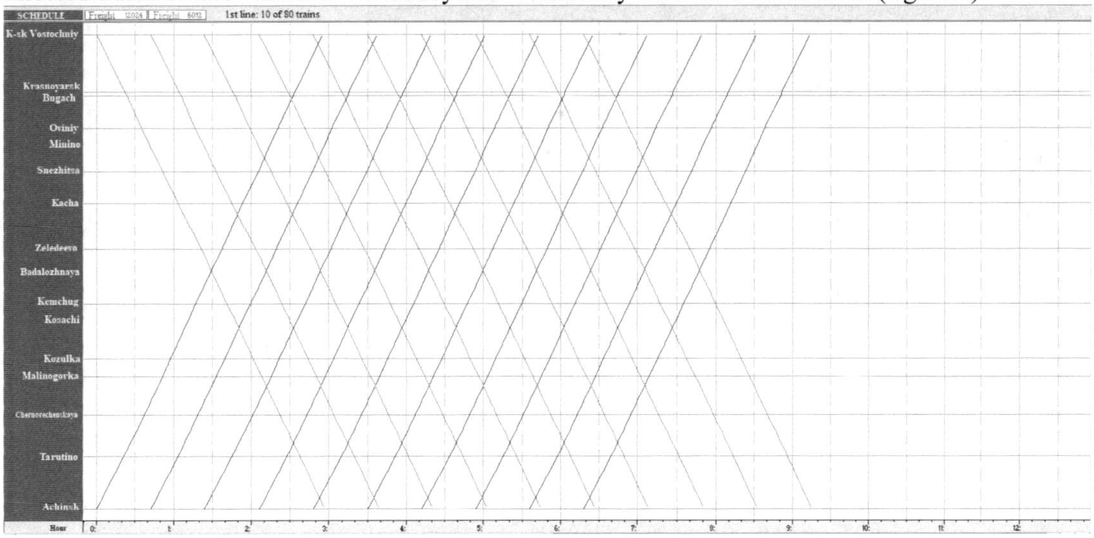

Figure 5. Batch schedule of connected trains of the first and second railway lines on the Achinsk – Krasnoyarsk-Vostochny section.

4. Conclusion

Analyzing the results obtained of the Achinsk-Krasnoyarsk-Vostochny section, it is clear that connected trains require a longer minimum inter-train interval to restore the traction network. With the same inter-train interval for two trains' types, the main operating mode characteristics of trains with a virtual coupling are better than those for connected trains. Using freight trains in a virtual coupling allows you to reduce the inter-train interval about 1.5-2 times.

References

[1] *Development strategy of the Russian Railways Holding for the period up to 2030* 2013 (Moscow: JSCo «RZD»)

[2] *Official website JSCo «RZD»*, available at: http://rzd.ru

[3] Rozentals E M, Buynova N V 2019 *Interval regulation: innovations and development prospects: thematic selection* (Krasnoyarsk: KrCNTIB)

[4] *Calculation rules of traction train operation* 1985 (Moscow: Transport)

[5] Olentsevich V A, Gozbenko V E, Kargopoltsev S K 2019 Security level assessment of complex transport systems with purpose of their competitiveness increase in the transport services market *Modern technology. System analysis. Modeling* **2** 250-259

[6] Olentsevich V A, Upyr R Y, Antipina A A 2020 Introduction of train movement interval regulation efficiency in the "virtual coupling" system on the section *Modern technology. System analysis. Modeling* **2 (66)** 182-189

[7] Osipov S I, Osipov S S 2000 *Fundamentals of train traction. Textbook for students of railway technical schools and colleges Tr-TA-M.* (UMK MPs of Russia)

[8] Rosenfeld V E, Isaev V P, Sidorov N N, Ozerov M I ed. Isaev I P 1995 *Theory of electric traction* (Moscow: Transport)

[9] Olentsevich V A, Upyr R Y, Gladkih A M 2020 Computational procedure for preparing the technical conditions for stowage and securing cargo in rail cars and containers *Journal of Physics: Conference Series* **1615** 012029 doi:10.1088/1742-6596/1615/1/012029 1

[10] *Management rules of the railways traction power network system of the Russian Federation TSE-462* 2007 (Moscow: Transport)

[11] Order of the Ministry of Transport of the Russian Federation No. 286 2010 *Railways technical operation rules of the Russian Federation*, available at: http://docs.cntd.ru/document/902256286.

[12] Cherepanov A V, Kutsy A P, Esaulenko A S 2020 Application of virtual coupling technology for high-mass trains *Young science of Siberia: electron. scientific journal* **2(8)**

[13] Ryabchenok N, Alekseeva T, Astrakhancev L, Astashkov N, Tikhomirov V 2020 Energy-Saving Driving of Heavy Trains *International Scientific Conference Energy Management of Municipal Facilities and Sustainable Energy Technologies EMMFT 2018. EMMFT-2018 2018. Advances in Intelligent Systems and Computing* Vol. **982** pp. 491-508 DOI: 10.1007/978-3-030-19756-8_47

[14] Gorbach R B, Rosentals E M 2018 *Digital technologies: a thematic selection* (Krasnoyarsk Krtsntib)

[15] Suslov K, Piskunova V, Gerasimov D, Ukolova E, Akhmetshin A, Lombardi P, Komarnicki P 2019 Development of the methodological basis of the simulation modelling of the multi-energy systems *E3S Web of Conferences* **124** 01049

[16] Tumanin A S 2019 *Technological process of the traffic management dispatching center of the Krasnoyarsk traffic management directorate – branch of JSCo "RZD"*

[17] Ministry of Railways of the Russian Federation 2000 *Railways technical operation rules of the Russian Federation* (Moscow: Techinform)

[18] Order of JSCo "RZD" No.1799p 2016 *On approval of instructions for organizing the circulation of increased mass and length freight trains on public railway lines of JSCo "RZD"*

IOP Conf. Series: Materials Science and Engineering 1151 (2021) 012039 doi:10.1088/1757-899X/1151/1/012039

Influence of spatial arrangement of wires on electromagnetic ecology near high-voltage power transmission lines at railway station

N V Buyakova[1], A V Kryukov[2,3], D A Seredkin[3], Le Van Thao[2] and K V Suslov[2]

[1] Angarsk State Technical University, 60 Tchaikovsky St., Angarsk, 665835, Russia
[2] Irkutsk National Research Technical University, 83, Lermontov St., Irkutsk, 664074, Russia
[3] Irkutsk State Transport University, 15, Chernyshevsky St., Irkutsk, 664074, Russia

E-mail: bn_900@mail.ru

Abstract: Based on computer modeling, we have considered the effect of a spatial arrangement of wires on electromagnetic pollution of environment on high-voltage electric transmission lines. This pollution is determined based on voltages of electric and magnetic fields. We have considered two variants of the arrangement of wires on power towers. The first version provides for the location of wires in a vertical plane and the second variant provides for the location of the wire at corners of a triangle. Three modes of the high-voltage power line operation were considered: 1) normal with a load at the receiving end of the power line 120 + *j90* MVA; 2) break of the phase A wire at the starting end of the power line; 3) two-phase short circuit. The simulation result in the Fazonord software package showed that the spatial arrangement of the wires significantly affects the maximum amplitudes of the electric and magnetic fields at a normalized height of 1.8 m. Moreover, the nature of dependence of the voltage on the x-axis coordinates directed perpendicular to the path of the analyzed power transmission section changes. The results of the simulation showed that under symmetrical load for a power line with a triangular arrangement of wires, the maximum value of the amplitude of the electric field show insignificant increase in comparison with high-voltage power lines with vertically arranged wires. The same indicator for the magnetic field shows increase of one and a half time.

1. Introduction

In modern conditions, electromagnetic pollution of the environment shows a considerable increase [1], which is characterized by the magnitude of intensity of the electromagnetic field (EMF). Fields near the high voltage overhead line [2–9] can generate interference that prevents the normal operation of electrical and electronic devices, as well as lead to serious accidents when operating on disconnected power lines or communications when exposed to induced voltage. Moreover, there is an adverse effect of EMF on the environment [10-14]. The spatial arrangement of the wires significantly affects the maximum amplitudes of the electric and magnetic fields at a standard height of 1.8 m. To study this effect, a computer modeling of the 220 kV power line was carried out with two options for the arrangement of wires on supports: in the vertical plane and at corners of the triangle.

Content from this work may be used under the terms of the Creative Commons Attribution 3.0 licence. Any further distribution of this work must maintain attribution to the author(s) and the title of the work, journal citation and DOI.
Published under licence by IOP Publishing Ltd

IOP Conf. Series: Materials Science and Engineering 1151 (2021) 012039 doi:10.1088/1757-899X/1151/1/012039

2. Modeling methods

Modeling using the Fazonord software package [15-17] was carried out in three stages:

1. Calculation of system operating states in phase coordinates based on which potential and current of all wires of high voltage overhead line were determined [3].
2. Calculation of vertical and horizontal components of intensity of electromagnetic and magnetic fields: $\dot{E}_x, \dot{E}_y, \dot{H}_x, \dot{H}_y$.
3. Calculation of amplitude values of intensities E_{max}, H_{max}, considering possible elliptical polarization of fields [3].

3. Results of modeling

An arrangement of the investigated electric network is shown in Figure 1. The scheme of the calculation model of the Fazonord software package is shown in Figure 2. The dependences of the components of electric and magnetic field intensities at a height of 1.8 meters on the x coordinate are shown in Figure 3 - 6.

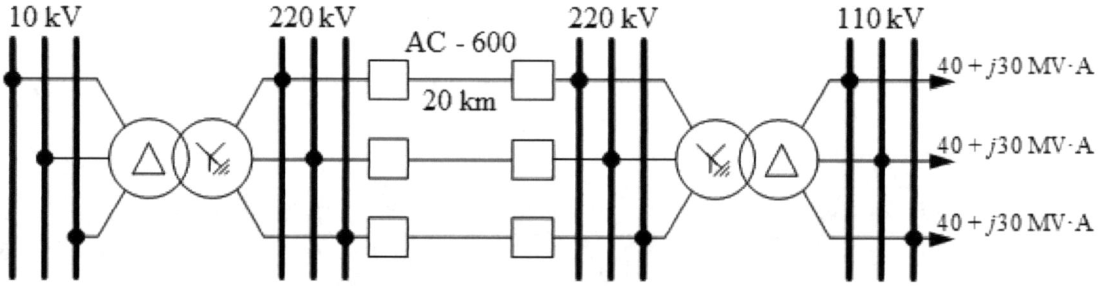

Figure 1. Scheme of electric circuit

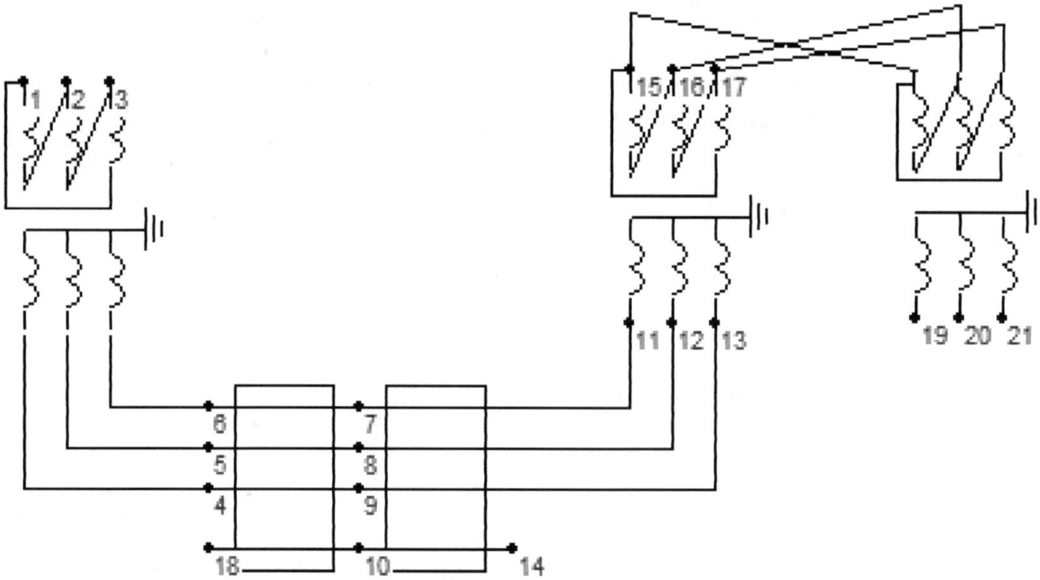

Figure 2. Arrangement of calculated model

IOP Conf. Series: Materials Science and Engineering 1151 (2021) 012039 doi:10.1088/1757-899X/1151/1/012039

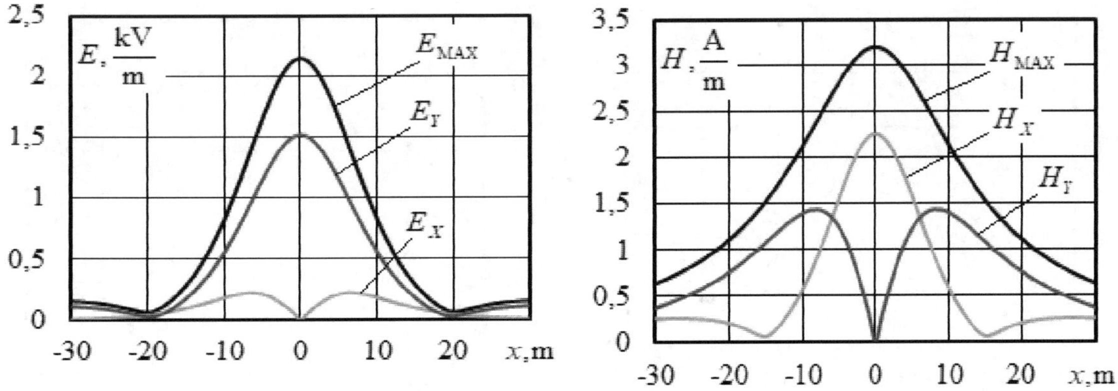

a) b)

Figure 3. Electric (a) and magnetic (b) fields at a height of 1.8 m at normal operation of high voltage overhead lines with a vertical arrangement of wires

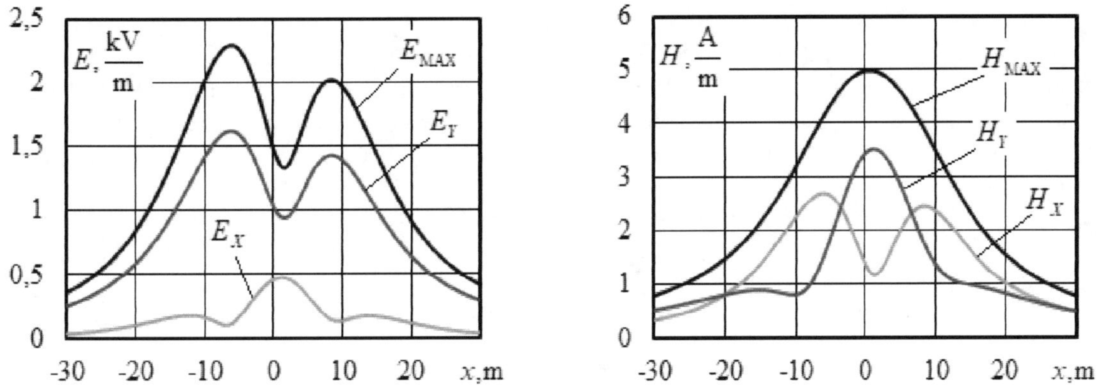

a) b)

Figure 4. Intensity of electric (a) and magnetic (b) fields at height of 1.8 m under normal operation mode of high voltage overhead line with triangular arrangement of wires

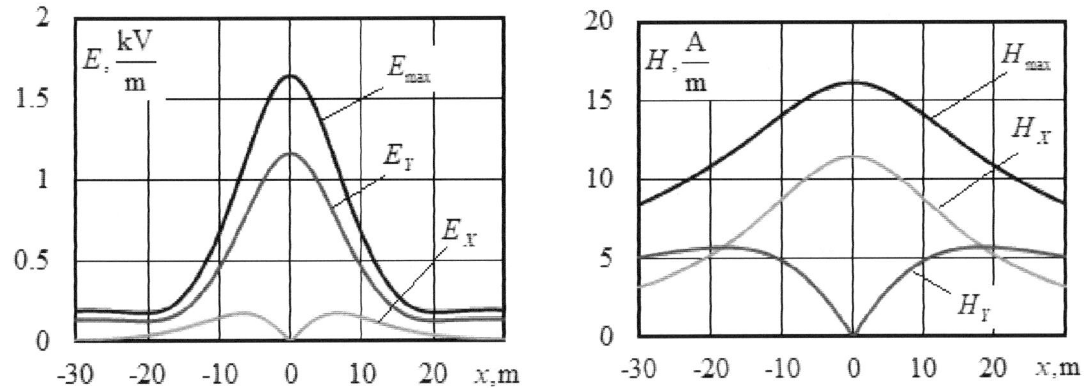

a) b)

Figure 5. Intensity of electric (a) and magnetic (b) fields at height of 1.8 upon loss of phase A of high voltage overhead lines with vertical arrangement of wires

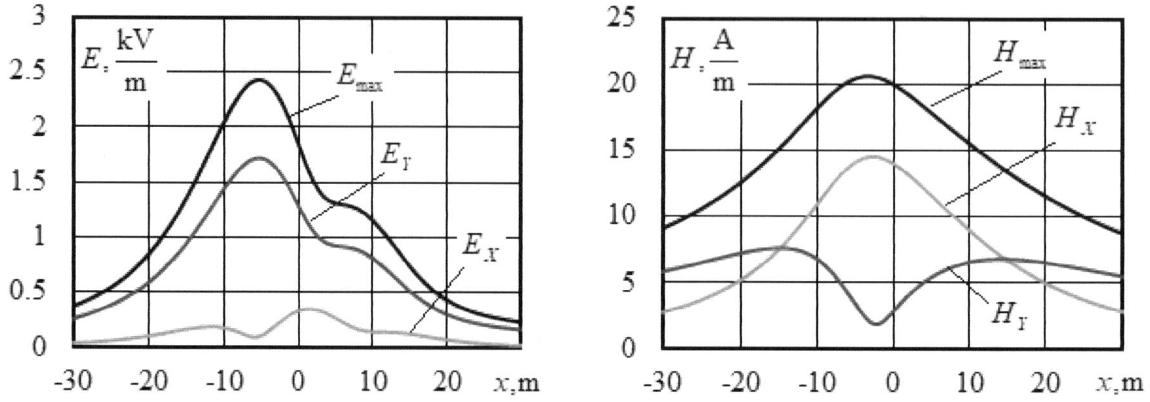

a) b)

Figure 6. Intensity of electric (a) and magnetic (b) fields at height of 1.8 upon loss of phase A of high voltage overhead lines with triangular arrangement of wires

4. Conclusions

Results of the simulation allow us to make following conclusions:

1. Under symmetrical load of high voltage overhead lines with a triangular arrangement of wires, the maximum value of the amplitude of the electric field increases slightly compared to high voltage overhead lines with vertically arranged wires. A similar indicator for the magnetic field increases one and a half times.

2. Upon loss of phase A, the intensity of electric field of high voltage overhead lines with a triangular arrangement of wires is 1.5 times higher than the value of high voltage overhead lines with a vertical arrangement, the intensity of magnetic field is 28%.

3. In the two-phase short-circuit mode, the intensity of electric and magnetic fields of high voltage overhead lines with a triangular arrangement of wires show the increase by 32% and 11%, respectively, in comparison with high voltage overhead lines with vertical arrangement of wires.

Acknowledgements

The studies were financially supported by the grant "Improving the quality of electric energy and electromagnetic safety in power supply systems of railways equipped with Smart Grid devices by applying methods and means of mathematical modeling based on phase coordinates.

References

[1] Spodobaev Yu M and Kubanov V P 2000 *Fundamentals of electromagnetic ecology* (Moscow: Radio and communications).

[2] Sidorov A I and Okrainskaya I S 2008 *Electromagnetic fields near electrical installations of extra-high voltage* (Chelyabinsk).

[3] Buyakova N, Zakaryukin V and Kryukov A 2018 Control of electromagnetic environment in smart traction power supply systems *ES3 Web of Conference 91,* 01009

[4] Braicu S F, Czumbil L, Stet D, Micu D D 2016 Evaluation of the Electric and Magnetic Field near High Voltage Power Lines *International Conference on Advancements of Medicine and Health Care through Technology; Cluj-Napoca, Romania* pp 141-146.

[5] Dahab A A, Amoura F K and Abu-Elhaija W S 2005 Comparison of Magnetic-Field Distribution of Noncompact and Compact Parallel Transmission-Line Configurations *IEEE Trans. On Power Delivery* **20(3)**.

[6] Filippopoulos G and Tsanakas D M 2005 Analytical Calculation of the Magnetic Field Produced by Electric Power Lines *IEEE Trans. On Power Delivery* **20(2)**

[7] Suslov K, Gerasimov D and Solodusha S 2015 Smart grid: Algorithms for control of active-adaptive network components, *IEEE Eindhoven PowerTech, PowerTech 2015,* 7232462
[8] Şteţ D, Micu D D, Ceclan A and Dărăbant L 2018 Study of Electromagnetic Interferences Between AC Systems and Metallic Structures *European Electromagnetics, Lausane, Switzerland,* July 21-25
[9] Sysoev I A, Kondrat'ev V V, Zimina T I and Karlina A I 2018 Simulation of the Energy States of Electrolyzers with Roasted Anodes at Elevated Currents *Metallurgist* **61(11-12)** 943-949
[10] Vanteyev A I 2017 Once again about induced voltages *Energetik* **8** 22-26.
[11] Visyashev A N, Klepikov S A, Sherbakov A V et al 2010 On safety of works performed on Overhead Power Lines subject to induced voltage *Energetik* **2** 18-22
[12] Shumanskaya E V and Goldobin D A 2018 Calculation of induced voltages on disconnected phases and parallel Overhead Power Lines using the modal wave channel method *Energetik* **3** 35-38
[13] Akhmetshin A, Mendeleev D and Marin G 2020 Improvement of electricity quality indicators in electric networks with voltage of 0.4-10 kV *International Russian Automation Conference (RusAutoCon), INSPEC,* DOI: 10.1109/RusAutoCon 49822.2020.9208158
[14] Bulatov Y N, Kryukov A V and Suslov K V 2017 Multi -agent technologies for control of distributed generation plants in the isolated power systems *Far East Journal of Electronics and Communications* **17(5)** 1197-1210
[15] Sivtsov A V, Yolkin K S, Kashlev I M and Karlina A I 2020 Processes in the Charge and Hearth Zones of Furnace Working Spaces and Problems in Controlling the Batch Dosing Mode during the Smelting of Industrial Silicon and High-Silicon Ferroalloys *Metallurgist* **64(5-6)** 396–403
[16] Soluyanov Y I, Fedotov A I, Soluyanov D Y and Akhmetshin A R 2020 Experimental research of electrical loads in residential and public buildings *IOP Conference Series: Materials Science and Engineering, The International Conference on Advances in Energy Industry and Power Generation (AdvEnGen-2020)* **860** 012026 DOI:10.1088/1757-899X/860/1/012026
[17] Nazarychev S A, Akhmetshin A R and Gaponenko S O Full compensation of reactive power in electric networks 0.4-10kV *Journal of Physics: Conference Series, XI Scientific Technical Conference on Low Temperature Plasma during the Deposition of Functional Coatings,* **1588** 012036 DOI: 10.1088/1742-6596/1588/1/012036

IOP Conf. Series: Materials Science and Engineering 1151 (2021) 012040 doi:10.1088/1757-899X/1151/1/012040

Development of the automatic control system of the blower motors of the diesel electric locomotive

T G Konovalova[1], I I Izotova[2], N P Astashkov[3] and A A Olentsevich[3]

[1] Irkutsk National Research Technical University, 83, Lermontova str., Irkutsk, 664074, Russia
[2] Baikal State University, 11, Lenin str., Irkutsk, 664003, Russia
[3] Irkutsk State Transport University, 15, Chernyshevskogo str., Irkutsk, 664074, Russia

E-mail: serafi1972@mail.ru

Abstract. The main tasks of the railway industry are focused on reducing the costs associated with the operation and repair of technological equipment. It is possible to solve them by the development of technical means that optimally take into account the natural and climatic, geographical and technical operating conditions. In these operating conditions, the development of locomotive traction involves the modernization of the existing fleet of traction rolling stock, improvement of the locomotives produced, the development of a paradigm-shifting platform of lines of innovative locomotives, the replacement of equipment by more reliable and modern one, and the automation of processes. Automation makes it easier to control the locomotive, reduce the risk of errors and ensure the most rational equipment operation modes. It was found that the damage to the electric parts of the electric motors accounted for more than 70 percent of the total number of faults. The purpose of this article is to develop a technical solution, the use of which will minimize the number of failures of auxiliary machines and locomotive equipment of this series. To implement this, it considers the main design features of electric motors, their connection diagram and the presence of protection elements to determine the trouble spots and justify the proposed technical solution.

1. Introduction

The railways include engineering facilities and various technical devices and tools to ensure timely complete and high-quality accommodation of needs of the population and the national economy in transportation. The uninterrupted and accident-free work of this kind of a multi-industry economy requires the interconnected and coordinated work of all its links. The automation of technological processes aimed at eliminating the so-called human factor – the main cause of emergencies – stands out among the set of measures.

In order to optimize the industry's performance, locomotive traction development is carried out in three main areas:

- modernization of the existing fleet of traction rolling stock;
- automation of individual operations of technological process;
- improving the produced locomotives in accordance with modern requirements, taking into account unification;
- development of a conceptually new platform for innovative locomotives [1-5].

Content from this work may be used under the terms of the Creative Commons Attribution 3.0 licence. Any further distribution of this work must maintain attribution to the author(s) and the title of the work, journal citation and DOI.
Published under licence by IOP Publishing Ltd

2. The reliability analysis of the 2TE116UM series of diesel electric locomotives

The main tasks of the railway industry are focused on reducing the costs associated with the operation and repair of technological equipment. It is possible to solve them by the development of technical means that optimally take into account the natural and climatic, geographical and technical operating conditions. According to the Russian Railways investment program, the renewal of the Ulan-Bator Railway locomotive fleet has begun with the 2TE116UM series diesel electric locomotive. Since 2014, the transportation process has been carried out by diesel electric locomotives of the 2TE116UD series with the American diesel GE. The emphasis on the 2TE116UM series is due to the good quality characteristics of the inventory fleet of locomotives in operation and its high degree of reliability (see figure 1) [6-10].

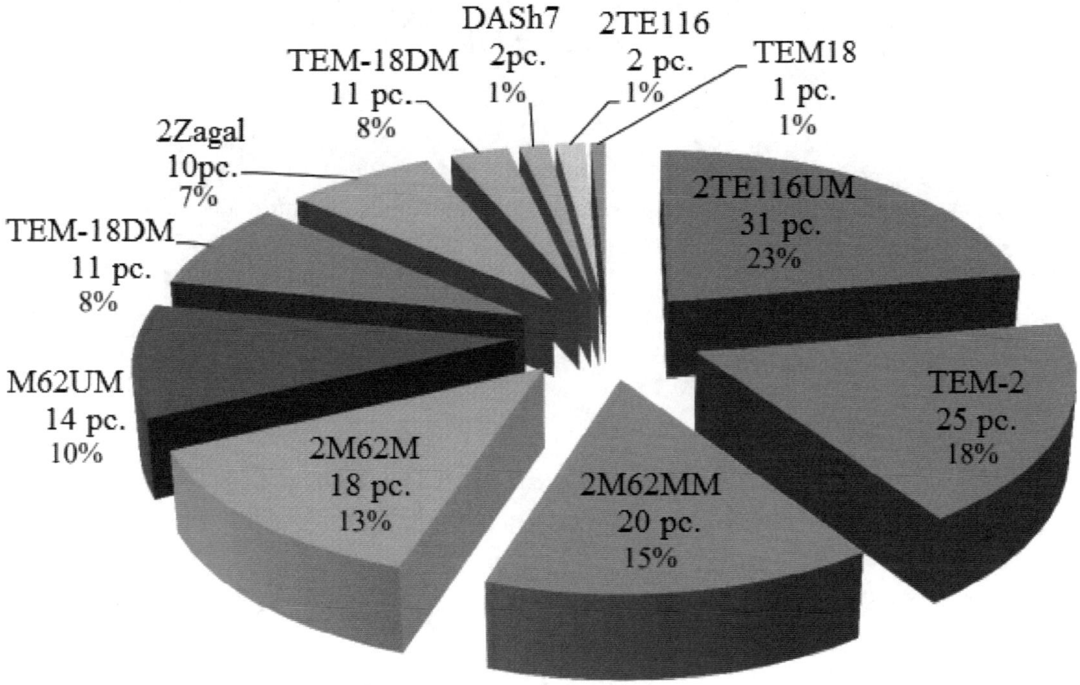

Figure 1. Inventory fleet of the diesel electric locomotives of the Ulan Bator Railway.

The analysis showed that a large proportion of failures can be attributed to electric machines (Figure 2). It was found that the damage to the electric parts of the considered electric motors for the analyzed period accounted for more than 70 percent of the total number of faults.

When the temperature of water or oil rises to the value of setting one of the sensors, it is triggered and switches on the corresponding electro-pneumatic valve PV and Contactor C (Figure 3). The proposed power diagram is presented in Figure 4.

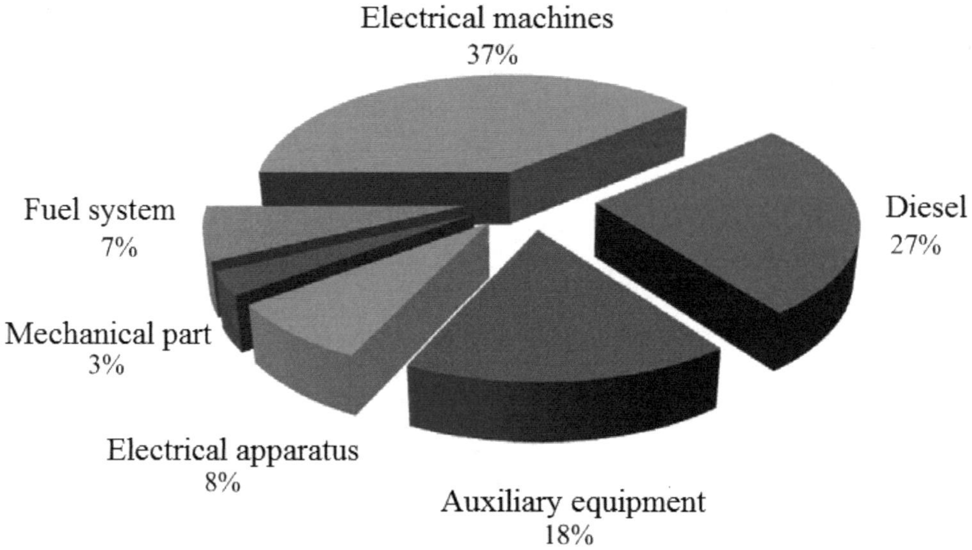

Figure 2. The reliability analysis of the 2TE116UM series of diesel electric locomotives for 2015-2019.

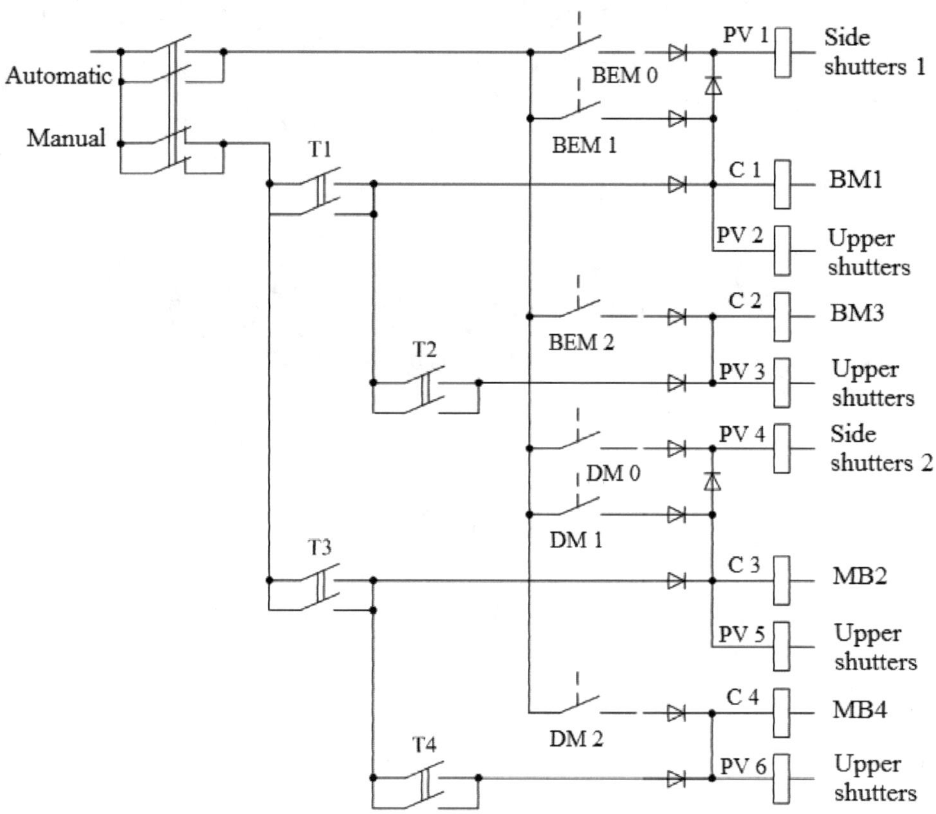

Figure 3. The schematic control diagram of the blower and the cooling device shutters of the 2TE116Um series diesel locomotive.

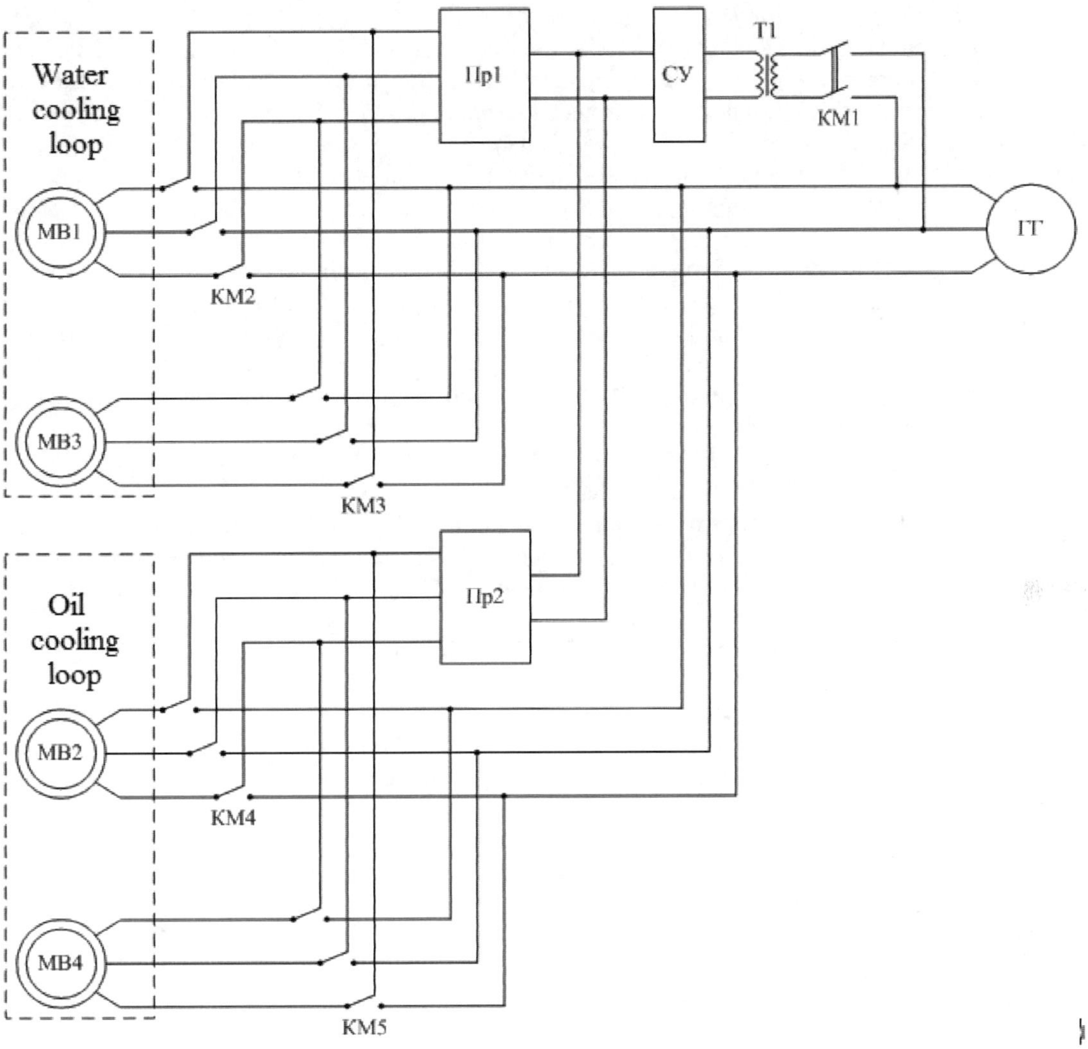

Figure 4. A modernized power diagram for blower motors' control of the diesel electric locomotive of the 2TE116UM series.

3. Conclusion

To practically implement the system proposed in the framework of a scientific article, it is necessary to review the technical parameters of semiconductor frequency converters of various manufacturers to assess the technical and economic efficiency and determine the required amount of capital investments. The effective factor in the implementation of the proposed system is to reduce the failure of the locomotive's blower motors, the inclusion of which is currently implemented in a direct way.

While substantiating the choice of the element base, in addition to the technical and economic aspects, it is necessary to take into account mass and dimension indicators, as the practical implementation of the proposed system should not clutter the locomotive body, and the installation place should fully correspond to the range of temperatures of the stable operation of the device.

References

[1] 2013 *The Development Strategy of Russian Railways Holding for the period until 2030* (Moscow: Russian Railways OAO).

[2] *The official website of the Russian Railways*, available at: http://rzd.ru
[3] Olentsevich V A, Konyukhov V Y, Olentsevich A A and Lysenko D A 2020 Efficiency of implementation of interval traffic regulation by the virtual coupling system on the section of the railway line in the framework of the digital railway project *Journal of Physics: Conference Series* **1661** 0121064
[4] Olentsevich V A, Gozbenko V E, Kargapoltsev S K and Kramynina G N 2019 Complex of organizational and technical and reconstructive measures aimed at improving the performance of the site based on the study of systemic links and patterns of the railway transport system *Modern technologies. System Analysis. Modeling* **3 (63)** 171-179
[5] Filonov S P, Gibalov A I, Bykovskii V E, et al. 1985 *Diesel electric locomotive 2TE116 – a design and maintenance manual. 2nd ed., revised and enlarged.* (Moscow: Transport Publ.)
[6] Sergeev S V, Yaitskikh A I and Yaitskikh I A 2020 Principles of diagnostics of the electric circuit of a diesel locomotive 2TE116U (UM) *Locomotive* **8(668)** 14-16
[7] Astrakhantsev L A and Astashkov N P 2012 Justification of the method of constructing an automated system for controlling the performance of motor fans on electric locomotives *Bulletin of Irkutsk State Technical University* **3 (62)** 90-95
[8] Gozbenko V E, Kargapoltsev S K, Karlina Yu I and Karlina A I 2019 Approximation of amplitude-frequency characteristics using equidistants *J. of Phys.: Conf. Ser.* **1384(1)** 012014
[9] Gozbenko V E, Kargapoltsev S K, Kuznetsov B O, Karlina A I and Karlina Yu I 2019 Determination of the principal coordinates in solving the problem of the vertical dynamics of the vehicle using the method of mathematical modeling *J. of Phys.: Conf. Ser.* **1333(5)** 052007
[10] Kuznetsov B O, Gozbenko V E, Kargapoltsev S K, Karlina Yu I and Karlina A I 2019 Dynamic vibration protection of the railway carriage *J. of Phys.: Conf. Ser.* **1333(5)** 052018
[11] Shtayger M G, Balanovskiy A E, Kargapoltsev S K, Gozbenko V E, Karlina A I, Karlina Yu I, Govorkov A S and Kuznetsov B O 2019 Investigation of macro and micro structures of compounds of high-strength rails implemented by contact butt welding using burning-off *IOP Conf. Ser.: Mater. Sci. Eng.* **560** 012190
[12] Balanovskiy A E, Shtayger M G, Karlina A I, Kargapoltsev S K, Gozbenko V E, Karlina Yu I, Govorkov A S and Kuznetsov B O 2019 Surface hardening of structural steel by cathode spot of welding arc *IOP Conf. Ser.: Mater. Sci. Eng.* **560** 012138
[13] Yolkin K S, Yolkin D K, Kolosov A D, Ivanov N A and Shtayger M G 2018 Technologies, which allow to reduce an impact of metal silicon production on the environment *IOP Conf. Ser.: Mater. Sci. Eng.* **411** 012028
[14] Ryabchenok N L, Alexeyeva T L and Astashkov N P 2013 Automatic Control system of motor fans, adaptive to the conditions of operation on electric locomotives *Modern technologies. System Analysis. Modeling* **4 (40)** 95-100
[15] Ryabchenok N, Alekseeva T, Astrakhancev L, Astashkov N and Tikhomirov V 2020 Energy-saving driving of heavy trains *Advances in Intelligent Systems and Computing* **982** 491-508
[16] Kuzmich V D, Rudnev V S and Frenkel S Ya 2005 *The theory of locomotive traction*. (Moscow: Marshrut Publ.)
[17] Poida A A, Khutoryanskii N M and Kononov V E *Diesel electric locomotives, mechanical equipment. Design and repair.*
[18] Mikhalchenko G S and Kashnikov V N 2006 *Theory and design of diesel electric locomotives* (Marshrut Publishing Route)
[19] Rakov V A 1979 *Locomotives and multiple units (1966-1975).* (Moscow: Transport Publ.).
[20] Konyuhov V Yu, Konstantinova M V and Gladkih A M 2019 Determination of restored units spectrum of equipment and development of the assembly unit repair method at industrial enterprises *J. of Phys.: Conf. Ser.* **1353** 012047
[21] Konyuhov V Yu, Gladkih A M, Galyautdinov I I and Shchadova E I 2020 Calculations of efficiency in implementing progressive mold forming methods *IOP Conf. Ser.: Mater. Sci. Eng.* **760(1)** 012027
[22] Konyuhov V Y, Gorban A V and Gladkih A M 2019 Investment in the improvement of

maintenance service efficiency of processing equipment of an industrial enterprise *J. of Phys.: Conf. Ser.* **1353(1)** 012102

[23] Rakov V.A. 1990 *Locomotives and multiple units of the railways of the Soviet Union. 1976-1985.* (Moscow: Transport Publ.).

[24] Voropai N, Ukolova E, Gerasimov D, Suslov K, Lombardi P and Komarnicki P 2019 Simulation approach to integrated energy systems study based on energy hub concept *IEEE Milan PowerTech* 8810666

[25] Ilyushin P and Suslov K 2019 Operation of automatic transfer switches in the networks with distributed generation *IEEE Milan PowerTech* 8810450

[26] Suslov K, Solonina N and Gerasimov D 2018 Assessment of an impact of power supply participants on power quality *Proc. of Int. Conf. on Harmonics and Quality of Power* pp 1-5

[27] Sobenin L A, Bakholdin V I, Sinchenko O V and Vorobyov A A 2004 *Design and repair of locomotives: A textbook for the basic vocational education.* (Moscow: Academy Publishing Center)

[28] Marchenko V S, Sergeev A A, Ivanchenko V T et al 2003 *Electric equipment of diesel electric locomotives: A reference book.* (Moscow: ICC "Akademkniga" Publ.)

IOP Conf. Series: Materials Science and Engineering 1151 (2021) 012041 doi:10.1088/1757-899X/1151/1/012041

Approaches to solving the resource supply problem of the railway company based on intellectual data analysis

O V Baryshnikova, V V Kukartsev, V V Ivanenko, V N Kernitskii, D K Gek and D V Rogova

Reshetnev Siberian State University of Science and Technology, 31, Krasnoyarsky Rabochy Av., Krasnoyarsk, 660037, Russia

E-mail: vlad_saa_2000@mail.ru

Abstract. The purchase of trade establishments is an important and complex process, on which companies spend a lot of resources in order to get the maximum benefit. There are many well-established methods for making a forecast of the required number of goods, but none of them can guarantee an accurate result, because in the matter of forecasting, there cannot be a method that produces a result with 100% accuracy. The object of our research is the process of creating a purchase forecast for trade enterprises. The subject is the analysis of the quality of the purchase forecast for retail outlets. To determine the quality of the purchase forecast of the retail outlets, it is necessary to understand how the purchasing process takes place, as well as highlight the factors that affect the forecast quality.

1. Introduction

Purchasing work is the backbone of commercial activities in trade. Commercial work essentially begins with it. To sell a product to a buyer (consumer) and make a profit, you need to have (own) the product.

Obviously, suboptimal procurement planning entails losses [1]. On the one hand, with excessive purchase of goods, freezing of funds occurs, there is a decrease in turnover, an overstocking of warehouse space and, consequently, a lack of space and under-purchase of other goods. The risk of losses from the expiration of the shelf life of the goods increases [2]. On the other hand, if there is insufficient or late purchase, the lack of goods on the shelf reduces sales and reduces customer loyalty [3, 4].

A specially trained person is engaged in forecasting purchases manually or using special software. It all depends on the size of the outlet and the criteria that must be taken into account.

Figure 1 shows the algorithm for delivering goods from the manufacturer to the end customer.

There are a lot of factors that influence purchases. These factors can be both external and internal. Table 1 provides a list of external and internal factors.

External factors are factors that cannot be influenced by a trading company, internal factors directly depend on the activities of the company.

This list includes obvious factors, some of which are used by many methods for forecasting, but the list contains factors that are aimed at the universality of the forecast. For example, a payment method.

Content from this work may be used under the terms of the Creative Commons Attribution 3.0 licence. Any further distribution of this work must maintain attribution to the author(s) and the title of the work, journal citation and DOI.
Published under licence by IOP Publishing Ltd

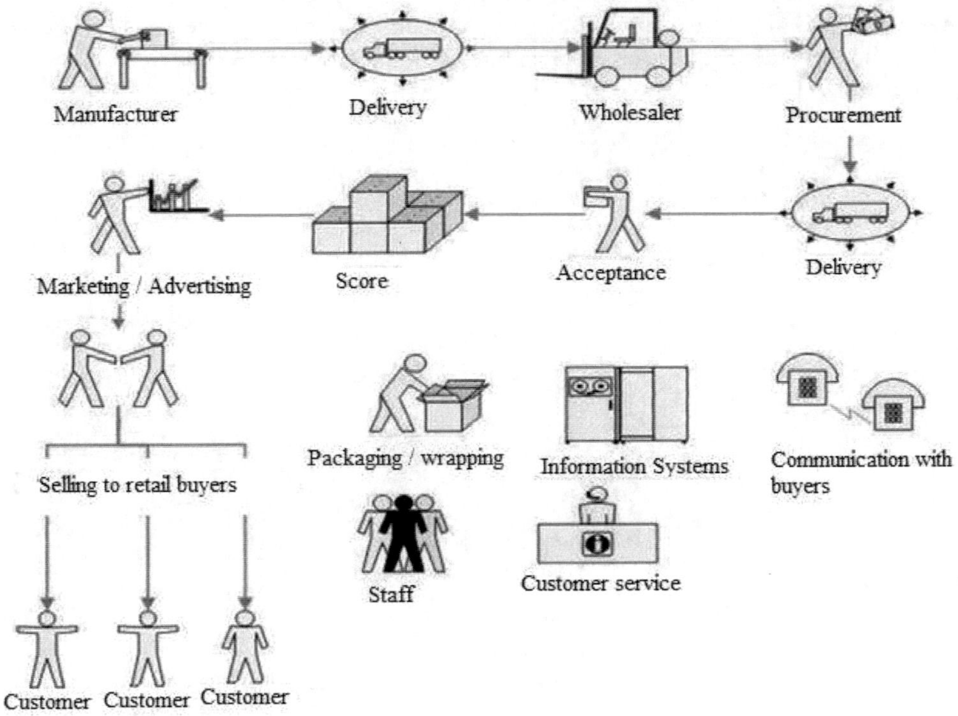

Figure 1. Life path of goods associated with a retail outlet.

Table 1. Parameters for the input layer of the neural network.

External factors	Internal factors
Customer demand. Indicator from 0 to 1	Sales volume for previous periods. The number of items sold
Supplier reliability. Indicator from 0 to 1	The number of unsold goods
Influence of competitors. Indicator from 0 to 1	The trend of the emergence of new products in a particular category. Index from 0 to 1
Seasonality of the product. Seasonal or non-seasonal item	Price for one unit of goods
Are there any delivery delays? Indicator from 0 to 1	There is payment by card (0 or 1)
	Product quality. Indicator from 0 to 1
	Is there a marketing campaign planned (0 or 1)

In Europe, buyers have no problems to carry cash with them, so the lack of card payment will not greatly reduce the income of the outlet, and in Russia people in large cities are accustomed to using a bank card, so if the outlet introduces a card payment system, it will have, in the near future, to buy more goods, because its sales will increase.

The importance of the above factors is obvious, but for practical use they must be converted into numerical values. This requires deciding which parameters are most important, how they are determined and whether they are continuous or discrete.

2. Artificial neural networks: essence, training and types

Artificial neural networks (ANN) are very popular to solve the prediction problems [5 - 7]. Figure 2 shows a graph of the growth in the number of publications in the Scopus citation system for the request "Neural networks" (neural networks).

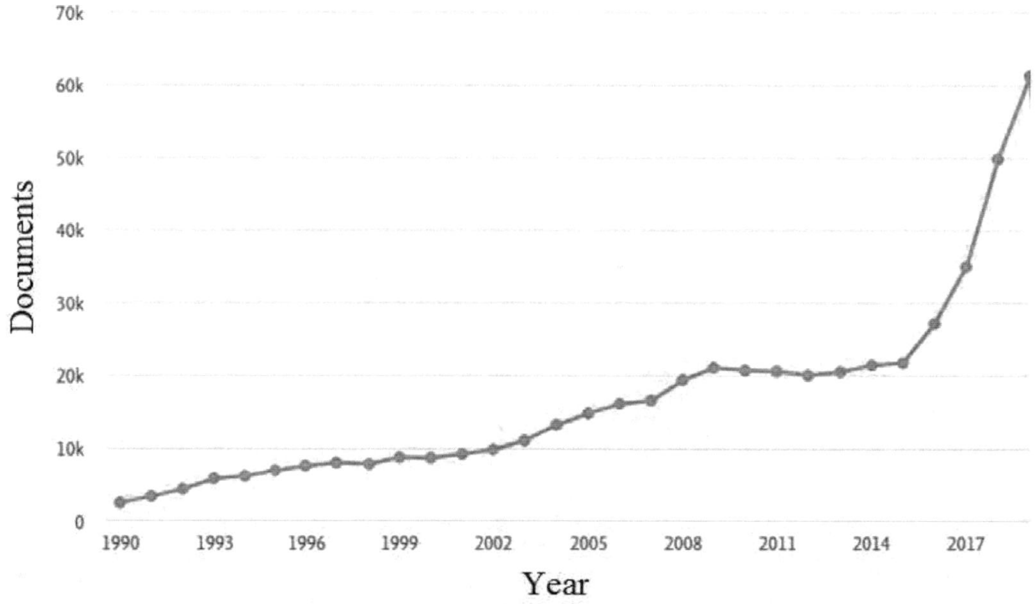

Figure 2. Graph of growing interest in the topic "Neural networks" (neural networks).

Interest in the topic only grows every year. To solve the regression problem, we need to choose the correct architecture of the neural network [8, 9].

The choice of the structure of the neural network is carried out in accordance with the characteristics and complexity of the problem [10]. Regression problems can be solved using different types of networks: multilayer perceptron, linear network, radial basis function and generalized regression network.

We based on a number of criteria for choosing a neural network architecture:

- Hardware limitations.
- If a more complex architecture shows similar results than a simpler one, then the choice is made in favor of a simpler one.

The recommendations of the developers of libraries for neural networks used in the developed program (TenserFlow, Keras) were also taken into account. The basic architecture model will be a multilayer perceptron. The main feature of the architecture is associated with the fact that the program uses a large number of hidden layers in order to reduce the number of elements in them. This principle is at the heart of deep learning.

One of the important advantages of the chosen model is that it will be able to process a large amount of data quickly enough, which for other models could become a problem faced by companies that are engaged in the purchase and retail of goods in huge quantities and over a long period.

Two hidden layers are justified by the fact that the developed product uses the non-linear activation function ReLu. Function mathematical notation:

$$f(x) = \begin{cases} x, if\ x > 0 \\ 0 \end{cases}. \quad (1)$$

Figure 3 shows the architecture of the neural network being developed.

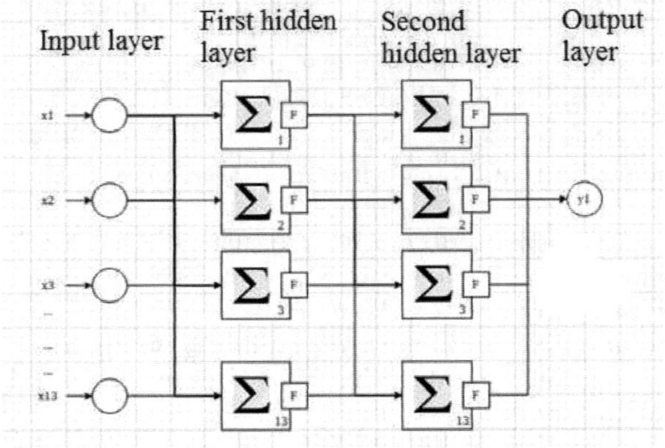

Figure 3. Neural network architecture.

The input layer is the 13 factors from Table 1; $\sum$ - adder; F - activator function.
Each new layer receives data with adjustments that were made in the previous layer.
At the output, we get a forecast.
The activation functions can be different, but for the chosen architecture the non-linear function ReLu will be used, the graph of which is shown in Figure 4.

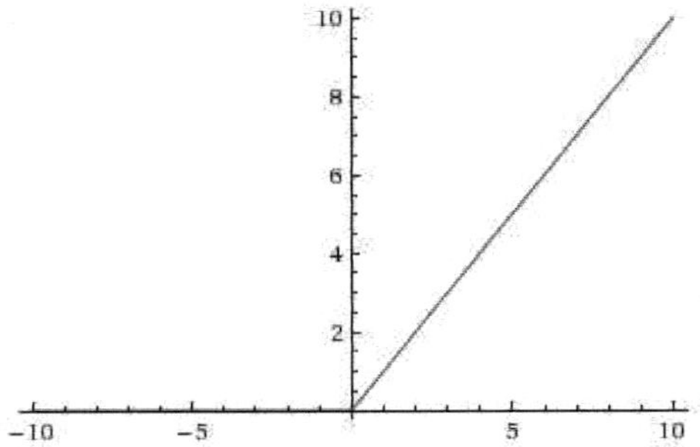

Figure 4. ReLu function graph.

We will calculate the learning error using the formulas for the mean square error S and the mean absolute error M:

$$S_n = \sqrt{\frac{1}{n-1}\sum_{i=1}^{n}(x_i - \bar{x})^2}, \quad (2)$$

$$M = \frac{1}{n}\sum_{i=1}^{n}\left|\frac{x_i - y_i}{x_i}\right|. \quad (3)$$

To visually look at the results of a neural network calculation, you need graphs. For this, you can use the MatLab library.
The activation and error counting functions, as well as the architecture, allow you to create a basic neural network for purchasing forecasting for retail enterprises. In the future, it may require modifications.
One of the most important tasks when creating a neural network is to correctly create the structure of the neural network. To do this, you need to make 4 steps:

- Select the activation function.
- Prepare initial data.

- Create a suitable neural network architecture.
- Select the error counting functions.

The main task is to predict the value of a product over a certain period, which is a regression task. Regression analysis is a set of statistical methods for studying the influence of one or more independent variables X_1, X_2,..., X_i on the dependent variable Y. The independent variables are otherwise called regressors or predictors, and the dependent variables are criterion variables. The most common type of regression analysis is linear regression, where a linear function is found that, according to certain mathematical criteria, it most closely matches the data. For example, the least squares method calculates a straight line (or hyperplane) whose sum of squares and the data is minimal.

When solving the regression problem, you will have to face the problems of data normalization, because the data will be in different ranges, which will increase their impact on the forecast. To do this, we consider deep neural networks.

Deep learning is understood as the field of machine learning devoted to the construction of nonlinear classification or regression models, the elements of superposition, which describe the corresponding level of data aggregation of features [11, 12]. A deep neural network (DNN) is an artificial neural network (ANN) with multiple layers between the input and output layers [13, 14]. A deep learning neural network aims to exclude the human factor from the forecasting process. Due to its high accuracy and low prediction errors, the neural network has advantages over physical prediction models [15].

3. Forecast quality estimation

To determine the quality of the forecast, different types of analyses are used, we will consider them below.

The most common measures of quality in regression problems are Mean Squared Error (MSE) and Mean Absolute Error (MAE).

The root-mean-square functional penalizes more for large deviations compared to the absolute mean, and therefore is more sensitive to outliers. When using any of these two functionals, it can be useful to analyze which objects make the greatest contribution to the total error - it is possible that an error was made on these objects when calculating features or a target value.

If there is no error, then it needs to be able to determine why there is such a large deviation. To do this, you must use the rules for dividing the product line into groups. To do this, we will use the standard XYZ analysis method.

XYZ analysis will be able to determine which goods should be paid special attention to when forecasting.

XYZ analysis is a tool that allows you to divide the assortment into groups, depending on the stability of sales and fluctuations in demand for products [16]. The purpose of the analysis is to predict the stability of certain objects of study, for example, the stability of sales of certain types of goods, fluctuations in the level of demand, etc.

XYZ analysis is carried out based on the following classification:

- X-parts have consistent and predictable demand patterns without significant fluctuations. They can be formalized, for example, using their small forecast error less than 10% of the real value.
- Y portions have lower forecast accuracy because their demand curve is subject to seasonal fluctuations or trends. The classification can be made based on the forecast error of their time series in about 10-30% of the real value.
- The Z part is characterized by particularly irregular variations. A high forecast error is possible with Z portions.

The use of XYZ analysis will make it possible to say with higher confidence whether it is worth purchasing goods for a retail outlet, based on the data obtained as a result of the neural network forecast.

XYZ analysis is often combined with ABC analysis. ABC analysis allows you to divide a large amount of data, based on their contribution to the profit of a trading enterprise, into three groups. This allows, first, to highlight the positions that are key. Second, we can concentrate on analyzing three groups instead of a large list, and work with positions within groups in a similar way [17]. These products fall into three categories as shown below:

- Type A parts include 10-20% of materials, which account for about 70-80% of the total profit. You should pay special attention to it.
- Type B parts are fewer valuable goods with a share of about 10-20% of the total assortment of goods and about 20-30% of the total profit of the enterprise.
- Type C part makes up the largest share of approximately 60-70% of the total assortment but bear only 5-10% of the total value.

First, an ABC analysis of goods is carried out by the amount of income received or by the amount of goods shipped for the entire accounting period (for example, for a year). Then an XYZ analysis of these products is carried out for the entire period (for example, monthly sales for the year). After that, the results are combined. Combining the two methods will create 9 product groups, which should further improve the limitation on the permissible forecast error.

4. Conclusion

Correctly organized wholesale purchases make it possible to form the necessary range of goods for the retail trade network to supply the population, to influence manufacturers of goods in accordance with the requirements of consumer demand, and also ensure the efficient operation of the trade enterprise. To do this, you need to know the methods for determining the quality of the forecast.

References

[1] Poston R and Grabski S 2001 Financial impacts of enterprise resource planning implementations *International Journal of Accounting Information Systems* **2(4)** 271-94

[2] Hunton J E, Lippincott B and Reck J L 2003 Enterprise resource planning systems: comparing firm performance of adopters and nonadopters *International Journal of Accounting information systems* **4(3)** 165-84

[3] Loh T C and Koh S C L 2004 Critical elements for a successful enterprise resource planning implementation in small-and medium-sized enterprises *International journal of production research* **42(17)** 3433-55

[4] Shi J J and Halpin D W 2003 Enterprise resource planning for construction business management *Journal of construction engineering and management* **129(2)** 214-21

[5] Milov A V, Tynchenko V S and Petrenko V E 2018 Algorithmic and software to identify errors in measuring equipment during the formation of permanent joints *In: 2018 Int. Multi-Conference on Industrial Engineering and Modern Technologies (FarEastCon)* pp. 1-5

[6] Sfidari E, Kadkhodaie-Ilkhchi A and Najjari S 2012 Comparison of intelligent and statistical clustering approaches to predicting total organic carbon using intelligent systems *Journal of Petroleum Science and Engineering* **86** 190-205

[7] Tynchenko V, Milov A, Tynchenko V, Bukhtoyarov V and Antamoshkin O 2018 Application of artificial neural networks for identification of non-normative errors in measuring instruments for controlling the induction soldering process *International Multidisciplinary Scientific GeoConference: SGEM* **18(2.1)** 117-24

[8] Subhadra K and Vikas B 2019 Neural network based intelligent system for predicting heart disease *International Journal of Innovative Technology and Exploring Engineering* **8(5)** 484-7

[9] Milov A V, Tynchenko V S, Kukartsev V V, Tynchenko V V and Bukhtoyarov V V 2018 Use of artificial neural networks to correct non-standard errors of measuring instruments when creating integral joints *Journal of Physics: Conference Series* **1118(1)** 012037
[10] Tynchenko V S, Tynchenko V V, Bukhtoyarov V V, Tynchenko S V and Petrovskyi E A 2016 The multi-objective optimization of complex objects neural network models *Indian Journal of Science and Technology* **9(29)** 99467
[11] Bakhteev O P, Popova M S and Strijov V V 2016 System and means of deep learning for classification problems *Systems and Means of Informatics* **26(2)** 4-22
[12] Kruglov V V and Borisov V V 2001 *Artificial neural networks. Theory and practice* (Moscow: Hotline-Telecom) 382
[13] Bengio Y 2009 Learning deep architectures for AI *Foundations and Trends in Machine Learning* **2(1)** 1-127
[14] Schmidhuber J 2015 Deep learning in neural networks: an overview *Neural Networks* **61** 85-117
[15] Haykin S 2006 *Neural networks: a comprehensive foundation* (Moscow: Williams) 1104
[16] Nowotynska I 2013 An Application of XYZ Analysis in Company Stock Management *In: Modern management review 2013 MMR* **XVIII** 7-17
[17] Pandya B and Thakkar H 2016 A review on inventory management control techniques: ABC-XYZ analysis *REST Journal on Emerging trends in Modelling and Manufacturing* **2** 15

IOP Conf. Series: Materials Science and Engineering 1151 (2021) 012042 doi:10.1088/1757-899X/1151/1/012042

Intelligent automation system of the railway company's purchasing activity based on artificial neural networks

D K Gek[1], V V Kukartsev[1,2], I V Markevich[1], D S Shalaeva[1], A S Yadrov[2] and D V Rogova[1]

[1]Reshetnev Siberian State University of Science and Technology, 31, Krasnoyarsky Rabochy Av., Krasnoyarsk, 660037, Russia
[2]Siberian Federal University, 79, Svobodny pr., Krasnoyarsk, 660041, Russia

E-mail: vlad_saa_2000@mail.ru

Abstract. A person is responsible for quality control of the purchase forecast of commercial enterprises, which entails a number of certain consequences. In this regard, we decided to create an intelligent system that will determine the quality of the forecast. Such a system will be a human assistant in deciding on the quality of the forecast. The system will also be able, if necessary, to indicate the necessary adjustments for its forecasting method. As part of this work, a software product will be created that will determine the quality of the created trade procurement plan, and, if necessary, help the user adjust the forecasting method used by the outlet. The aim of the work is to create an intelligent procurement system that will generate its forecast using a neural network, and then determine the forecast error using XYZ analysis.

1. Introduction

To understand what will happen as a result of development, it is necessary to define an intelligent system.

An intelligent system (IS) is a technical or software system capable of solving problems traditionally considered creative [1 - 3], belonging to a specific subject area, the knowledge about which is stored in the memory of such a system [4 - 6]. The structure of an intelligent system includes three main blocks - a knowledge base, a mechanism for choosing solutions and a program interface [7 - 9].

The knowledge base is an important component of an intelligent system. The most famous class of such programs is expert systems [10 - 12]. They are designed to search for ways to solve problems from a certain subject area, based on knowledge base records and a user description of the situation [13 - 15].

Simple knowledge bases can be used to create expert data storage systems in an organization: documentation, manuals, technical articles. The main purpose of creating such databases is to help less experienced people find an already existing description of a way to solve a problem [3, 7, 16].

The decision-making mechanism is a generalized procedure for finding a solution to a problem, which, based on the knowledge base and in accordance with the user's information needs, builds a chain of reasoning (logically related units of knowledge) leading to a specific result [1, 5].

The user interface provides the transfer of information by the user of the application and the software and hardware complex [6, 8, 11]. The user enters the necessary data, which the system processes, and then, using the interface, displays information that is understandable to the user.

Predicting the purchase of an item is a regression problem. In a regression problem, the sample usually looks like:

Content from this work may be used under the terms of the Creative Commons Attribution 3.0 licence. Any further distribution of this work must maintain attribution to the author(s) and the title of the work, journal citation and DOI.
Published under licence by IOP Publishing Ltd

$$S = \{(x_1, y_1), \dots, (x_i, y_i)\} \subseteq R^n \times R. \quad (1)$$

The answers are real numbers (maybe also vectors from real numbers). The regression task is to predict the values of this function at other points using the known values of y_i of some unknown function at given points x_i ($i = 1, \dots, n$).

The creation of a procurement plan was and remains an urgent problem; therefore, there are currently many products that are engaged in the creation of products. The most successful companies charge a lot of money for their product, but the main amount goes to the fact that they have to apply an individual approach to each client.

The modern market is very changeable, in order to create the most accurate product, they conduct a point-of-sale analysis to find the most accurate parameters that the system will use in forecasting.

Supply of goods is a very difficult regression task, the solution of which depends on many factors.

The simplest and most accessible methods can only indicate a trend when creating a forecast. This conclusion was made thanks to a review of software products. Results of the survey are given below.

2. Review of programs and methods for forecasting purchases

In the course of analyzing the market for ready-made software products, a large number of ready-made software solutions were found for creating a purchase forecast for a commercial enterprise by various methods. Some ready-made solutions are Excel tools for inventory management, BUY © ER system, inventory management system on Access, analytical Internet service for inventory management and procurement planning, procurement assistant, KORUS - Inventory management, SCM: Purchase planning, FORECAST & SUPPLY, KAIS AutoStandard, oremax.

Such a large number of programs prove that the topic of forecasting is in demand, and also that there is no universal program, because these programs have many differences, starting from the price and ending with the forecasting method.

Each method has its own percentage of forecast error. There are applications in this list that, just like the system being developed, use neural networks to create a forecast. Other methods worth noting are OLAP cubes, trendline, least squares, etc.

The more and more varied data a trading enterprise has, the more difficult it is to forecast. Simple methods are not capable of making accurate predictions with large and varied amounts of data, and programs with more modern methods are very expensive.

One of the most popular and complex forecasting methods from the point of view of implementation is a neural network. The developed system generates its forecast using an artificial neural network.

To implement the neural network of the project, the Python programming language was used due to the large availability of ready-made libraries for creating neural networks. For the basis, an open library from Google - TensforFlow and an open-source library for deep neural networks Keras were chosen.

To create the application interface, QtDesigner will be used - a cross-platform free environment for developing graphical interfaces (GUI) programs using the Qt library, a wrapper for the Tkinter Python PySimpleGUI API. The MatLab library for Python will be used to create graphs. First, you need to download the data for training the neural network. The data was previously digitized. After loading, the data needs to be transformed for more accurate neural network operation. We replace all data types with numeric so that the neural network does not have problems during data normalization. Baseline data is the sale of supermarket goods for a certain period.

Input data is entered into the program, after that they are transformed, unnecessary columns are removed, and the products are grouped. Part of the input transformed data is shown in Figure 1.

IOP Conf. Series: Materials Science and Engineering 1151 (2021) 012042 doi:10.1088/1757-899X/1151/1/012042

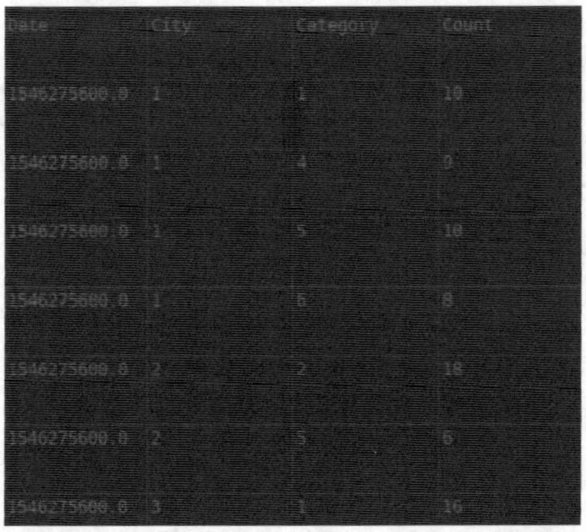

Date	City	Category	Count
1546275600.0	1	1	10
1546275600.0	1	4	9
1546275600.0	1	5	10
1546275600.0	1	6	8
1546275600.0	2	2	18
1546275600.0	2	5	6
1546275600.0	3	1	16

Figure 1. Part of the transformed input data for training the neural network.

After loading, output part of the data, as well as the description of the input data, the next step is to normalize the input data. Before normalization, it is necessary to split the DataSet into training and test samples.

Input normalization is a process in which all input data goes through a process of "flattening", i.e. reduction to a certain interval, for example [0,1] or [-1,1] [20]. Normalization can significantly increase the convergence rate of the neural network learning algorithm. If you do not perform normalization, then the input data will have an additional effect on the neuron, which can lead to incorrect decisions. In other words, you cannot compare quantities of different orders.

For normalization, the Min-Max function was used:

$$X' = \frac{X - X_{min}}{X_{max} - X_{min}}. \quad (2)$$

We separate target values or "labels" from features. Now we need to create a neural network model. Our model will be sequential, each layer will be the input for a new layer. Next there is the training of the neural network directly.

Figure 2 shows the indicators of the mean square error and the mean absolute error when training a neural network.

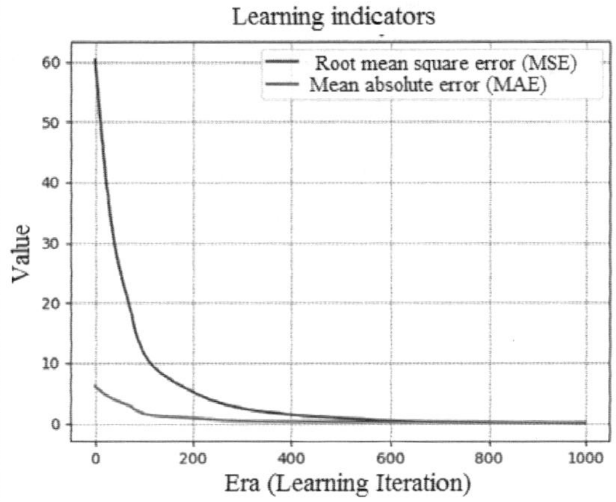

Figure 2. Error indicators when training a neural network.

After testing, we need to check the created neural network against test data. The result of the neural network forecast is shown in Figure 3.

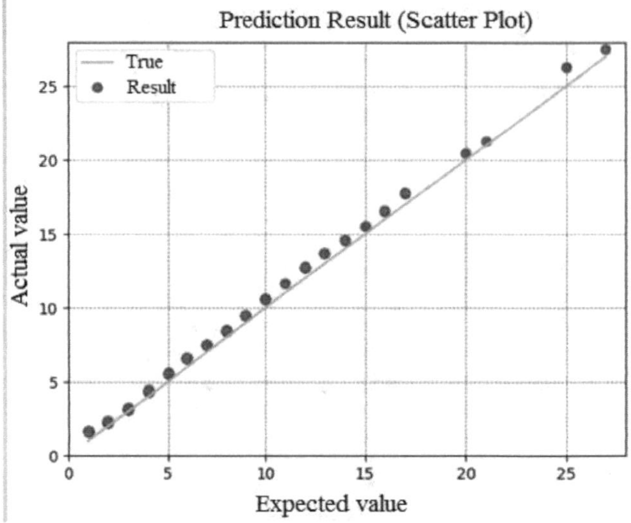

Figure 3. Scatter plot.

The prediction result is very accurate, now you need to look at the graph of the distribution of prediction errors, which is shown in Figure 4.

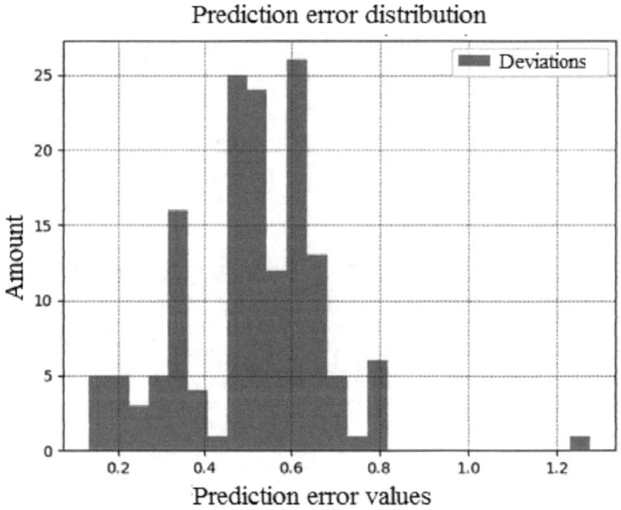

Figure 4. Prediction error distribution.

3. XYZ Analysis Module Implementation

To understand the work of XYZ analysis, it is necessary to introduce two concepts of the deviation of the variation series and the coefficient of variation.

The variation series deviation is the standard deviation of a product over a known period. The coefficient of variation is the stability of product sales over a known period. It is calculated using the formula:

$$v = \frac{\sigma}{x} \times 100\%, \tag{3}$$

where σ is the standard deviation; x is the average number of items.

After all calculations, our products are divided into categories X, Y, Z. The neural network made a forecast, the system finds deviations between real data and those received from the neural network. If the result falls within the specified scope of the product category, then the system can recommend the user to purchase the product based on the neural network forecast.

For example, a standard deviation of 100 can have a product with an average monthly sale of 200 or 20,000. In the first case, the significance of monthly fluctuations will be 50%, in the other case - 0.5%. It is obvious that the sales of the second product are much more stable and, as a result, more predictable, which means that the forecast for them needs to be as accurate as possible. The first product is poorly predictable, so the system can only suggest the sales trend for product number 1.

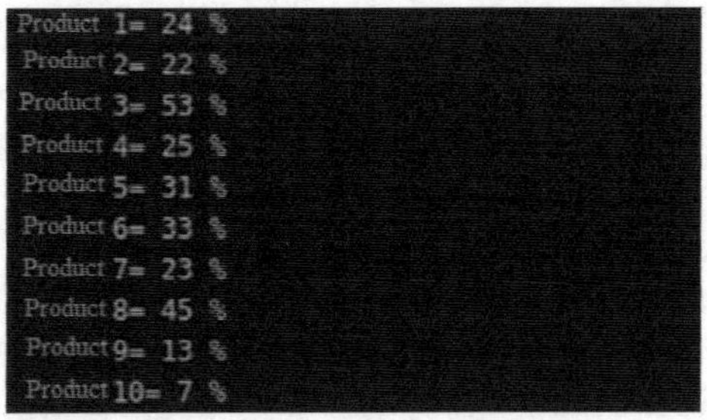

Figure 5. Product variation coefficient.

Figure 5 shows the coefficient of variation for each product from the original data.

Now you need to design your user interface layouts. To get started, the user must download the initial data for creating a forecast, as well as data on the forecast results of the custom forecasting method.

Figure 6 shows the layout of the application start window.

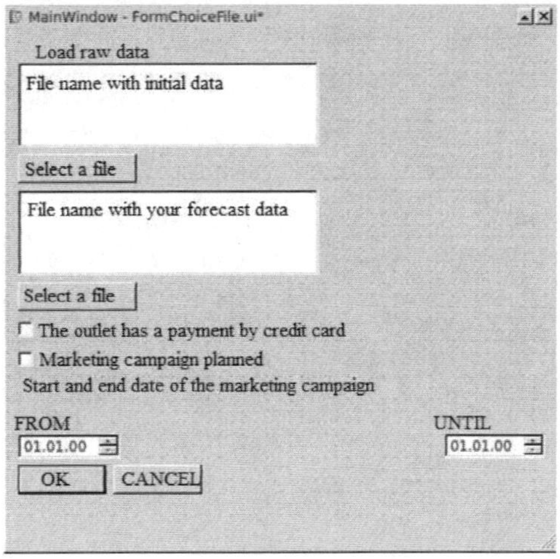

Figure 6. Data loading window prototype for application operation.

Next, the user should look at the forecast for each of the selected products, which have already been divided into groups thanks to the XYZ analysis.

Figure 7 shows the layout of the application window, where the user can see the distribution of the product by groups X, Y, Z.

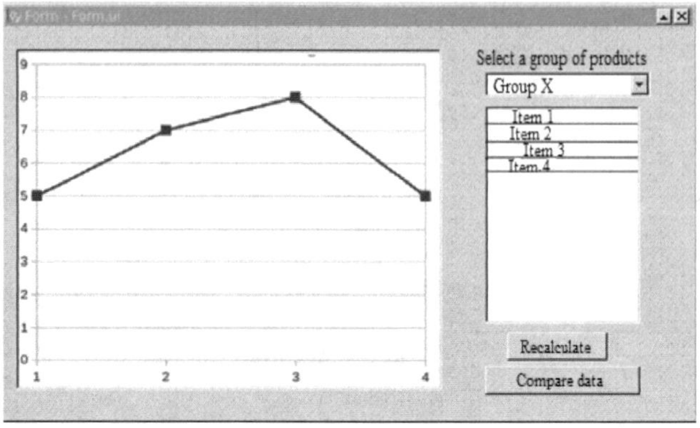

Figure 7. Layout of the "Forecasting Result" window.

Further, the system should produce a result with a recommendation about the quality of the forecast and the forecasting method used. The layout of the application window, where the system did not find deviations in errors, is shown in Figure 8.

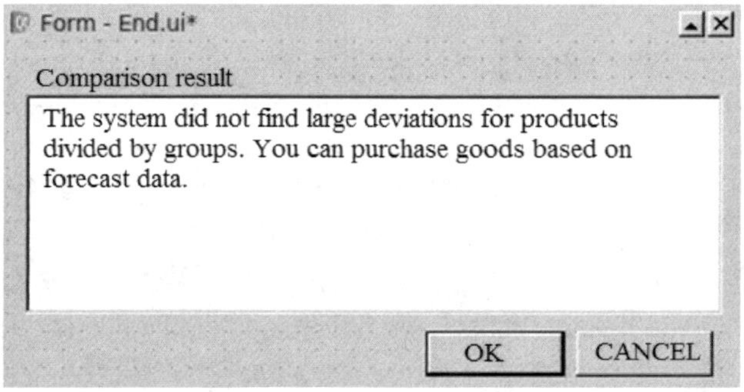

Figure 8. Example of a "Successful Result" result.

Figure 9 shows the possible results when the system found an error in the forecast, as well as an example of the result when the system advises to change the forecasting method for a given outlet.

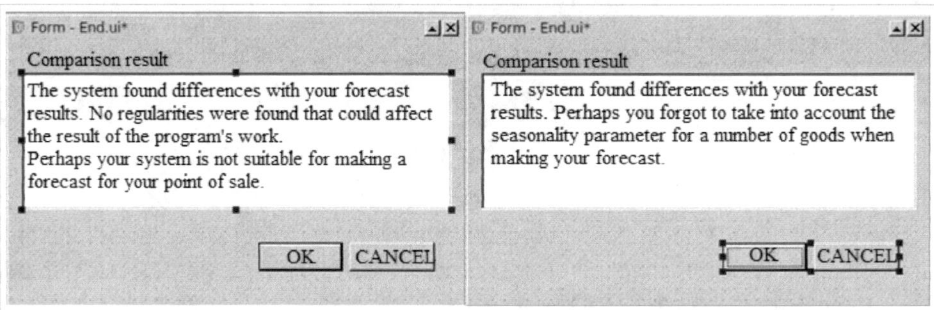

Figure 9. An example of a "negative" result output layout.

4. Experimental study

For ease of use and to increase the speed of the program, the layouts of the application have been slightly modified. The test launch was performed on the data from Figure 1. After starting the program, the user sees the start screen of the program, which is shown in Figure 10.

IOP Conf. Series: Materials Science and Engineering 1151 (2021) 012042 doi:10.1088/1757-899X/1151/1/012042

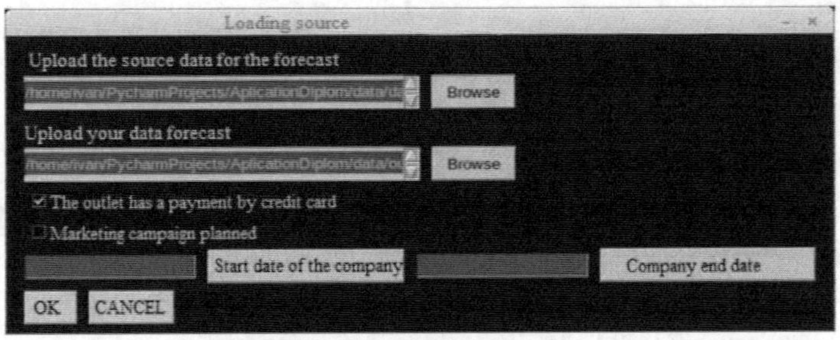

Figure 10. Application start screen.

One needs to download the initial data for the developed network, as well as the data of the available forecast. If necessary, one needs to indicate the date of the marketing campaign.

Then the neural network is trained, and also creates a forecast.

Before creating a forecast, the system divides products into 3 groups according to the XYZ analysis algorithm. The result of data splitting is shown in Figure 11.

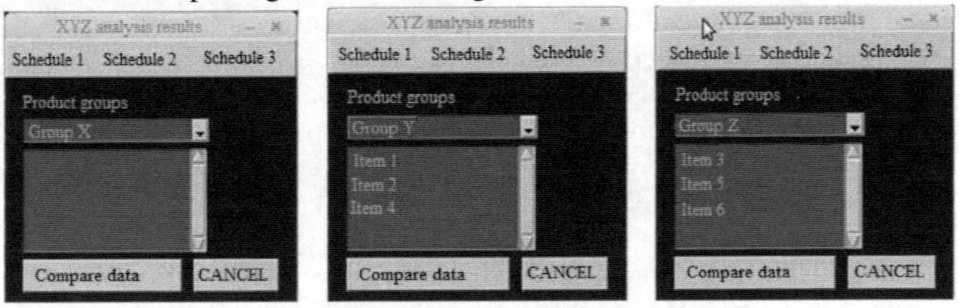

Figure 11. Application start screen.

The user can view the graphs of the error and the forecast result by clicking on the menu items.

Then the system analyzes the forecast and initial data and gives out the problems found. Figure 12 shows the result of analyzing the program when running on test data.

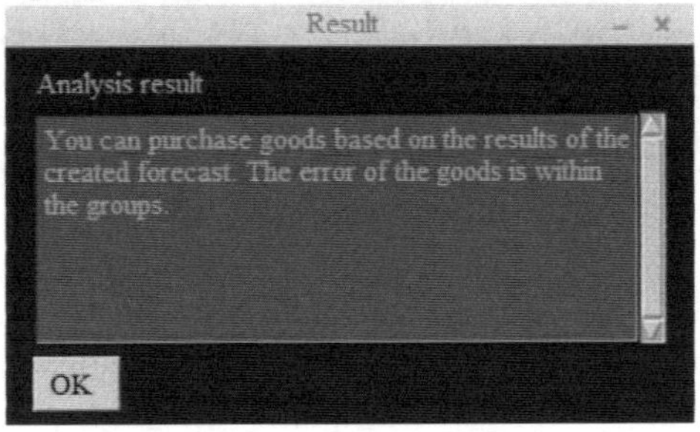

Figure 12. Program analysis result.

As a result, the system did not find any problems in the created forecast, so the user can order goods according to the system forecast.

Now it is necessary to take stock of the research and evaluate the created program.

5. Conclusion

After conducting market research of programs for creating a procurement forecast, improvements in programs and methods can be noted, but there are still problems that do not allow creating the most accurate forecast.

Until now, a person has to make a decision about the quality of the created forecast and whether it is worth purchasing goods according to the created forecast.

The developed program should help outlets in making this decision.

References

[1] Meystel A M and Albus J S 1997 *Intelligent Systems* (Architecture: Design and Control)
[2] Grosan C and Abraham A 2011 *Intelligent systems* (Berlin: Springer)
[3] Rudas I J and Fodor J 2008 Intelligent systems International *Journal of Computers, Communications & Control* **3(3)** 132-8
[4] Tynchenko V S, Tynchenko V V, Bukhtoyarov V V, Tynchenko S V and Petrovskyi E A 2016 The multi-objective optimization of complex objects neural network models *Indian Journal of Science and Technology* **9(29)** 99467
[5] Milov A V, Tynchenko V S and Petrenko V E 2018 Algorithmic and software to identify errors in measuring equipment during the formation of permanent joints *In: 2018 Int. Multi-Conference on Industrial Engineering and Modern Technologies (FarEastCon)* pp. 1-5
[6] Hayes-Roth B 1995 An architecture for adaptive intelligent systems *Artificial intelligence* **72(1-2)** 329-65
[7] Aronson J E, Liang T P and MacCarthy R V 2005 *Decision support systems and intelligent systems* (Upper Saddle River, NJ, USA: Pearson Prentice-Hall)
[8] Stephanopoulos G and Han C 1996 Intelligent systems in process engineering: A review *Computers & Chemical Engineering* **20(6-7)** 743-91
[9] Tynchenko V, Milov A, Tynchenko V, Bukhtoyarov V and Antamoshkin O 2018 Application of artificial neural networks for identification of non-normative errors in measuring instruments for controlling the induction soldering process *International Multidisciplinary Scientific GeoConference: SGEM* **18(2.1)** 117-24
[10] Bukhtoyarov V V, Tynchenko V S, Petrovskiy E A, Tynchenko S V and Milov A V 2019 Intelligently informed control over the process variables of oil and gas equipment maintenance *International Review of Automatic Control* **12(2)** 59-66
[11] Gregor S and Benbasat I 1999 Explanations from intelligent systems: Theoretical foundations and implications for practice *MIS quarterly* 497-530
[12] Abraham A 2003 *Intelligent systems: Architectures and perspectives Recent advances in intelligent paradigms and applications* (Physica: Heidelberg) 1-35
[13] Sfidari E, Kadkhodaie-Ilkhchi A and Najjari S 2012 Comparison of intelligent and statistical clustering approaches to predicting total organic carbon using intelligent systems *Journal of Petroleum Science and Engineering* **86** 190-205
[14] Bukhtoyarov V V, Tynchenko V S, Petrovsky E A, Kukartsev V V and Kuklina A I 2018 Evolutionary method for automated design of models of vortex flowmeters transformation function *Journal of Physics: Conference Series* **1118(1)** 012041
[15] Subhadra K and Vikas B 2019 Neural network based intelligent system for predicting heart disease *International Journal of Innovative Technology and Exploring Engineering* **8(5)** 484-7
[16] Milov A V, Tynchenko V S, Kukartsev V V, Tynchenko V V and Bukhtoyarov V V 2018 Use of artificial neural networks to correct non-standard errors of measuring instruments when creating integral joints *Journal of Physics: Conference Series* **1118(1)** 012037

An intelligent approach for creation of control systems for road sections with pedestrian crossings

V S Tynchenko[1,2], A S Bondarev[1], S V Apanasenko[1], D V Rogova[1], A S Yadrov[2] and S A Moskvin[2,3]

[1]Reshetnev Siberian State University of Science and Technology, 31, Krasnoyarsky Rabochy Av., Krasnoyarsk, 660037, Russia
[2]Siberian Federal University, 79, Svobodny pr., Krasnoyarsk, 660041, Russia
[3]State Institution - Krasnoyarsk Regional Branch of the Social Insurance Fund of the Russian Federation, 117, Krasnoyarsky Rabochy Av., Krasnoyarsk, 660010, Russia

E-mail: vadimond@mail.ru

Abstract. The article discusses the development of a system for recognizing people at a pedestrian crossing. Such recognition system includes a trained classifier and two sets of images taken from an open database containing images of city streets from outdoor cameras. Some shortcomings that can be easily solved with the help of the developed recognition system are analyzed. We also studied common methods of detection, for example, the Viola-Jones method, and their advantages in the operation of the system. A comparison of all the considered methods is carried out and the most suitable one for the set goal is selected and the method of constructing a classifier is described separately. This system is becoming more and more relevant with the development of the technological capabilities of urban life and improves its quality due to its smooth operation and good performance indicators.

1. Introduction

In large cities, in practice, centralized traffic light control systems are used, taking into account the developed cycles and phases of automobile and pedestrian traffic, for the development of which empirically collected information on flows and density of each type of traffic is used [1 - 3].

Taking into account the development of machine learning, which provides a wide range of tools for working with data in real time, this direction is a promising direction for implementation into existing systems or their expansion, allowing one to optimize or automate existing solutions [4 - 6].

Next, the existing solution to the task of adjusting the pedestrian phase will be considered, and the requirements for the system being developed will be described.

2. Recognition system for pedestrian crossing people

At the moment, there is a solution to the problem of adjusting the pedestrian phase in the form of pedestrian crossing sections (PCS): in the immediate vicinity of the pedestrian crossing. A button is installed with a connection to the controller, when pressed, the pedestrian traffic light switches to a green signal, a road traffic light – to a red signal. After a given amount of time, the traffic lights switch back. An example of such a PCS is shown in Figure 1.

Content from this work may be used under the terms of the Creative Commons Attribution 3.0 licence. Any further distribution of this work must maintain attribution to the author(s) and the title of the work, journal citation and DOI.
Published under licence by IOP Publishing Ltd

IOP Conf. Series: Materials Science and Engineering 1151 (2021) 012043 doi:10.1088/1757-899X/1151/1/012043

Figure 1. Example of PCS.

Despite the simplicity of this solution, it has significant drawbacks, which are successfully solved by the developed recognition system.

First, the button design is unreliable. The buttons themselves are firm enough to press. Moreover, due to frequent use, vandalism and dampness, the PCS begins to work intermittently, leading either to the impossibility of switching when pressed, or to the constant inclusion of the pedestrian phase.

The system does not require direct pedestrian intervention, so the service life of the equipment will be higher than the corresponding PCS indicator. This also applies to acts of vandalism: the surveillance camera is out of reach for a person without special technical means, so the likelihood of equipment breakdown in this way will be much lower.

Second, PCSs do not allow maintaining the cycle length and the same number of phases, since the controllers do not support the phase selection mode according to the principle "either one or the other".

Likewise, PCS is incompatible with traffic lights in coordinated control, when they are controlled remotely and turn on simultaneously or with a shift according to the "green wave" principle. When working in remote mode, the technical control device intercepts control of the controller installed directly on the road, including phases according to the specified personal rules. This relationship is one-way, describing a top-down relationship. At the same time, in order to be included in the nearest window, communication must be two-way, which is currently not technically implemented.

The developed system has the prospect of expanding to a network of interconnected controllers transmitting information to each other, or implementation into existing coordination systems. This will allow you to individually and centrally adjust the adjustment of the pedestrian phases for each traffic situation, both for the local controller system and for each traffic situation, as well as in accordance with the existing algorithms of the coordination systems with their expansion.

3. Description of system requirements

The system being developed is faced with the task of recognizing a human figure in a video fragment in the context of an urban environment, that is, it must be able to distinguish people from all other objects of the environment [7 - 9].

It is understood that the system can both solve the problem of detecting a human figure and fix the fact of the presence of a pedestrian in the frame for a certain amount of time to transmit information either to a traffic light or to a local or central traffic controller [10].

The recognition system requires a video surveillance camera or other video recorder above the pedestrian crossing area [11]. The device should be installed at a distance that allows the human figure to be clearly distinguished.

In addition to the reader, hardware is required to operate the system itself. It is possible to implement the developed solution into the existing coordination systems.

An important initial step is the selection of the best solution for the classification problem. The chosen method must meet the specified requirements, that is, the final decision based on the analysis results must be based on the specifics of the subject area.

This section provides an overview of common detection methods and indicates the selected development tools. Moreover, simple indicators of the efficiency of the classifier are described for further evaluation of the results obtained at the end of training.

Object recognition is a scientific discipline that aims to classify objects into several categories or classes. The recognition task is to classify the considered new, unexplored object to any given class [12]. Thus, in this context, object recognition is one of the varieties of classification of identification; that is, assigning an unambiguous name to the object in question. In other words, obtaining characteristic points on the image will allow further classification of the object under consideration.

If we consider the problem of finding objects on a digital video fragment, then it can be divided into three subtasks:

- Detection: selection of a fragment in the image that may contain the desired object.
- Recognition: determination of the found object to a certain class.
- Tracking: The found object is fixed and its position is tracked.

These tasks are reduced to the construction of an image classifier.

For the system being developed, the first two subtasks are considered taking into account the tasks assigned to the system.

Two of the most common object recognition methods are discussed below: convolutional neural networks and the Viola-Jones method.

The convolutional neural network is a special neural network architecture proposed in 1988 and aimed at efficient recognition of objects and patterns. The architecture of such neural networks models some of the features of the visual cortex of the brain: there are some simple cells that respond to straight lines at different angles, and complex ones, the work of which is associated with the launch of a specific group of simple cells.

This architecture received its name in connection with the use of the convolution operation, the essence of which is element-wise multiplication of each image fragment by the convolution kernel and then writing the result to the corresponding position of the output image (Figure 2).

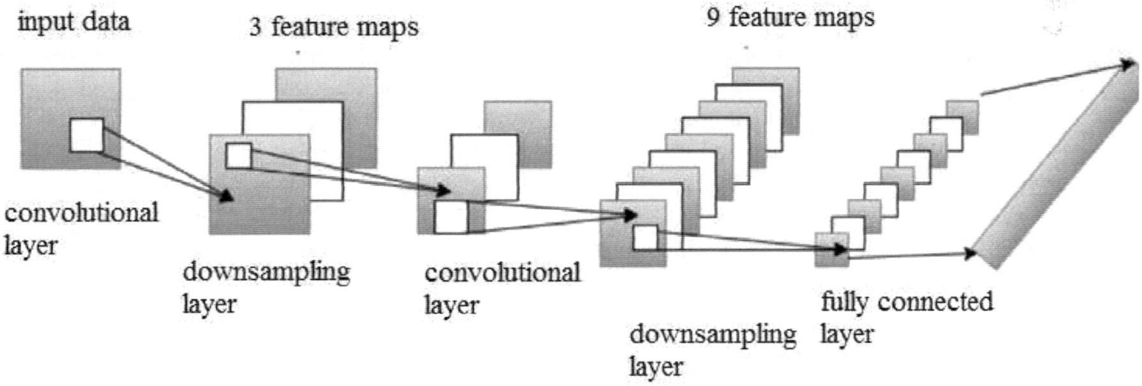

Figure 2. Convolutional neural network architecture.

The method under consideration uses three types of layers: convolution, downsampling (pulling), and a fully connected layer.

The convolution operator forms the basis of the convolutional layer of the network. This layer consists of three levels:

- Creation of a set of linear activations by performing one or more parallel convolutions.
- Detection, that is, the application of a non-linear activation function to all linear activations.
- Spatial aggregation to modify the output for transmission to the next network layer.

IOP Conf. Series: Materials Science and Engineering 1151 (2021) 012043 doi:10.1088/1757-899X/1151/1/012043

A layer consists of a set of cores, and its task is to calculate the convolution of the output image from the previous layer using this set and add an offset corresponding to the core at each iteration. The essence of these processes is the addition and scaling of the input image parameters (pixels).

The pulling layer is a non-linear compaction of the feature map. The layer receives at the output separate fragments of the image and combines them into one value. There are various ways of aggregation, usually the maximum function is used. This operation can significantly reduce the spatial volume of the image and avoid overfitting.

After several passes of image convolution and compaction using the pooling operation, the system rebuilds from a specific pixel grid to more abstract feature maps, as a rule, on each next layer, the number of channels increases and the image dimension in each channel decreases. Thus, by the end, there is a large set of channels storing a small amount of data, which are interpreted as the most abstract concepts identified from the original image.

The simplest and most common learning method is supervised learning - back propagation.

Convolutional neural networks have the following advantages:

- Solving problems with unknown patterns.
- Relative resistance to rotation and shift of the recognized image.
- Resistance to noise in the input data.
- Potentially fast work.
- Fault tolerance with hardware implementation of a neural network.

The Viola-Jones method was developed and presented in 2001, but it is still effective for finding objects in images and video sequences in real time. Initially, its task was to recognize faces, but the scope of possible application of the algorithm is quite wide.

A simplified diagram of the algorithm is shown in Figure 3. At the output of the algorithm, a set of found objects at different scales are given. As mentioned earlier, the classifier is trained with a teacher, i.e., there is a training sample and information indicating which of the training images contain the desired object.

The fundamental idea of the algorithm is the extraction of image features and the subsequent training of the algorithm on them. As features for the recognition algorithm, the authors proposed Haar features based on Haar wavelets.

The Haar trait is a set of adjacent rectangular regions, which are divided into two types. There are a lot of possible Haar features (various combinations of regions of different widths and heights with different positions in the image). In the standard Viola-Jones method, the features presented in Figure 4, called Haar primitives, are used.

To obtain the value of a certain feature for the image under consideration, it is necessary to add the brightness of the pixels in both groups of rectangular regions separately, and then subtract the second from the first resulting sum. The resulting difference is the value of a particular Haar feature for the image under consideration.

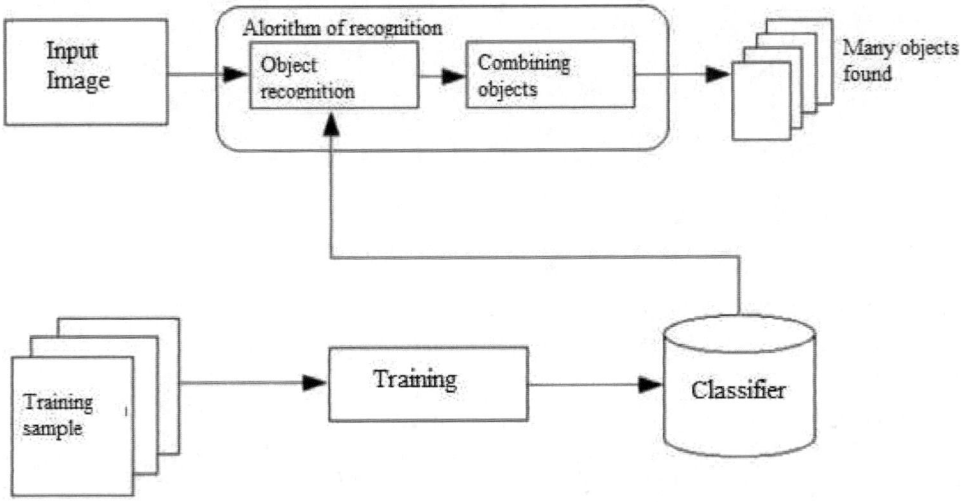

Figure 3. Generalized recognition scheme in the Viola-Jones algorithm.

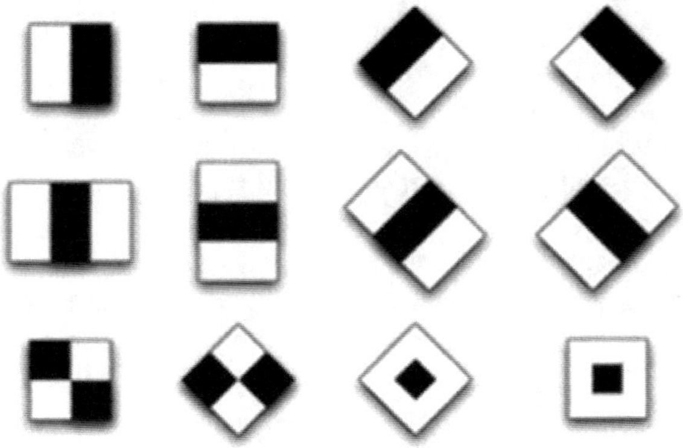

Figure 4. Haar primitives.

Calculating the value of each Haar feature for an image requires many addition operations. For quick calculation of the total brightness, the calculation operations are performed using integral representations of images.

In the integral form of the image, the value of each pixel is the sum of the brightness of that pixel and all the pixels that are above and to the left of it (in case the pixel with unit coordinates is located in the upper left corner of the image). By translating the image into an integral form, you can calculate the values of Haar features for it, without performing the summation of all the required brightness values each time anew. It is enough to calculate the sum of the brightness for each rectangular area using the properties of the integral shape of the image.

Boosting is a set of methods that improve the accuracy of analytical models. Boosting literally means "strengthening" of "weak" models - it is a procedure for sequentially building a composition of machine learning algorithms, when each next algorithm seeks to compensate for the shortcomings of the composition of all previous algorithms. The idea of boosting was proposed by Robert Shapir in the late 90s, when it was necessary to find a solution to the problem of obtaining one good one, having many slightly different learning algorithms.

Thus, upon completion of training, the cascade shown in figure 5 will be obtained, consisting of a chain of classifiers (layers) trained using the boosting procedure.

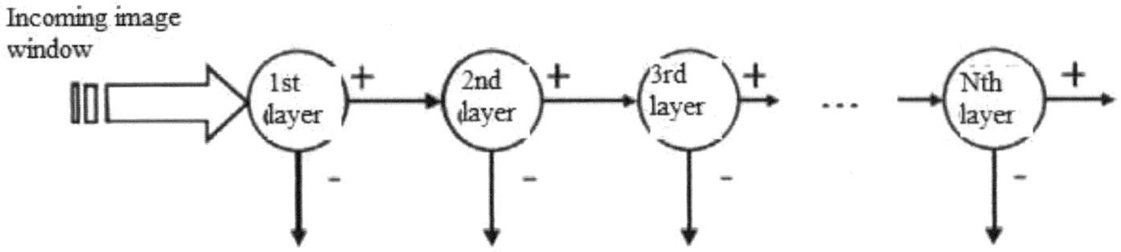

Figure 5. Cascade structure.

The Viola-Jones method has the following advantages:

- The ability to detect a large number of objects in the original image.
- Good speed of work.
- The possibility of training the classifier on the location of any object.
- Checking a feature at a certain position in the image takes linear time due to the use of the integral representation of the image.
- A large number of implementations in various open-source libraries and in various programming languages.
- A small number of false positives when searching for an object in the image.

Based on the results of the analysis of the considered methods, we can say that both methods are suitable for solving the problem of classifying pedestrians in a video stream.

Both algorithms are fast, with high detection rates. Both the convolutional neural network and the Viola-Jones method require a lot of time and resources in the learning process, so this factor, like the speed of work, cannot be considered as decisive.

However, one of the advantages of the Viola-Jones method is the low rate of false positives, which is a critical feature in the work being done. It is assumed that a security camera or other video recording device will have a static field of view. Thus, in case of false classification of a third-party relatively static object (plant, road sign, etc.), the resulting classifier will have no practical application: a signal will be transmitted on an ongoing basis to switch the pedestrian green traffic light in the absence of a pedestrian, which corresponds to normal operation controllers at the moment without implementing the system.

For the above reason, the Viola-Jones method was chosen for the developed system.

4. Conclusion

As a result of this work, a review of development tools for control systems for road sections with pedestrian crossings was carried out. To implement the classifier, the methods of object detection were considered. The Viola-Jones method was selected for the most suitable for the specifics of the task at hand, and a method for constructing a classifier based on the AdaBoost algorithm was described. The considered classifier is offered to systems for regulating traffic sections with pedestrian crossings.

References

[1] Datola G, Bottero M and De Angelis E 2020 Addressing Social Inclusion Within Urban Resilience: A System Dynamics Approach *In: INTERNATIONAL SYMPOSIUM: New Metropolitan Perspectives* pp 510-9

[2] Dong C Z, Bas S and Catbas F N 2020 A portable monitoring approach using cameras and computer vision for bridge load rating in smart cities *Journal of Civil Structural Health Monitoring* **10(5)** 1001-21

[3] Oc T and Tiesdell S 1998 City centre management and safer city centres: approaches in Coventry and Nottingham *Cities* **15(2)** 85-103

[4] Obedin A V, Soroka E O, Kukartsev V V, Mikhalev A S, Tynchenko V S, Semenova E I and Bashmur K A 2019 The developing program system of social monitoring of road improvement and urban infrastructure *Journal of Physics: Conference Series 1399(5)* 055021

[5] Oc T and Tiesdell S 1999 The fortress, the panoptic, the regulatory and the animated: Planning and urban design approaches to safer city centres *Landscape Research* **24(3)** 265-86

[6] Ballesteros J, Rahman M, Carbunar B and Rishe N 2012 Safe cities. A participatory sensing approach *In: 37th Annual IEEE Conference on Local Computer Networks* pp 626-34

[7] Tubaishat A and Elsayed R 2017 Safe city: Introducing a flawless model and implementation guidelines to reduce crime rate and response time *In: Proceedings of the 29th International Business Information Management Association Conference - Education Excellence and Innovation Management through Vision 2020: From Regional Development Sustainability to Global Economic Growth* pp 474-85

[8] Li Y H, Zhang X D, Feng Z C and Luan Z H 2014 Research on the Key Equipment Performance of Stage Video Surveillance and Scheduling System Based on IP Network *Applied Mechanics and Materials* **644** 957-60

[9] Idimechev I A, Mikhalev A S, Goponenko A A, Stavenko S S, Kukartsev V V and Tynchenko V S 2020 Comparative analysis of background subtraction algorithms in the task of recognizing and identifying people in a video stream *Journal of Physics: Conference Series* **1661(1)** 012041

[10] Zeng X, Guo H and Hu W 2019 Design and Implementation of Shipping Video Surveillance Equipment Based on Raspberry Pi *In: 2019 IEEE International Conference on Computational Science and Engineering (CSE) and IEEE International Conference on Embedded and Ubiquitous Computing (EUC)* pp 66-70

[11] Pan X, Hu H, Xu J and Li M 2020 Research on Video Surveillance Robot Based on SSH Reverse Tunnel Technology *In: 2020 3rd International Conference on Advanced Electronic Materials, Computers and Software Engineering (AEMCSE)* pp 298-302

[12] Kukartsev V V, Mikhalev A S, Tarasevich A V, Tynchenko V S, Ogol A R and Khramkov V V 2019 Selection of appropriate architecture and parameters of neural network for images recognition and classification *Journal of Physics: Conference Series* **1399(3)** 033105

IOP Conf. Series: Materials Science and Engineering 1151 (2021) 012044 doi:10.1088/1757-899X/1151/1/012044

Use of the multi-factor regression model for temperature control in structurally isolated rooms of the railway stations

V S Tynchenko[1], I V Markevich[1], A R Ogol[1], O V Baryshnikova[1], D V Rogova[1] and O N Kotova[2]

[1] Reshetnev Siberian State University of Science and Technology, 31, Krasnoyarsky Rabochy Av., Krasnoyarsk, 660037, Russian Federation
[2] Kemerovo State University, 6, Krasnaya St., Kemerovo, 650000, Russian Federation

E-mail: vadimond@mail.ru

Abstract. The article is devoted to the analysis and construction of a multivariate regression model. The mode of operation of generalized heat points is considered. The positive and negative aspects of using heat points to maintain the temperature at the desired level have been identified. The analysis of the multivariate model is carried out. The input data for the generated model has been defined. The factors that affect the temperature in structurally isolated rooms have been identified. A multifactorial regression model of temperature control in heated rooms has been built, with the help of which it is possible to determine how much heat needs to be supplied to a particular structurally isolated room, which will allow more efficient use of resources to maintain the temperature at the required level.

1. Introduction

Maintaining the temperature in various rooms at a certain level is one of the most important tasks [1 - 3]. For example, heat enters the apartments after the average daily temperature is kept at +8 and below degrees for five days in a row. But the final decision on the beginning of the heating season is made by the executive authorities [4, 5].

Starting up a central heating system is a gradual process. After opening the valves at CHPPs, boiler houses and other heat sources, heat enters the rooms for several days, and it takes time to obtain the necessary heat supply parameters.

The living quarters use both local and central heating. The heating mode in isolated rooms of the railway stations is most often not chosen by residents. It depends on if the heaters installed by the builders support energy from a common heat source and do not have the ability to be regulated. But, at the same time, the supply of heat to rooms and premises often occurs in strange modes, which forces the owners of apartments and premises to change radiators, install adjusting devices on them and form additional heating circuits.

Based on this, the purpose of this article is to develop a multifactorial model for temperature control in isolated rooms of the railway stations during the heating season, which will lead to a decrease in resource use. To achieve this goal, it is necessary to solve the following tasks:

- To identify factors influencing the temperature in living quarters.
- To develop a multifactor model of temperature control in isolated rooms of the railway stations during the heating season.

Content from this work may be used under the terms of the Creative Commons Attribution 3.0 licence. Any further distribution of this work must maintain attribution to the author(s) and the title of the work, journal citation and DOI.
Published under licence by IOP Publishing Ltd

2. Materials and methods

In the modern world, in practice, used in construction, central water heating systems are used. A fairly large part of the housing stock is heated by central networks. Central heating provides a more uniform indoor climate and shows great demand and efficiency, despite the active introduction and use of such sources to maintain a favorable microclimate at home [6]. Central heating is most often used in railway stations. The uninterrupted operation and efficiency of centralized networks at home depends on high-quality assembly and proper operation of all constituent elements.

The main difference between central heating and local heating is that there is no pollution by combustion products and fuel, more convenient and reliable control. The delivery of the coolant is carried out through the main pipelines. This is a complex ramified engineering system that provides heat energy to many objects and is dispersed over a large area.

For the modern practice of multi-storey construction, central heating is a necessary element of the sanitary engineering of buildings.

The central heating system has the following advantages:

- Use of inexpensive fuels.
- Regular monitoring of performance and technical condition by special services.
- Use of environmentally friendly equipment.
- Ease of use.
- Ensuring a more uniform thermal regime in the premises during the day.
- Lack of pollution in the premises by combustion products.
- More convenient, reliable and safe management.
- No labor costs by the population for heating the premises.

Among the disadvantages of such a heating scheme for the railway stations, the following should be noted:

- Operation of the system according to a strict seasonal schedule.
- Impossibility of individual regulation of the temperature of heating devices.
- Frequent pressure drops in the system.
- Significant heat loss during transportation and heating.
- High cost of equipment and its installation.

Due to the above disadvantages, it was decided to develop a multifactor model for temperature control in isolated rooms of the railway stations.

A decision was made to conduct a regression analysis in order to establish the degree of influence of independent variables on the dependent variable. To identify dependencies, a generally accepted linear regression equation will be used [7 - 9].

In the regression equation, Y means the variable, the influence of factors on which we are trying to study. In our case, this is the number of buyers. The x-value is the various factors that affect the variable.

To build a model, you must specify the following input data:

- Purpose of the isolated rooms of the railway stations (residential / non-residential).
- Number of storeys.
- Floor height.
- Height of the window.
- Outside air temperature text equal to the average temperature of the coldest five-day period.
- Average temperature of the heating season (tht).
- Duration of the heating season (zht).
- Maximum of the average wind speeds in terms of points for January V.
- Heat carrier source.
- Heat carrier parameters: in the heating network and in the heating system.
- Differential pressure at the subscriber input.

For calculations, we will use the following parameters: railway station, 5 floors, floor height - 3 meters, window height - 1.5 meters, text = -31°C, tht = -4 °C, zht = 221 days, V = 5.5 m / s, heat carrier source - city heating networks, heat carrier parameters: in the heating network - 130-70 °C, in the heating system - 95-70 °C, pressure drop at the subscriber input = 60 kPA.

3. Experimental study and discussion

The problem under study can be solved using a multivariate regression model, which will allow you to build and test the relationship between one resulting variable and several independent factors that influence it [10, 11].

In general terms, the multivariate regression equation can be written as follows:

$$Y = \hat{f}(\overrightarrow{X}) = f(\vec{X}) + e, \tag{1}$$

where $(\bar{X})$ = X1, X2,... Xn – independent explanatory variables; e - random variable, an additional residual term that reflects the influence of random errors, measurement features and actions that affect the resulting variable, and other explanatory variables not included in the equation. It is also called resentment or remnant. This random component can be considered a random forecast error Y for a given value X.

Table 1. Data for regression.

Apartment	X0	X1	X2	X3	X4	Y
101	1020	20	22	18,91	1	1834
102	490	20	20	20,13	1	1316
103	430	10	20	16,47	1	1106
104	470	20	20	18,3	1	1221
105	490	20	20	20,13	1	1316
106	490	20	20	20,13	1	1316
107	470	20	20	18,3	1	1221
108	430	10	20	16,47	1	1106
109	490	20	20	20,13	1	1316
110	1070	70	20	18,91	1	1884
111	1070	70	20	18,91	1	1884
112	490	20	20	20,13	1	1316
113	430	10	20	16,47	1	1106
114	470	20	20	18,3	1	1221
115	470	20	20	18,3	1	1221
116	490	20	20	20,13	1	1316
117	490	20	20	20,13	1	1316
118	470	20	20	18,3	1	1221
119	470	20	20	18,3	1	1221
120	430	10	20	16,47	1	1106
121	490	20	20	20,13	1	1316
122	1020	20	20	18,91	1	1834
201	880	20	20	18,91	2	1694
202	350	20	20	20,13	2	1176
203	310	10	20	16,47	2	986
204	350	20	20	18,3	2	1101
205	340	20	20	20,13	2	1166
206	350	20	20	20,13	2	1176
207	350	20	20	18,3	2	1101
208	310	10	20	16,47	2	986
209	350	20	20	20,13	2	1176
210	930	70	20	18,91	2	1744
211	620	70	20	18,91	2	1434
212	350	20	20	20,13	2	1176

The value of the required load on heating devices in each apartment is taken as the dependent variable Y.

Explanatory variables:

- X0 is the value of the main heat losses.
- X1 - value of additional heat losses.
- X2 - required temperature in isolated rooms.
- X3 - room area.
- X4 - floor.

Table 1 presents some of the input data for constructing the regression.

Tables 2 - 4 show the performed regression calculations.

Table 2. Regression statistics.

Multiple R	1,00
R- square	0,999
Normalized R- square	1,00
Standard error	5,88
Observations	110

Table 3. Analysis of variance.

	df	SS	MS	F	Significance F
Regression	5	7087520,10	1417504,02	40983,8012	1,3725E-169
Balance	104	3597,04	34,58693384		
Total	109	7091117,14			

Table 4. Regression coefficients.

	Odds	Standard error	t- statistics	P- Value	Lower 95%	Upper 95%
Y- crossing	-50,07	47,34	-1,06	0,29	-143,95	43,81
Variable X1	1,05	0,00	299,27	0,00	1,04	1,05
Variable X2	1,31	0,05	27,70	0,00	1,22	1,40
Variable X3	2,55	2,36	1,08	0,28	-2,13	7,23
Variable X4	39,90	0,44	89,81	0,00	39,02	40,78
Variable X5	-0,65	0,40	-1,62	0,11	-1,44	0,14

One of the main indicators of regression is the R-squared, which indicates the quality of the model [12]. In our case, this ratio is 0.999 or about 99%. This is an acceptable level of quality. Dependency less than 0.5 is bad.

The Y-intersection shows what will be Y if all factors in the model are equal to 0. In our case, this value is -50.07.

- 1.05 - coefficient that shows the importance of the influence of the factor X0 on Y, that is, the main heat losses within this model affect the heat load with a weight of only 1.05.
- 1.31 - coefficient that shows that additional heat loss within the given model affects the heat load with a weight of 1.31.
- 2.55 - coefficient that shows that the required temperature in the apartment within the given model affects the heat load with a weight of 2.55.
- 39.9 is a coefficient that shows that the area of an apartment within a given model affects the heat load with a weight of 39.9, ie. is the most significant factor in the model.
- -0.65 is a coefficient that shows that a floor within a given model affects the heat load with a weight of -0.65. A negative value of the coefficient means that an inverse relationship is established between the factor and the dependent variable, i.e., the higher the floor, the lower the value of the heat load on electrical appliances.

Using these coefficients, you can build a regression model and it will look like this:

$$Y = 1{,}05 * X0 + 1{,}31 * X1 + 2{,}55 * X2 + 39{,}9 * X3 - 0{,}65 * X4 - 50{,}07. \quad (2)$$

Using a regression model, you can make predictions for this, you need to substitute the resulting coefficients instead of X into the model.

Table 5 shows a fragment of the forecast calculation.

Table 5. Forecast.

Apartment	X0	X1	X2	X3	X4	Y
101	1020	20	22	18,91	1	1832
102	490	20	20	20,13	1	1321
103	430	10	20	16,47	1	1110
104	470	20	20	18,3	1	1228
105	490	20	20	20,13	1	1321
106	490	20	20	20,13	1	1321
107	470	20	20	18,3	1	1228
108	430	10	20	16,47	1	1110
109	490	20	20	20,13	1	1321
110	1070	70	20	18,91	1	1893
111	1070	70	20	18,91	1	1893
112	490	20	20	20,13	1	1321
113	430	10	20	16,47	1	1110
114	470	20	20	18,3	1	1228
115	470	20	20	18,3	1	1228
116	490	20	20	20,13	1	1321
117	490	20	20	20,13	1	1321
118	470	20	20	18,3	1	1228
119	470	20	20	18,3	1	1228
120	430	10	20	16,47	1	1110
121	490	20	20	20,13	1	1321
122	1020	20	20	18,91	1	1827
201	880	20	20	18,91	2	1680
202	350	20	20	20,13	2	1174

Looking at the table, you can see that the discrepancy between the actual data and the forecast is not significant.

4. Conclusion

As a result of this work, the positive and negative aspects of using heat points to maintain the temperature at the desired level have been identified. The analysis of the multivariate model was carried out, while the input data for the generated model has been defined. The factors that affect the temperature in structurally isolated rooms have been identified.

A multifactorial regression model of temperature control in heated rooms has been built, with the help of which it is possible to determine how much heat needs to be supplied to a particular structurally isolated room. It allows one to use the resources more efficiently to maintain the temperature in isolated rooms of the railway station at the required level.

References

[1] Ferdyn-Grygierek J 2014 Indoor environment quality in the museum building and its effect on heating and cooling demand *Energy and Buildings* **85** 32-44

[2] Sodha M S, Sharma A K, Singh S P, Bansal N K and Kumar A 1985 Evaluation of an earth—air tunnel system for cooling/heating of a hospital complex *Building and Environment* **20(2)** 115-22

[3] Stryjakowska-Sekulska M, Piotraszewska-Pajak A, Szyszka A, Nowicki M and Filipiak M 2007 Microbiological quality of indoor air in university rooms *Polish Journal of Environmental Studies* **16(4)** 623
[4] Wan G H, Chung F F and Tang C S 2011 Long-term surveillance of air quality in medical center operating rooms *American journal of infection control* **39(4)** 302-8
[5] Ng S F, Oo C S, Loh K H, Lim P Y, Chan Y H and Ong B C 2003 A comparative study of three warming interventions to determine the most effective in maintaining perioperative normothermia *Anesthesia & Analgesia* **96(1)** 171-6
[6] Moriske H J, Drews M, Ebert G, Menk G, Scheller C, Schöndube M and Konieczny L 1996 Indoor air pollution by different heating systems: coal burning, open fireplace and central heating *Toxicology letters* **88(1-3)** 349-54
[7] Bukhtoyarov V V, Tynchenko V S, Petrovsky E A, Dokshanin S G and Kukartsev V V 2019 Research of methods for design of regression models of oil and gas refinery technological units *IOP Conference Series: Materials Science and Engineering* **537(4)** 042078
[8] Chamberlain G 1982 Multivariate regression models for panel data *Journal of econometrics* **18(1)** 5-46
[9] Boyko A A, Kukartsev V V, Tynchenko V S, Korpacheva L N, Dzhioeva N N, Rozhkova A V and Aponasenko S V 2020 Using linear regression with the least squares method to determine the parameters of the Solow model *Journal of Physics: Conference Series* **1582(1)** 012016
[10] Eerola T, Lartillot O and Toiviainen P 2009 Prediction of Multidimensional Emotional Ratings in Music from Audio Using Multivariate Regression Models *In: ISMIR* pp 621-6
[11] Azen R and Budescu D V 2006 Comparing predictors in multivariate regression models: An extension of dominance analysis *Journal of Educational and Behavioral Statistics* **31(2)** 157-80
[12] Santiago C B, Guo J Y and Sigman M S 2018 Predictive and mechanistic multivariate linear regression models for reaction development *Chemical science* **9(9)** 2398-412

AUTHOR INDEX